Radiation Damage in Materials

Radiation Damage in Materials: Helium Effects

Special Issue Editors

Yongqiang Wang
Khalid Hattar

MDPI • Basel • Beijing • Wuhan • Barcelona • Belgrade • Manchester • Tokyo • Cluj • Tianjin

Special Issue Editors

Yongqiang Wang
Los Alamos National Laboratory
USA

Khalid Hattar
Sandia National Laboratories
USA

Editorial Office
MDPI
St. Alban-Anlage 66
4052 Basel, Switzerland

This is a reprint of articles from the Special Issue published online in the open access journal *Materials* (ISSN 1996-1944) (available at: https://www.mdpi.com/journal/materials/special_issues/ Radiation-Damage).

For citation purposes, cite each article independently as indicated on the article page online and as indicated below:

LastName, A.A.; LastName, B.B.; LastName, C.C. Article Title. *Journal Name* **Year**, *Article Number*, Page Range.

ISBN 978-3-03936-362-9 (Pbk)
ISBN 978-3-03936-363-6 (PDF)

Cover image courtesy of Hyosim Kim.

Contents

About the Special Issue Editors . . . . **vii**

Yongqiang Wang and Khalid Hattar
Special Issue: Radiation Damage in Materials—Helium Effects
Reprinted from: *Materials* **2020**, *13*, 2143, doi:10.3390/ma13092143 . . . . **1**

Luis Sandoval, Danny Perez, Blas P. Uberuaga and ArthurF. Voter
An Overview of Recent Standard and Accelerated MolecularDynamics Simulations of Helium
Behavior in Tungsten
Reprinted from: *Materials* **2019**, *12*, 2500, doi:10.3390/ma12162500 . . . . **5**

Chunping Xu, Wenjun Wang
Simulation Study of Helium Effect on the Microstructure of Nanocrystalline Body-Centered
Cubic Iron
Reprinted from: *Materials* **2019**, *12*, 91, doi:10.3390/ma12010091 . . . . **25**

Shi-Hao Li, Jing-Ting Li and Wei-Zhong Han
Radiation-Induced Helium Bubbles in Metals
Reprinted from: *Materials* **2019**, *12*, 1036, doi:10.3390/ma12071036 . . . . **39**

**Osman El Atwani, Kaan Unal, William Streit Cunningham, Saryu Fensin, Jonathan Hinks,
Graeme Greaves and Stuart Maloy**
In-Situ Helium Implantation and TEM Investigation of Radiation Tolerance to Helium Bubble
Damage in Equiaxed Nanocrystalline Tungsten and Ultrafine Tungsten-TiC Alloy
Reprinted from: *Materials* **2020**, *13*, 794, doi:10.3390/ma13030794 . . . . **71**

**Hyosim Kim, Tianyao Wang, Jonathan G. Gigax, Shigeharu Ukai, Frank A. Garner and
Lin Shao**
Effect of Helium on Dispersoid Evolution under Self-Ion Irradiation in A Dual-Phase 12Cr
Oxide-Dispersion-Strengthened Alloy
Reprinted from: *Materials* **2019**, *12*, 3343, doi:10.3390/ma12203343 . . . . **81**

Feifei Zhang, Lynn Boatner, Yanwen Zhang, Di Chen, Yongqiang Wang and Lumin Wang
Swelling and Helium Bubble Morphology in a Cryogenically Treated FeCrNi Alloy with
Martensitic Transformation and Reversion after Helium Implantation
Reprinted from: *Materials* **2019**, *12*, 2821, doi:10.3390/ma12172821 . . . . **95**

Cuncai Fan, Zhongxia Shang, Tongjun Niu, Jin Li, Haiyan Wang and Xinghang Zhang
Dual Beam In Situ Radiation Studies of Nanocrystalline Cu
Reprinted from: *Materials* **2019**, *12*, 2721, doi:10.3390/ma12172721 . . . . **111**

**Caitlin A. Taylor, Samuel Briggs, Graeme Greaves, Anthony Monterrosa, Emily Aradi,
Joshua D. Sugar, David B. Robinson, Khalid Hattar and Jonathan A. Hinks**
Investigating Helium Bubble Nucleation and Growth through Simultaneous In-Situ Cryogenic,
Ion Implantation, and Environmental Transmission Electron Microscopy
Reprinted from: *Materials* **2019**, *12*, 2618, doi:10.3390/ma12162628 . . . . **125**

Wenfan Yang, Jingyu Pang, Shijian Zheng, Jian Wang, Xinghang Zhang and Xiuliang Ma
Interface Effects on He Ion Irradiation in Nanostructured Materials
Reprinted from: *Materials* **2019**, *12*, 2639, doi:10.3390/ma12162639 . . . . **135**

Huaqiang Chen, Jinlong Du, Yanxia Liang, Peipei Wang, Jinchi Huang, Jian Zhang, Yunbiao Zhao, Xingjun Wang, Xianfeng Zhang, Yuehui Wang, George A. Stanciu and Engang Fu
Comparison of Vacancy Sink Efficiency of Cu/V and Cu/Nb Interfaces by the Shared Cu Layer
Reprinted from: *Materials* **2019**, *12*, 2628, doi:10.3390/ma12162628 . . . . **157**

Qing Su, Tianyao Wang, Jonathan Gigax, Lin Shao and Michael Nastasi
Resistance to Helium Bubble Formation in Amorphous SiOC/Crystalline Fe Nanocomposite
Reprinted from: *Materials* **2019**, *12*, 93, doi:10.3390/ma12010093 . . . . **165**

Saryu Fensin, David Jones, Daniel Martinez, Calvin Lear and Jeremy Payton
The Role of Helium on Ejecta Production in Copper
Reprinted from: *Materials* **2020**, *13*, 1270, doi:10.3390/ma13061270 . . . . **175**

About the Special Issue Editors

Yongqiang Wang currently serves as Director of Ion Beam Materials Laboratory and Team Leader for Radiation Science Experimental Team at Materials Science and Technology Division, Los Alamos National Laboratory, Los Alamos, New Mexico, USA. He received his B.S. M.S. and Ph.D. in nuclear physics and technology from Lanzhou University in China. He has over 30 years of experience using energetic ion beams in materials research, including radiation damage effects, ion beam modification of materials, and ion beam analysis. He is a member of the International Committee for Ion Beam Analysis and a member of the Advisory Editorial Board for Nuclear Instruments and Methods in Physics Research: Beam Interactions with Materials and Atoms. He also co-chairs the Biennial International Conference Series on Application of Accelerators in Research and Industry. He has co-authored more than 350 publications, including two books: Handbook of Modern Ion Beam Materials Analysis (Second Edition) and Ion Beam Analysis: Fundamentals and Applications.

Khalid Hattar is a Principle Member of the Technical Staff in the Ion Beam Lab and Center for Integrated Nanotechnologies at Sandia National Laboratories in Albuquerque, New Mexico, USA. He received a B.S. in Chemical Engineering with an emphasis in Materials Science from the University of California, Santa Barbara, in 2003, and a Ph.D. in Materials Science and Engineering from the University of Illinois, Urbana-Champaign, in 2009 under the guidance of Professor Ian M. Robertson. He specializes in determining the property–microstructure relationship for a variety of structural and functional materials through in-situ electron microscopy techniques in various extreme environments, as well as tailoring local properties of materials through ion beam modification. He has co-authored more than 100 publications, including three book chapters.

Editorial

Special Issue: Radiation Damage in Materials—Helium Effects

Yongqiang Wang [1],* and Khalid Hattar [2],*

[1]   Materials Science and Technology Division, Los Alamos National Laboratory, Los Alamos, NM 87545, USA
[2]   Center for Integrated Nanotechnologies, Sandia National Laboratories, Albuquerque, NM 87185, USA
*   Correspondence: yqwang@lanl.gov (Y.W.); khattar@sandia.gov (K.H.)

Received: 30 April 2020; Accepted: 5 May 2020; Published: 6 May 2020

Abstract: Despite its scarcity in terrestrial life, helium effects on microstructure evolution and thermo-mechanical properties can have a significant impact on the operation and lifetime of applications, including: advanced structural steels in fast fission reactors, plasma facing and structural materials in fusion devices, spallation neutron target designs, energetic alpha emissions in actinides, helium precipitation in tritium-containing materials, and nuclear waste materials. The small size of a helium atom combined with its near insolubility in almost every solid makes the helium–solid interaction extremely complex over multiple length and time scales. This *Special Issue*, "Radiation Damage in Materials—Helium Effects", contains review articles and full-length papers on new irradiation material research activities and novel material ideas using experimental and/or modeling approaches. These studies elucidate the interactions of helium with various extreme environments and tailored nanostructures, as well as their impact on microstructural evolution and material properties.

Keywords: helium; in situ transmission electron microscopy (TEM); ion beam modification (IBM); extreme environments; molecular dynamic (MD) simulation; nanostructure stability

---

Understanding radiation damage effects in materials, used in various irradiation environments, has been an ongoing challenge since the Manhattan Project, more than 75 years ago. The complexity stems from the fundamental particle–solid interactions and the subsequent damage recovery dynamics after collision cascades, which involves a range of both spatial (Å to m) and temporal (ps to decades) length scales. Adding to this complexity are the transmuted impurities that are unavoidable from accompanying nuclear processes, such as (neutron, alpha) reactions and their interactions with both intrinsic and extrinsic defects through damage recovery and defect evolution processes [1]. Helium (He) is one such impurity that plays an important and unique role in controlling the microstructure and properties of materials. Although abundant in the universe, He is a rare terrestrial resource with even greater scarcity in solid matter due in part to its virtually zero solubility in any material systems [2]. The ultra-low solubility forces He atoms to self-precipitate into small He bubbles that become nucleation sites for further void growth under radiation induced vacancy supersaturations, resulting in material swelling and high temperature He embrittlement, as well as surface blistering under low energy and high flux He bombardment at elevated temperatures. Because it is the large voids (not the small bubbles) that contribute to the detrimental effects, two general approaches have been adopted over the years to mitigate the bubble to void transition [3,4]:

(1)   Maximize the critical radius at which bubbles transform into voids, for example, by reducing the vacancy supersaturation; and

(2)   Increase the number of stable bubbles by maximizing the number of their nucleation sites (e.g., as in nanoferritic alloys [5,6]), which reduces the He flux to individual bubbles for any given He implantation rate.

Obviously, both approaches do not prevent the formation and eventual linkup of voids, but merely delay it with a hope that other degradation mechanisms become life-limiting. More recently, a totally different approach was proposed, in which carefully engineered semicoherent metal nanolayers were found to alter the fundamental growth trajectory of He-precipitates from the traditional equiaxed growth of 3D nanobubbles into a preferential formation of elongated 1D nanochannels [7]. These 1D He nanochannels that once interconnected to form a 3D network have a potential to outgas He outside of the material in Operando when the material is continuously being implanted with He particles, and thus, effectively reduce the net He particle flux received by the material. During these studies, He implantation has emerged as a useful tool for understanding complex and diverse environments, ranging from solar winds in space [8] to 'nanofuzz' formation in fusion energy systems [9].

This *Special Issue*, "Radiation Damage in Materials—Helium Effects", includes three review articles and nine full length papers (12 publications in total) on new irradiation material research activities and novel material ideas focused on understanding He effects on microstructure evolution and the subsequent properties. The research, originating from 24 different institutions in five different countries (USA, China, UK, Romania, and Japan), utilizes both experimental and modeling approaches to explore the complex interaction of He in a wide range of metallic-based microstructures and compositions (10 in total). These compositions included four ferrous-based systems (Fe, Fe/SiOC, FeCrNi, and Fe12Cr), three copper-based systems (Cu, Cu/V, and Cu/Nb), two tungsten-based systems (W and W-TiC), and palladium (Pd). The impact of the research ranged from fundamental shock wave physics questions such as the role of He impurity on ejecta production in dynamic materials to elucidating the nuclear engineering candidacy of certain alloys and processing routes for advanced fission and fusion reactor concepts.

The *Special Issue* starts with two modeling papers exploring the fundamental nuances of helium–solid interactions, which in turn permit a greater understanding of the physics and the development of more reliable models predicting the response of He-containing materials. The first article by the team at Los Alamos reviews the rapid developments in the Accelerated Molecular Dynamic (AMD) simulations as applied to the interaction of He in W, which is important for the success of current and future applications in the area of magnetic confinement fusion [10]. This is followed up by a detailed MD simulation by Xu et al., that explores the role of He generation on both grain boundary stability and crack growth in BCC-Fe. This modeling effort also takes advantage of recent advancements in computational code to produce simulated X-Ray Diffraction (XRD) patterns that permit rapid experimental validation [11].

The following six papers nicely demonstrate the complex He defects formed depending greatly on the underlying microstructure and the nuances of the radiation environment. The current understanding of He evolution in solids is well presented in the review "Radiation-Induced Helium Bubbles in Metals" [12]. The nuanced importance of composition and nanostructure for both fusion and Generation IV fission relevant materials are highlighted in the papers by El Atwani et al. exploring equiaxed nanocrystalline W and ultrafine grained W-TiC Alloy [13], by Kim et al.'s study of dual-phase 12Cr oxide-dispersion-strengthened alloy [14], and by Zhang et al. examining swelling and He bubble morphology in a cryogenically treated FeCrNi alloy with martensitic transformation and reversion after He implantation [15]. Two studies in model metal systems (Cu and Pd) noted the impact of radiation environments by comparing various sequential and concurrent heavy ion irradiation and He implantation conditions and by controlling the irradiation temperature. The study on nanocrystalline Cu by the team at Purdue University [16] suggests that He bubbles at grain boundaries and grain interiors may retard grain coarsening. The work on Pd hydriding behavior by the Sandia and University of Huddersfield team [17] utilized hydrogen over-pressure during in situ TEM observation in order to effectively mimic tritium-decay-induced He-3 precipitates in Pd. In both model and application alloys, it is clear that the evolution of He can vary greatly as a function of alloy composition and microstructure, as well as the details of the radiation environment.

Utilizing the understanding of He evolution at the time, it was proposed and demonstrated in Cu/Nb in 2005 that nanolayered materials with tailored layer thickness and interface structures could greatly decrease the large scale and catastrophic failure in He implanted metals [18]. In the last 15 years, the type of interfaces and layer thickness that have been He implanted has exponentially expanded with a range of engineered microstructures. These research activities are fortunately reviewed in "Interface Effects on He Ion Irradiation in Nanostructured Materials" [19]. This field continues to grow with new results from Chen et al. showing a clever triple layer approach to demonstrate the difference between Cu/V and Cu/Nb interfaces vacancy sink efficiency [20]. Pushing the community away from just metallic nanolayers, the team from the University of Nebraska and Texas A&M University has explored the He evolution in a ceramic/metal (SiOC/Crystalline Fe) nanolayer system [21]. These combined studies show the rapid growth in the study of nanolayers for radiation tolerance and the exciting new directions that are still left to be explored.

The final paper in this *Special Issue* by S. Fensin et al. deviated somewhat from the traditional He effects in materials, where He-induced defects affect microstructural evolution, which further impacts material properties and performance. Instead, this paper used a clever experimental design, utilizing the Richtmyer–Meshkov Instability (RMI) technique to determine "The Role of Helium on Ejecta Production in Copper" [22] and demonstrated a new area of interest in dynamic materials research, where He-doped microstructures are found to directly influence the surface material ejection behavior under shock wave extreme conditions.

Taken together, these studies show a vibrant and still evolving simulation and experimental research community exploring the unique impact He can have on solid matrixes. Despite the scarcity of He on earth, we expect studies exploring the interaction of He in matter will increase as the accessibility of He implantation capabilities via the He Ion Microscope (HIM) [23] becomes a common commercial tool at most research institutes, while simultaneously, the demand for such studies from advanced fission and fusion nuclear reactor, space exploration, actinide research, and nuclear waste communities continues to increase.

Author Contributions: Conceptualization, Y.W.; solicitation of authors, Y.W. and K.H.; writing—original draft preparation, K.H.; writing—review and editing, Y.W. and K.H.; All authors have read and agreed to the published version of the manuscript.

Funding: K.H. was supported by the DOE Office of Basic Energy Science, Materials Science and Engineering Division. This work was performed, in part, at the Center for Integrated Nanotechnologies, an Office of Science User Facility operated for the U.S. Department of Energy (DOE) Office of Science. Los Alamos National Laboratory, an affirmative action equal opportunity employer, is managed by Triad National Security, LLC for the U.S. Department of Energy's NNSA, under contract 89233218CNA000001. Sandia National Laboratories is a multimission laboratory managed and operated by National Technology & Engineering Solutions of Sandia, LLC, a wholly owned subsidiary of Honeywell International, Inc., for the U.S. DOE's National Nuclear Security Administration under contract DE-NA-0003525. The views expressed in the article do not necessarily represent the views of the U.S. DOE or the United States Government.

Conflicts of Interest: The authors declare no conflict of interest.

References

1. Was, G.S. Ion beam modification of metals: Compositional and microstructural changes. *Prog. Surf. Sci.* **1989**, *32*, 211–332. [CrossRef]
2. Moore, C.A.; Esfandiari, B. Geochemistry and Geology of Helium. *Adv. Geophys.* **1971**, *15*, 1–57. [CrossRef]
3. Mansur, L.; Lee, E.; Maziasz, P.; Rowcliffe, A. Control of helium effects in irradiated materials based on theory and experiment. *J. Nucl. Mater.* **1986**, *141*, 633–646. [CrossRef]
4. Odette, G.; Stoller, R. A theoretical assessment of the effect of microchemical, microstructural and environmental mechanisms on swelling incubation in austenitic stainless steels. *J. Nucl. Mater.* **1984**, *122*, 514–519. [CrossRef]
5. Odette, G.; Alinger, M.; Wirth, B. Recent Developments in Irradiation-Resistant Steels. *Annu. Rev. Mater. Res.* **2008**, *38*, 471–503. [CrossRef]

6. Kim, I.-S.; Hunn, J.D.; Hashimoto, N.; Larson, D.; Maziasz, P.; Miyahara, K.; Lee, E. Defect and void evolution in oxide dispersion strengthened ferritic steels under 3.2 MeV Fe$^+$ ion irradiation with simultaneous helium injection. *J. Nucl. Mater.* **2000**, *280*, 264–274. [CrossRef]

7. Chen, D.; Li, N.; Yuryev, D.; Baldwin, J.K.; Wang, Y.; Demkowicz, M. Self-organization of helium precipitates into elongated channels within metal nanolayers. *Sci. Adv.* **2017**, *3*, eaao2710. [CrossRef]

8. Johnson, J.R.; Swindle, T.D.; Lucey, P.G. Estimated solar wind-implanted helium-3 distribution on the Moon. *Geophys. Res. Lett.* **1999**, *26*, 385–388. [CrossRef]

9. Das, S. Recent advances in characterising irradiation damage in tungsten for fusion power. *SN Appl. Sci.* **2019**, *1*, 1614. [CrossRef]

10. Sandoval, L.; Perez, D.; Uberuaga, B.P.; Voter, A.F. An Overview of Recent Standard and Accelerated Molecular Dynamics Simulations of Helium Behavior in Tungsten. *Materials* **2019**, *12*, 2500. [CrossRef]

11. Xu, C.; Wang, W. Simulation Study of Helium Effect on the Microstructure of Nanocrystalline Body-Centered Cubic Iron. *Materials* **2018**, *12*, 91. [CrossRef] [PubMed]

12. Li, S.-H.; Li, J.-T.; Han, W.-Z. Radiation-Induced Helium Bubbles in Metals. *Materials* **2019**, *12*, 1036. [CrossRef] [PubMed]

13. El-Atwani, O.; Unal, K.; Cunningham, W.; Fensin, S.; Hinks, J.; Greaves, G.; Maloy, S. In-Situ Helium Implantation and TEM Investigation of Radiation Tolerance to Helium Bubble Damage in Equiaxed Nanocrystalline Tungsten and Ultrafine Tungsten-TiC Alloy. *Materials* **2020**, *13*, 794. [CrossRef]

14. Kim, H.; Wang, T.; Gigax, J.G.; Ukai, S.; Garner, F.; Shao, L. Effect of Helium on Dispersoid Evolution under Self-Ion Irradiation in A Dual-Phase 12Cr Oxide-Dispersion-Strengthened Alloy. *Materials* **2019**, *12*, 3343. [CrossRef] [PubMed]

15. Zhang, F.; Boatner, L.A.; Zhang, Y.; Chen, D.; Wang, Y.; Wang, L. Swelling and Helium Bubble Morphology in a Cryogenically Treated FeCrNi Alloy with Martensitic Transformation and Reversion after Helium Implantation. *Materials* **2019**, *12*, 2821. [CrossRef]

16. Fan, C.; Shang, Z.; Niu, T.; Li, J.; Wang, H.; Zhang, X. Dual Beam In Situ Radiation Studies of Nanocrystalline Cu. *Materials* **2019**, *12*, 2721. [CrossRef] [PubMed]

17. Taylor, C.A.; Briggs, S.A.; Greaves, G.; Monterrosa, A.; Aradi, E.; Sugar, J.D.; Robinson, D.B.; Hattar, K.; Hinks, J. Investigating Helium Bubble Nucleation and Growth through Simultaneous In-Situ Cryogenic, Ion Implantation, and Environmental Transmission Electron Microscopy. *Materials* **2019**, *12*, 2618. [CrossRef]

18. Höchbauer, T.; Misra, A.; Hattar, K.; Hoagland, R.G. Influence of interfaces on the storage of ion-implanted He in multilayered metallic composites. *J. Appl. Phys.* **2005**, *98*, 123516. [CrossRef]

19. Yang, W.; Pang, J.; Zheng, S.; Wang, J.; Zhang, X.; Ma, X. Interface Effects on He Ion Irradiation in Nanostructured Materials. *Materials* **2019**, *12*, 2639. [CrossRef]

20. Chen, H.; Du, J.; Liang, Y.; Wang, P.; Huang, J.; Zhang, J.; Zhao, Y.; Wang, X.; Zhang, X.; Wang, Y.; et al. Comparison of Vacancy Sink Efficiency of Cu/V and Cu/Nb Interfaces by the Shared Cu Layer. *Materials* **2019**, *12*, 2628. [CrossRef]

21. Su, Q.; Wang, T.; Gigax, J.; Shao, L.; Nastasi, M. Resistance to Helium Bubble Formation in Amorphous SiOC/Crystalline Fe Nanocomposite. *Materials* **2018**, *12*, 93. [CrossRef] [PubMed]

22. Fensin, S.; Jones, D.; Martinez, D.; Lear, C.; Payton, J. The Role of Helium on Ejecta Production in Copper. *Materials* **2020**, *13*, 1270. [CrossRef] [PubMed]

23. Fox, D.S.; Chen, Y.; Faulkner, C.C.; Zhang, H. Nano-structuring, surface and bulk modification with a focused helium ion beam. *Beilstein J. Nanotechnol.* **2012**, *3*, 579–585. [CrossRef] [PubMed]

Review

An Overview of Recent Standard and Accelerated Molecular Dynamics Simulations of Helium Behavior in Tungsten

Luis Sandoval [1], **Danny Perez** [1], **Blas P. Uberuaga** [2] **and Arthur F. Voter** [1,*]

[1] Theoretical Division T-1, Los Alamos National Laboratory, Los Alamos, NM 87545, USA
[2] Materials Science and Technology Division MST-8, Los Alamos National Laboratory, Los Alamos, NM 87545, USA
* Correspondence: afv@lanl.gov

Received: 1 July 2019; Accepted: 2 August 2019; Published: 7 August 2019

Abstract: One of the most critical challenges for the successful adoption of nuclear fusion power corresponds to plasma-facing materials. Due to its favorable properties in this context (low sputtering yield, high thermal conductivity, high melting point, among others), tungsten is a leading candidate material. Nevertheless, tungsten is affected by the plasma and fusion byproducts. Irradiation by helium nuclei, in particular, strongly modifies the surface structure by a synergy of processes, whose origin is the nucleation and growth of helium bubbles. In this review, we present recent advances in the understanding of helium effects in tungsten from a simulational approach based on accelerated molecular dynamics, which emphasizes the use of realistic parameters, as are expected in experimental and operational fusion power conditions.

Keywords: helium bubbles; tungsten; nucleation and growth

1. Introduction

Conditions expected in the divertor of the world's largest tokamak under construction, ITER ("The Way" in Latin, formerly known as International Thermonuclear Experimental Reactor), include the low-energy (≤ 100 eV) helium nuclei impact on a high-temperature tungsten surface (~ 1000 K) [1]. No tungsten defects are created upon impact because the maximum energy transferable from the collision is significantly below the tungsten displacement threshold [2]. The helium atoms diffuse in the tungsten matrix, eventually forming helium clusters with lower diffusion rates. At a critical size, a helium cluster converts into a practically immobile (under Molecular Dynamics (MD) time scales) entity, composed of the helium cluster and a tungsten Frenkel pair [3]. This helium (nano-)bubble is able to collect additional helium clusters, triggering the nucleation of additional Frenkel pairs. As a result, the bubble grows and the tungsten interstitials form a dislocation line pinned to the bubble, which eventually detaches as a loop, effectively displacing tungsten atoms in the matrix [4–6].

Recent simulation results, which are the focus of the present review, have shown that the process of the nucleation and growth of He bubbles in W exhibits extremely rich dynamics, involving competing mechanisms defined in an ample interval of time scales, which should be studied by considering advanced atomistic techniques beyond the standard MD approach. In doing so, we mostly focus on work from our own group where so-called Accelerated MD (AMD) techniques have been applied to the problem. This manuscript is therefore not meant to be an exhaustive review of the vast literature in the field, but to summarize our research effort over the last few years. First of all, we briefly review in Section 2 some results concerning the reflection and implantation of He atoms in W, which give information about the initial distribution of diffusing He atoms as a function of the impact energy, angle of incidence, surface orientation and temperature. In Section 3, the kinetics of He clusters is considered, with particular attention on the mobility of small clusters and the phenomenon of trap-mutation, as well as the mobility of small vacancy/helium complexes. The nucleation and growth of He bubbles in W is the focus of Section 4. In Section 5, we present some results from recent studies on the interaction between He clusters and He bubbles. Interaction between He bubbles and defects, specifically grain boundaries, is reviewed in Section 6. Finally, in Section 7, we discuss the advantages of the AMD techniques, as compared to standard MD, which have allowed us to gain fundamental insight into the nucleation and growth of helium bubbles in tungsten.

2. Reflection and Implantation of He Atoms

For the sake of completeness, we first discuss the reflection and implantation processes of He atoms in W. In a fusion reactor, the plasma irradiation flux is composed of deuterium, tritium, and helium. Contrary to the other two species, which may form a deposited layer, incoming He atoms are either reflected upon impact or implanted below the W surface. As a first approach to the implantation problem, we can ignore the interaction with deuterium and tritium, in order to focus on the behavior of He atoms.

Borovikov et al. [7] studied this topic via MD simulations, significantly extending the scope of previous computational efforts [8] by considering the effect of temperature in the substrate (300 K, 1000 K, and 1500 K), incidence energy E_i ($\leq$100 eV), deposition angles θ_i (0–75°), and substrate surface orientation ((100), (110), and (310)). The interaction between W atoms was determined by an Ackland–Thetford potential [9], modified at short distances by Juslin and Wirth [10]. He–W interactions were obtained from Juslin and Wirth [10], while the He–He potential corresponded to the one used by Beck [11] modified at short distances by Morishita et al. [12]. Incidentally, these potentials were also used to obtain all results discussed in the following sections. For each set of parameters considered, 1000 non-accumulative He impact simulations were performed, generating the statistics. The main focus was the determination of the reflection coefficient R (ratio of reflected He atoms to the total number of impacting He atoms), the energy reflection coefficient R_E (ratio of the kinetic energy of the reflected He atom to the incident kinetic energy of the He atom), and the average projected range L (implantation depth normal to the surface) for the implanted He atoms.

Given a substrate temperature and orientation and a deposition angle $\theta_i \leq 75°$, Borovikov et al. [7] showed that an increase of impact energy ($E_i > 10$ eV), in general, implies a decrease of R and R_E and an increase of L. For $\theta_i = 75°$, all the atoms were reflected. Low impact energies ($E_i \leq 10$ eV), even for small deposition angles, resulted in a reflection coefficient equal to one. They also observed that the dependence of particle/energy reflection coefficients with the incidence angle was non-monotonic, with a minimum at specific angles (e.g., $\theta_i \sim 30$–45° for a surface orientation of (100)), a behavior attributed to channeling effects. At low temperatures ($\sim$300 K), L increases substantially at $\theta_i = 0°$ due to the high probability of channeling events [13]. If channeling is absent, the implantation depth distributions can be well described by a Gaussian distribution [13], which, at low impact energies, is truncated at the surface, as shown in Figure 1. More recent works have extended these results by considering additional surface orientations [14], the effects of pre-existing bubbles in W surfaces [15], and curvature effects in W nanoparticles [16].

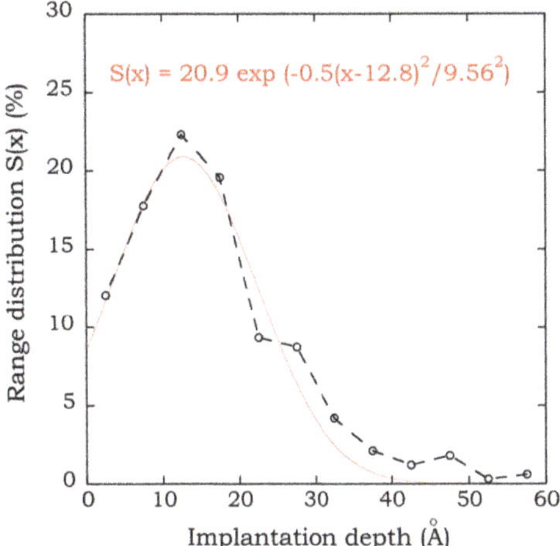

Figure 1. (Color online) He-implantation depth distribution for a (low) impact energy $E_i = 80$ eV, at a deposition angle $\theta_i = 0°$, on a W (100) surface equilibrated at 1000 K. A fit to a Gaussian distribution truncated at the surface is also shown. Taken from Borovikov et al. [7].

3. Kinetics of He Clusters

Once individual He atoms are implanted into the W matrix, their evolution is dictated by diffusion and clustering. In this section, we highlight some recent results concerning the evolution of interstitial He.

3.1. Mobility of Small He Clusters and Trap Mutation

As a He atom diffuses through the lattice, it either encounters traps (point defects, dislocations, etc.) or other He atoms. Clustering is favored because of the binding between He atoms generated by the elastic interactions caused by the repulsion between the metal atoms and He [17]. At a certain size, a growing He cluster can force the emission of W interstitials (crowdions), nucleating vacancies (V) that accommodate the He cluster. This process is known as trap mutation [18] or self-trapping [3].

Using conventional MD, Temperature-Accelerated Dynamics (TAD) [19], Statistical Temperature (STMD) [20], and multicanonical MD [21,22], Perez et al. [23] determined important thermodynamic and kinetic parameters describing the diffusion and transformation of small He clusters in W. They specifically investigated the behavior of He$_N$ clusters with N ranging from 2–7 He atoms, in an interval of temperatures relevant to nuclear fusion applications centered around 1000 K.

In order to find the most important low-temperature diffusion pathways, Perez et al. [23] performed TAD with low and high target temperatures of 300 K and 600 K, respectively. For a given cluster with size N, their simulations started with the lowest energy structure found for the cluster with size $N - 1$ plus one additional He interstitial. Once the two species encountered one another, forming a He cluster with size N, its diffusion was simulated. For each cluster, the lowest energy diffusion pathway, as well as the lowest energy structure was identified. With the exception of the case corresponding to $N = 5$, the ground state structure for all the clusters was found to be composed of He atoms located in tetrahedral interstices. For $N = 5$, the fifth He atom occupies an octahedral interstice; note that there are 24 tetrahedral and 18 octahedral positions within the BCC unit cell. As shown in Figure 2, the strength of the binding of the He cluster increases with the cluster size. If we consider the change of energy after removing a He atom from the clusters, the case $N = 5$ corresponds to a local maximum. Density Functional Theory (DFT) calculations [24] show a similar behavior, but with higher values, approximately 30%, as compared to the classical potential used in this work.

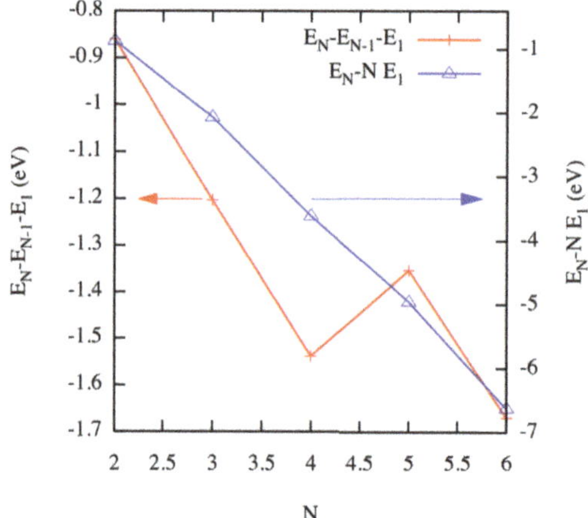

Figure 2. (Color online) Energy change after removing single atoms from a He cluster of size N ($E_N - E_{N-1} - E_1$; red, left axis) and after complete fragmentation into N single He atoms ($E_N - NE_1$; blue, right axis). These values correspond to $T = 0$ K. Taken from Perez et al. [23].

Using STMD and multicanonical simulations, Perez et al. [23] obtained finite-temperature effects on the stability of the clusters, specifically the canonical distributions of cluster compositions for all temperatures, which allowed determining the probabilities $p_Q(T)$ of finding certain cluster compositions Q. As an example, the case corresponding to $N = 4$ is shown in Figure 3. For temperatures $T \leq 2500$ K, the most notable cluster configuration is a cluster containing all four He atoms. With significant probability, single He atoms start to be observed at temperatures above 1500 K. On the other hand, the equilibrium probability of single He atoms becomes vanishingly small at low temperatures. Furthermore, using free energy data, Perez et al. [23] showed that the temperature at which He atoms separate from the clusters, or at which clusters are completely fragmented, grows with N, in accordance with the results obtained at $T = 0$ K. Figure 3 also provides a picture of the most probable dissociation scenario as a function of temperature. For instance, approximately between 1500 K and 2700 K, the four-atom cluster most likely transforms to a three-atom cluster plus a dissociated He atom, although the additional dissociation of He atoms starts to be more significant above 2000 K, probably involving intermediate steps. Additional insight is gained when we consider the volume-independent free energy of the clusters, which provides transition points for a given transformation.

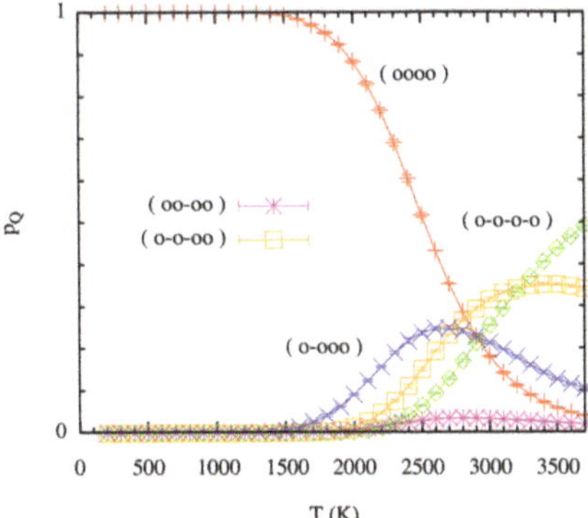

Figure 3. (Color online) Probability of finding a certain cluster distribution $p_Q(T)$ as a function of temperature for the case involving four He atoms ($N = 4$). Notation: (o–o–o–o) = four single atoms; (o–o–oo) = two single atoms plus a two-atom cluster; (oo–oo) = two two-atom clusters; (o–ooo) = one single atoms plus a three-atom cluster; (oooo) = one four-atom cluster. Taken from Perez et al. [23].

Available transition pathways are characterized via the Nudged Elastic Band (NEB) method [25], which is an inherent step in TAD. This analysis was performed by Perez et al. [23] for He cluster sizes ranging from $N = 1$–$N = 6$. For instance, for $N = 1$, the diffusion pathway goes from a tetrahedral interstice (lowest energy), through an octahedral interstice (saddle point), to another tetrahedral interstice position. The corresponding energy barrier was 0.15 eV. Note that this value is almost one order of magnitude higher than the one obtained via ab initio calculations [26], a result attributed to the limitations of the interatomic potential. The diffusion pathways become more complex (additional intermediate minima) as the He cluster size increases. As an example, Figure 4 shows the lowest energy migration pathway for $N = 4$.

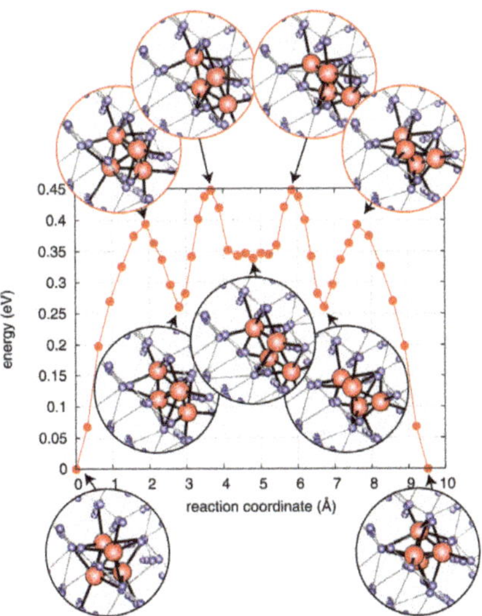

Figure 4. (Color online) Minimum Energy Path (MEP) for a He cluster containing four atoms. Small (blue) and large (red) spheres correspond to W and He, respectively. Taken from Perez et al. [23].

The corresponding migration energy [23] as a function of cluster size is presented in Figure 5. Coinciding with the unusual structure of the clusters highlighted previously, the cluster with $N = 5$ He atoms exhibits a very low diffusion barrier, indicating that its mobility is significantly higher.

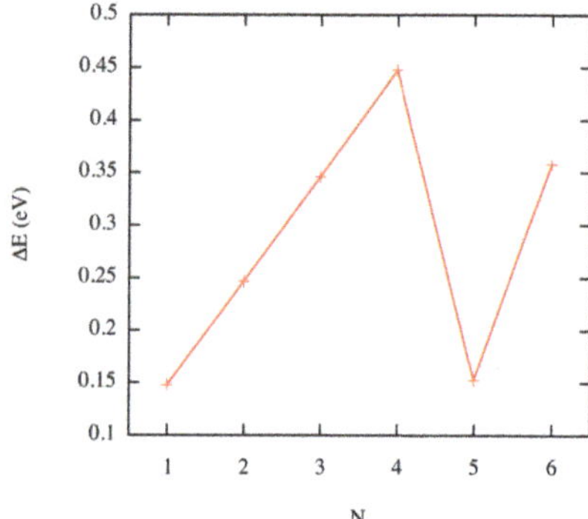

Figure 5. (Color online) Migration energies as a function of the He cluster size. Taken from Perez et al. [23].

The diffusivity of the different clusters, as a function of temperature, was obtained from the time-dependent Mean Squared Displacement (MSD) via a linear fit. As an example, in Figure 6, we show the results corresponding to $N = 4$ (similar behavior is seen for $N = 1,6$). For the cluster sizes considered in this study, the main result was the departure from a standard Arrhenius behavior. Additional insight is gained by considering what Perez et al. [23] termed Superbasin Harmonic Transition State Theory (SB-HTST), from which a generalized Arrhenius expression is obtained with an activation energy that becomes a function of temperature. In very good agreement with the MD results, SB-HTST predicts the downward bending of the slope of the Arrhenius curve as the temperature rises, due to the increase in the sampling frequency of higher-lying energy basins.

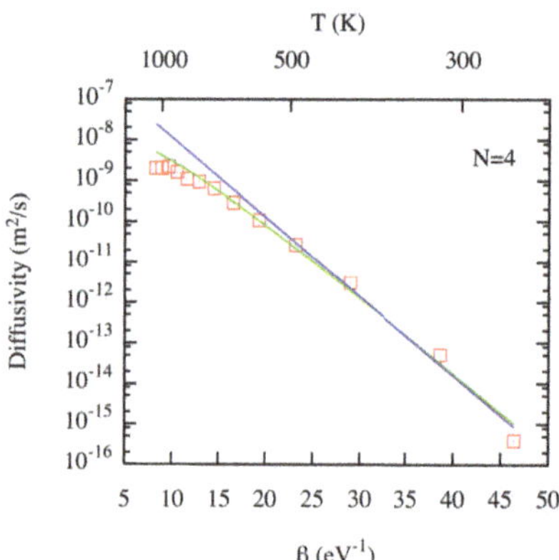

Figure 6. (Color online) Diffusivity as a function of $\beta = \frac{1}{k_B T}$ for a He cluster with size $N = 4$. Notation: red squares: MD results; blue line: Harmonic Transition State Theory (HTST); green line: Superbasin (SB)-HTST. Taken from Perez et al. [23].

On the other hand, even if the formation of He clusters is energetically favorable, configurational entropy effects favor isolated He atoms at low concentrations and high temperatures, which motivates the consideration of breakup reactions. After a careful definition of bound and unbound radii,

Perez et al. [23] directly determined the breakup rate of He clusters in a range of temperatures between 1000 K and 1500 K using MD. Their results indicated a fast decrease of the breakup rate (and an increase of activation barriers) as the He cluster size increases, which is shown in Figure 7, in agreement with the fact that the binding energy per He atom increases with the cluster size.

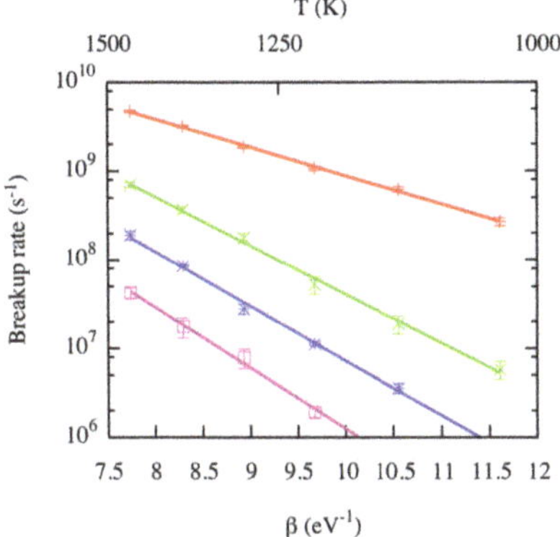

Figure 7. (Color online) Breakup rate as a function of β. Red $+$: $N = 2$; green $\times$: $N = 3$; blue $*$: $N = 4$; pink $\square$: $N = 5$. Lines correspond to Arrhenius fits. Taken from Perez et al. [23].

Concerning the trap mutation process, Perez et al. [23] also obtained the mutation rates as a function of temperature, which are shown in Figure 8. The mutation rate increases sharply with size, with a clear Arrhenius behavior over the range of temperatures considered. Other MD calculations [24] also exhibit an Arrhenius behavior, although the values reported for the characteristic times were approximately two orders of magnitude higher. As expected, the larger the cluster, the higher the mutation rate. Further compounding this effect is the fact that mutation is reversible at smaller N, i.e., that the W vacancy and W interstitial can recombine, restoring the He atoms to interstitial positions.

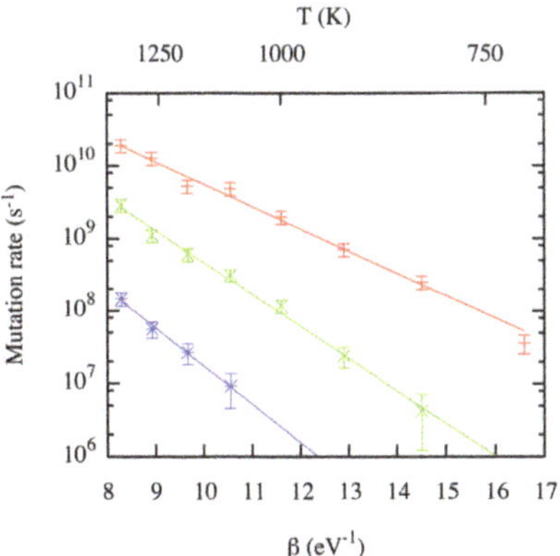

Figure 8. (Color online) Mutation rate as a function of β. Red $+$: $N = 7$; green $\times$: $N = 6$; blue $*$: $N = 5$. Lines corresponds to Arrhenius fits. Taken from Perez et al. [23].

The results presented in this section provide a comprehensive characterization of the kinetics of small interstitial He clusters in bulk W. The behavior close to surfaces has been also studied via MD simulations [27,28] and DFT calculations [29]. This detailed analysis of He cluster diffusion, breakup, and mutation into He bubbles can feed, e.g., cluster dynamics [30,31] or kinetic Monte Carlo [32,33] models able to describe the performance of W at the mesoscale. As an example, by recording the cluster pressure and using an object kinetic Monte Carlo code, Valles et al. [33] showed that the elastic strain energy in the interior of the grains of nanocrystalline W was significantly lower than the one calculated for monocrystalline W, indicating that nanocrystalline W has a better mechanical response under He irradiation.

3.2. Mobility of Small Vacancy/Helium Complexes

A further step in understanding the He damage in W is the study of the kinetics of small vacancy/helium ($V_N He_M$) complexes. After implantation, He atoms diffuse and form clusters, as discussed in the previous section. Eventually, these He clusters are capable of self-trapping via the nucleation of a Frenkel pair (W interstitial/vacancy recombination is prevented by He atoms filling the vacancy) or of binding to preexisting W vacancies. Perez et al. [34] performed AMD simulations via the Parallel trajectory Splicing (ParSplice) method to study the motion of the complexes involving one or two W vacancies, characterizing their diffusivity.

A first striking observation is the effect of a single He atom on the mobility of W vacancies. Perez et al. [34] found that adding a He atom to a single W vacancy ($N = 1$) causes the diffusivity to change by at least four orders of magnitude, from $\sim 10^{-12}$ m^2/s to a value below $\sim 10^{-16}$ m^2/s (with 90% confidence). Note that previous DFT calculations [35] provided explicit values for the change in migration energy (from 1.81 eV–4.83 eV), which explains the immobilization of the $V_1 He_1$ complex on ms timescales. Similar values were obtained for two and three He atoms, as shown in Figure 9.

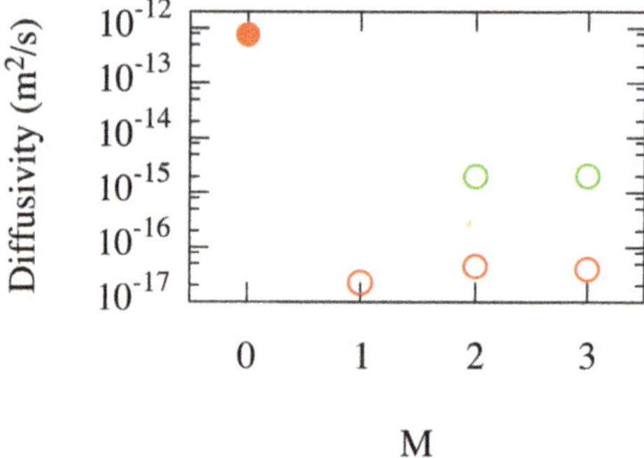

Figure 9. (Color online) Diffusivity of low He content $V_N He_M$ complexes at $T = 1000$ K. N and M denote the number of W vacancies and He atoms, respectively. Red: $N = 1$; green: $N = 2$. The filled circle symbol corresponds to a single W vacancy without He atoms. Open symbols are upper bounds at a 90% confidence level for cases where the complexes did not diffuse on accessible simulation timescales. Taken from Perez et al. [34].

A di-vacancy ($N = 2$), on the other hand, is weakly bounded. Over long times, the dimer breaks apart. In this case, the single He atom immobilizes just one vacancy. However, over longer time scales, the dimer may reform, eventually allowing the He atom to jump to the other vacancy. An additional He atom, as seen in Figure 9, immobilizes the dimer over tens of μs time scales.

An increase of the M-to-N ratio leads to He bubble growth via Frenkel pair nucleation [36,37]. Perez et al. [34] highlighted that the Frenkel pair nucleation process is reversible. Using ParSplice simulations, Perez et al. [34] found that the rates at which nucleation and annihilation (untrapping) occur are quite sensitive to He content: as M increases, the nucleation rate increases, while the annihilation rate decreases, as He pressure favors nucleation, but counteracts annihilation. Similar results for He in Fe have been obtained by Gao et al. [38] also using an AMD approach.

Annihilation is not restricted to the reverse process of the last nucleation event, but instead, many variants are possible due to the diffusion of W interstitial around the vacancies. Sequences of such nucleation/interstitial migration/annihilation events give rise to net migration of the cluster. Figure 10 shows that the diffusivity of complexes depends sensitively on the He content, through the dependence of the nucleation and annihilation rates.

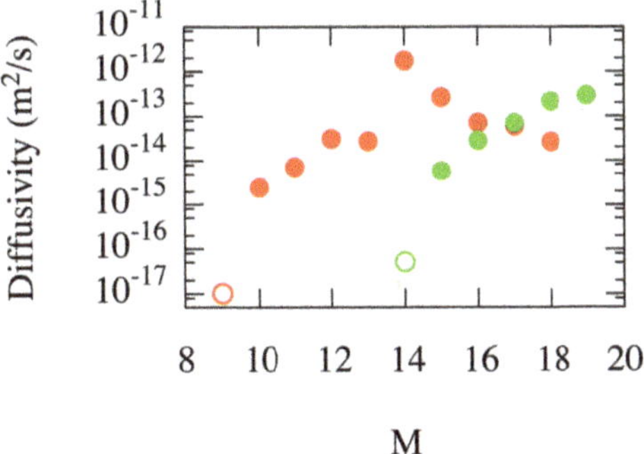

Figure 10. (Color online) Diffusivity of high He content $V_N He_M$ complexes at $T = 1000$ K as estimated from kinetic Monte Carlo simulations informed by ParSplice simulations. Red: $N = 1$; green: $N = 2$. Filled symbols correspond to estimates of the diffusivity, while open symbols are upper bounds at a 90% confidence level for cases where the complexes did not diffuse on accessible simulation timescales. Taken from Perez et al. [34].

In order to evaluate the importance of considering the mobility of such complexes, Perez et al. [34] performed mesoscale cluster dynamics simulations using the Xolotl model of microstructural evolution [30]. Figure 11 shows the overall retention of He atoms within the W wall when the diffusion of He/vacancy complexes is either allowed or forbidden. The results clearly showed that the mobility of these complexes provides an efficient outgassing pathway for He. These results also demonstrate that incorrect conclusions can be reached from standard MD simulations, as their limited time-scale restricts the set of events that can be observed. Surface effects on the stability of helium-vacancy complexes, on the other hand, have been recently studied via DFT calculations [39].

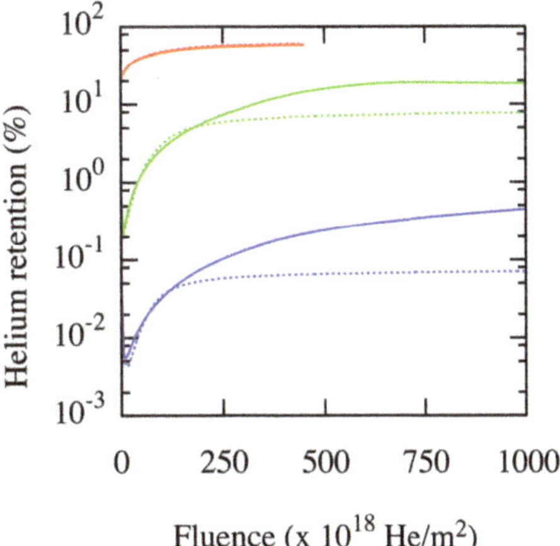

Figure 11. (Color online) He retention as a function of fluence. Red: 4×10^{25} He/m^2/s; green: 4×10^{23} He/m^2/s; blue: 4×10^{22} He/m^2/s. Continuous and dashed lines correspond to immobile and mobile complexes, respectively. Taken from Perez et al. [34].

4. Nucleation and Growth of He Bubbles

In this section, we focus on the growth of He bubbles after self-trapping, whereby additional He atoms are captured by helium/vacancy complexes, driving the nucleation of additional Frenkel pairs and the formation of dislocation lines around the bubbles.

Sandoval et al. [37] studied the growth of isolated He bubbles using AMD simulations in order to isolate the effect of the growth rate on the microstructural evolution. Starting with a W vacancy filled with eight He atoms, located 1.9 nm below the surface of the material, Sandoval et al. [37] carefully inserted He atoms at constant time intervals (the diffusion process of He atoms and their capture by the bubble were not explicitly considered). The growth rates spanned six orders of magnitude, from 10^{12}–2×10^6 He s^{-1}, which can be associated with He fluxes in the range of 10^{30}–10^{24} He m^{-2}s^{-1}, the last one being on the order of magnitude of fluxes expected at ITER [40].

Successive incorporation of He atoms in a bubble increases its pressure, which is the driving force for nucleation of Frenkel pairs, as has been shown by standard MD simulations [36]. The nucleation of Frenkel pairs increases the bubble size and partially releases the pressure in the bubble. Prismatic $\langle 111 \rangle$ dislocation loops, formed by the aggregation of W interstitials, are emitted from the bubble. Sandoval et al. [37] observed clear kinetic differences as a function of the growth rate. For instance, the first detected event (the nucleation of a Frenkel pair) required fewer He atoms at slower growth rates, as shown in Figure 12a.

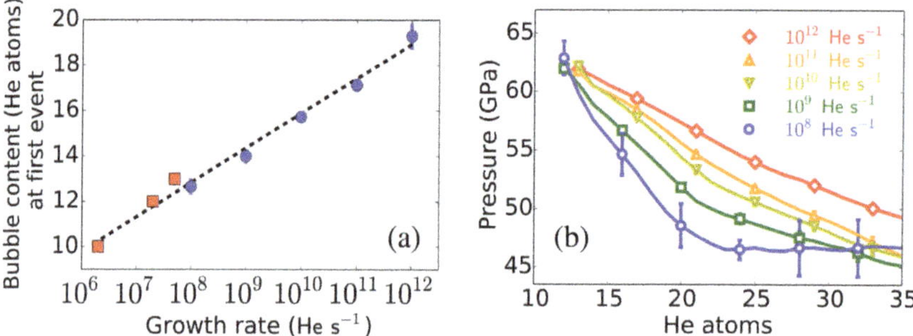

Figure 12. (Color online). Bubble growth as a function of the He insertion rate. (**a**) Average content of He atoms in the bubble at the time of the first detected event. Points with no error bars (red squares) are obtained from a single Parallel Replica Dynamics (ParRep) simulation. Points corresponding to rates $\geq 10^{11}$ He s^{-1} were obtained via standard MD simulations. (**b**) Average pressure in the He bubble as a function of the He content and growth rate. Taken from Sandoval et al. [37].

After nucleation of a Frenkel pair, the new W vacancy increases the bubble volume, while the W interstitial becomes a $\langle 111 \rangle$ crowdion tangent to the bubble. If the growth rate is slow enough, the crowdion is able to move around the bubble. Simulations show that subsequent interstitial emissions (driven by the constant addition of He) are most likely to occur in the neighborhood of interstitials already decorating the bubble, hence forming an incipient $\langle 111 \rangle$ dislocation line. For shallow bubbles, these dislocations elastically interact with the W surface, which drives their motion to the side of the bubble closer to the surface. This in turn favors subsequent interstitial emission from the surface-facing side of the bubble, which causes a preferential growth towards the surface. The dislocation arc eventually detaches from the bubble forming a $\langle 111 \rangle$ dislocation loop, which glides to the surface displacing W atoms and increasing the surface roughness. Note that, in general, as the growth rate is lowered, the probability of nucleating a new Frenkel pair before the next insertion increases, leading to lower pressure because of smaller helium/vacancy ratios, as seen in Figure 12b. This allows the definition of slow and fast growth regimes based on the speed at which W interstitials (crowdions) can diffuse around the bubble. At fast growth rates, typical of the ones commonly used in standard MD simulations, the emission of W interstitials is fast compared to their diffusion around the bubble. Therefore, dislocation lines grow where they are first nucleated, generating a more isotropic growth, delaying the bursting point as compared to the slow growth regime where emitted interstitials have time to diffuse around the bubble, facilitating the interaction with the surface and leading to a more directional growth process.

The growth of deeper and bigger He bubbles was studied by Sandoval et al. [41], following the approach described previously, where MD and AMD simulations were performed to explore the role of the growth rate, whose values spanned six orders of magnitude (10^6–10^{12} H s^{-1}). In this case, the bubble was created 6 nm below the surfaces. As observed in Sandoval et al. [37] for shallow bubbles, fast growth rates lead to higher pressures than slow growth rates. Between successive insertions, at slow growth rates, the system is able to explore its phase-space more thoroughly, eventually activating relaxation mechanisms such as the nucleation of Frenkel pairs, which allows for the release of the pressure. In addition, a remarkable difference is observed between simulations with different growth rates, corresponding to the number of coexisting dislocation lines attached to the bubble. As is shown in Figure 13, multiple dislocations were observed at fast growth rates, in contrast to only one at slower rates. This was again attributed to the insufficient time for the interstitials to reorganize around the bubble when the interstitial emission rate was fast compared to their diffusion rate.

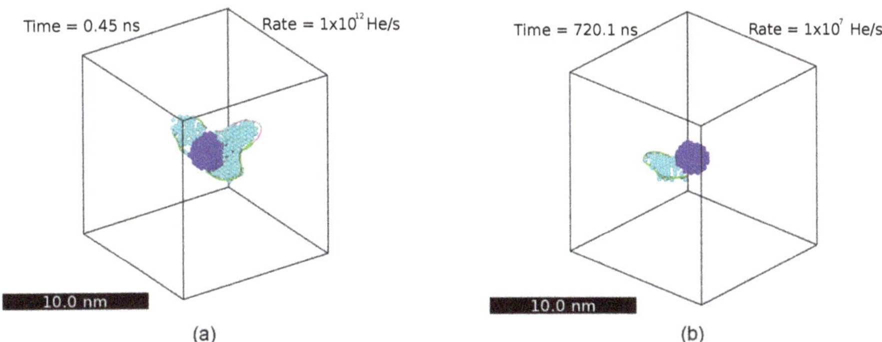

Figure 13. (Color online) Representative snapshots highlighting the dislocation loops attached to a growing He bubble for (**a**) fast and (**b**) slow growth rates. Dark and light spheres correspond to W vacancies and W interstitials, respectively. The detection of point defects is performed by the Wigner–Seitz defect analysis tool implemented in OVITO [42]. Taken from Sandoval et al. [41].

Using a Parallel Replica Dynamics (ParRep) [43] simulation, Sandoval et al. [41] showed that the nature of the dislocation structure around a growing bubble strongly depends on the growth rate. Specifically, they took, as the initial ParRep configuration, a dislocation structure composed of many arcs obtained from an MD simulation of a bubble growing at a fast rate. The entangled dislocation lines evolved to a single dislocation via a reorganization of W interstitials, which was then released from the bubble as a single loop.

Recent works have advanced this knowledge by studying the loop-punching mechanism for large bubbles at low temperatures (300 K) [44], the energetics and kinetics of He bubble growth for insertion rates in the fast growth regime ($\sim$30 ps between He insertions) [45], the lifetimes of non-growing He bubbles close to W surfaces [46], and the effect of strain fields [47]. MD and AMD results describing the bubble bursting process have also been incorporated into the cluster dynamics model Xolotl [31].

Including the diffusion process of He clusters to study the growth process of He bubbles adds interesting aspects to the simulation results, as shown by Sandoval et al. [48]. He monomers and dimers were introduced in a simulation box thermalized at different temperatures in the range [600, 2000] K. After nucleation of an initial He bubble and a W interstitial (crowdion) via self-trapping, incoming He clusters were allowed to diffuse freely and interact with them. The simulations showed that incoming He atoms strongly interact with the crowdion structure developed around a bubble, either impeding their incorporation into the bubble or even trapping them. If an He atom remains trapped in the crowdion structure for a sufficiently long time, it can eventually be joined by a newly-inserted He atom, forming a dimer. Sandoval et al. [48] also showed that the interaction with a He dimer is stronger, as compared to single He atoms. The simulations revealed that if the He-dimer, already trapped in the crowdion structure, captured two additional He atoms, the nucleation of a new He bubble via self-trapping was highly probable. Subsequently, as new bubbles grow, more crowdions are generated, which, in turn, can capture additional He atoms. A network of He (nano-)bubbles is formed, which is strongly dependent on the temperature, as shown in Figure 14. High temperatures reduce the probability that He atoms would remain trapped by the crowdions long enough to facilitate the nucleation of new He bubbles. Furthermore, at low He fluxes the trapped clusters would likely have time to de-trap from the crowdion structures, which affects the formation of the network.

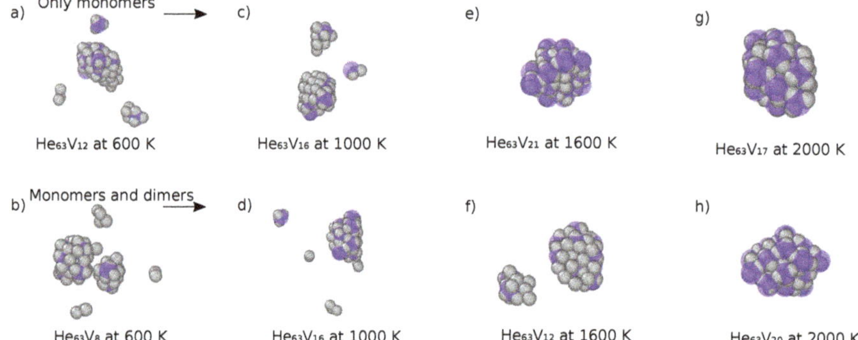

Figure 14. (Color online) Formation of He bubble networks. The top row shows snapshots after 55 He monomers have been inserted at a rate of 1×10^9 He s^{-1}. The bottom row shows snapshots after insertion of 49 He monomers, at the same rate, plus three He dimers inserted at $t = 10, 30,$ and 50 ns. Gray and blue spheres denote He atoms and W vacancies, respectively. Taken from Sandoval et al. [48].

5. Interaction between He-Clusters and He-Bubbles

Now, we extend the observations considered in the previous section to include the interaction between pre-existing bubbles and He clusters. Perez et al. [49] characterized the thermodynamics and kinetics of these interactions, showing that, in addition to attracting clusters, He bubbles also enhance the probability of trap-mutation in their neighborhood. In Figure 15a we show the simulation setup used by Perez et al. [49]. In a cubic simulation box containing 16,000 W atoms, a rhombic dodecahedral bubble with (110) facets was constructed by creating a void of 175 W vacancies, which was subsequently filled with 481 He atoms (in the original paper [49], the bubble geometry used was incorrectly described). The system was equilibrated at a temperature of $T = 1000$ K. Previous MD simulations indicated that the internal pressure was 26 GPa, a typical value for a growing bubble that has just released a dislocation loop. In addition to the large bubble, individual He$_N$ ($N = 1$–6) clusters are introduced in the simulation cell.

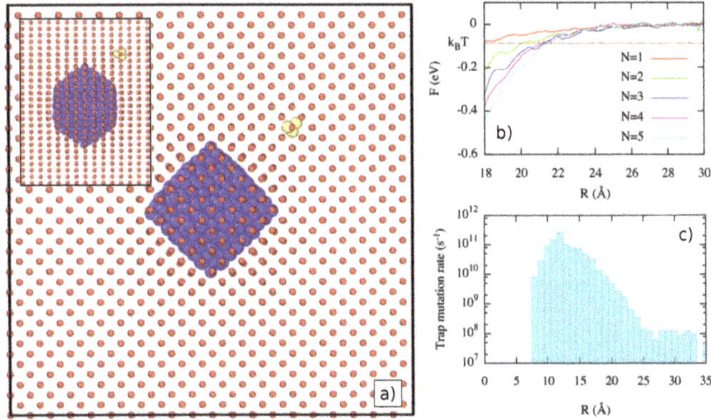

Figure 15. (Color online) Interaction of He clusters with He bubbles. (a) Simulation setup: W, substitutional He, and interstitial He atoms are shown as red, blue, and yellow spheres, respectively. The plot axes are aligned with [100] directions. An He cluster containing three atoms is placed in [110] orientation relative to the center of the bubble. The inset corresponds to a view from a [110] direction. (b) Free energy of He$_N$ clusters as a function of the distance to the center of the bubble. (c) Radially-averaged trap mutation rates as a function of the distance to the center of the bubble for $N = 5$. Taken from Perez et al. [49].

Using a direct histogram method, Perez et al. [49] obtained the free energy of He clusters at different positions around the bubble. The free energy gradually decreases at short range, as seen in Figure 15b, while the elastic interaction becomes stronger as the He cluster size increases. The capture distance goes from ~18 Å for $N = 1$–~22 Å for $N = 6$. Furthermore, the He cluster–bubble interaction presents a notable angular dependence because of the elastic anisotropy. He clusters are channeled along [100] directions, as approaches along [111] directions are strongly disfavored.

Although the incorporation of He clusters into existing bubbles is thermodynamically favored, there also are mechanisms that might obstruct the bubble growth. One of them is trap-mutation, where the cluster becomes a satellite nanobubble. Perez et al. [49] computed the trap mutation rates for He$_N$ clusters as a function of the position around the bubble. Their results clearly showed an enhancement of the trap mutation rate for distances going from 8 Å to 25 Å. Figure 15c shows the case corresponding to $N = 5$, where the asymptotic behavior to bulk values at long distances is clear, while at short distance, the trap-mutation rate increases by almost three orders of magnitude. Much like the interaction with the dislocations above, strains due to the surrounding microstructure can enhance the probability of self-trapping.

Using AMD simulations, Perez et al. [49] determined the role of the satellite nanobubbles in the bubble growth process. One interaction mechanism observed corresponds to the annihilation of the satellites, for $N = 1$ or 2 and short distances, by means of a W interstitial emitted by the bubble via a Frenkel pair nucleation. The nanobubble becomes a mobile interstitial cluster again, which is easily absorbed by the bubble. At longer distances, the annihilation of W vacancies in satellite bubbles can be activated when $\langle 111 \rangle$ dislocation loops emitted by the bubble impinge upon the nanobubble; this mechanism is more significant for small N. For larger values of N, we observed that transfer of He atoms to the bubble preceded annihilation. In particular, it was observed that, for $N \in (4, 5, 6)$, dislocation loops and nanobubbles were transiently bound, which facilitated the transfer of He to the bubble.

6. Interaction between He-Clusters/Bubbles and Defects

We have so far considered the nucleation and growth of He bubbles in initially pristine crystal lattices or in lattices with a controlled configuration of vacancies where He atoms have been carefully introduced. All crystal defects (vacancies, interstitials, and dislocations) were generated in the process of nucleation and growth. In this section, we highlight some recent results obtained via AMD simulations, which consider pre-existing defects, in particular Grain Boundaries (GBs).

Recently, Liu et al. [50] studied the process of He-bubble nucleation and growth in the neighborhood of a $\Sigma 5[100](310)$ tilt GB. This GB possesses a well-ordered structure, stable at temperatures typical of a fusion environment (1000 K in that work), as shown in Figure 16.

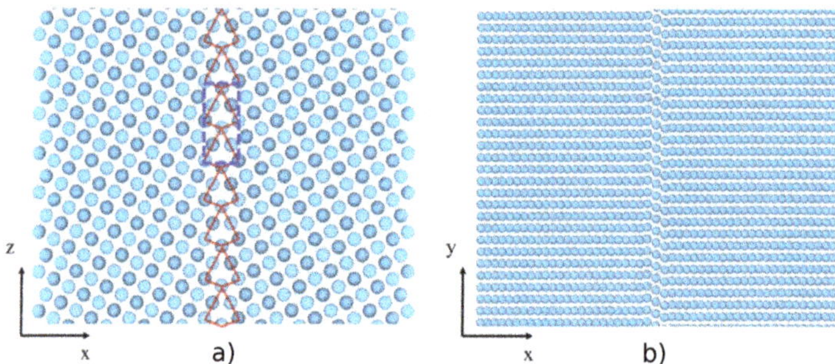

Figure 16. (Color online) Structure of the $\Sigma 5[100](310)$ tilt grain boundary in W. (**a**) Along the tilt axis. (**b**) along $[130]_{grain\,1}/[310]_{grain\,2}$. Taken from Liu et al. [50].

Liu et al. [50] used a simulation supercell for ParRep simulations initially containing 39,200 atoms, with periodic boundary conditions applied along the y and z directions. Similar to Sandoval et al. [37], eight He atoms were placed inside a pre-existing vacancy at the center of the GB plane. New He atoms were inserted inside the bubble at a rate of 1 He atom/10 ns, a value that corresponds to the slow growth regime in bulk conditions [37], close to what is expected at ITER for surface GBs [40]. During growth, W interstitials were emitted within the GB, and some of them escaped away from the nanobubble. As the bubble grew, it was observed that self-interstitial-atom loops were not formed, contrary to the behavior in the bulk, due to the strong binding of W interstitials to the GB plane. Eventually, a halo of W interstitials was formed around the bubble, trapped at the GB, as shown in Figure 17, corresponding to the final snapshot of the ParRep simulation.

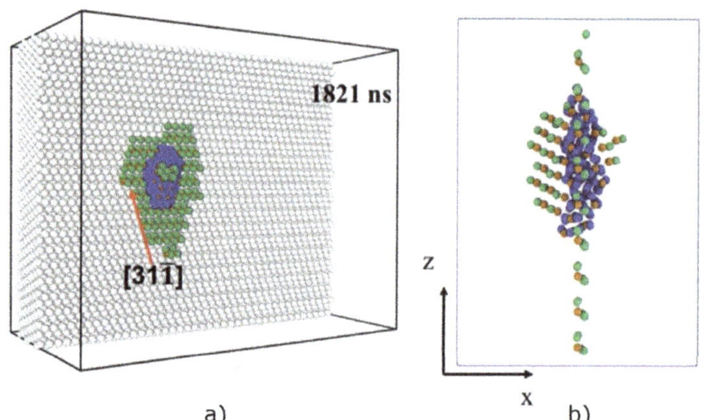

Figure 17. (Color online) Final atom configuration of a ParRep simulation of the He growth process of a bubble pre-nucleated at the grain boundary. (**a**) Perspective view showing the boundary of one of the grains. (**b**) Side view showing only the defects. White, orange, green, and blue spheres denote non-defective W atoms, W vacancies, W interstitials, and He atoms, respectively. Taken from Liu et al. [50].

In a second set of simulations, Liu et al. [50] evaluated the effect of the W interstitial halo on the arrival of a diffusing He atom. Snapshots corresponding to these simulations are shown in Figure 18. It was observed that the W halo attracts the incoming He atom, but prevents it from reaching the bubble, which suggests that the W halo may limit the diffusion of He atoms into the bubble, affecting the bubble growth process. This hypothesis was further supported by directly computing the diffusion barriers for He using climbing-image nudged elastic band calculations [25,51].

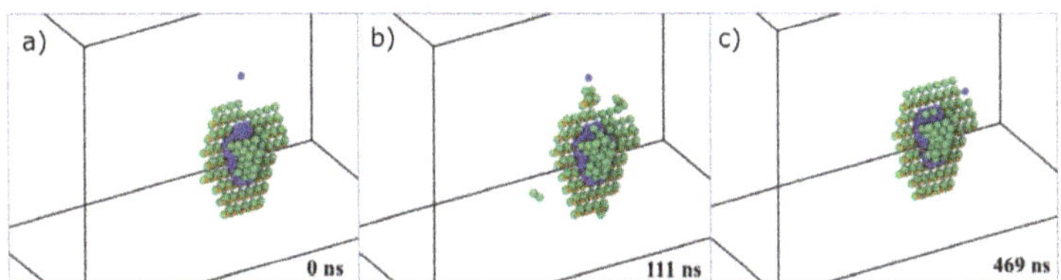

Figure 18. (Color online) Snapshots of the temporal evolution of the interaction of an excess He atom and a He bubble, both located at the grain boundary, obtained via ParRep simulations. (**a**) 0 ns, (**b**) 111 ns, and (**c**) 469 ns. White, orange, green, and blue spheres denote non-defective W atoms, W vacancies, W interstitials, and He atoms, respectively. Taken from Liu et al. [50].

In a last set of ParRep simulations, Liu et al. [50] introduced He atoms randomly into the GB at a rate of one He atom/100 ns. Inserting He atoms directly into the GB plane is supported by the fact that He atoms introduced into bulk W quickly migrate to the GB, as shown by recent simulations [27,30,52], where a sink segregation strength has been defined and calculated for some symmetric tilt grain boundaries, as well as for a few surface orientations. As in the bulk, He atoms migrating in the GB eventually encounter other He atoms, trap-mutating and nucleating bubbles. Again, contrary to the bulk case, interstitials emitted during the bubble growth process cannot escape, as previously described. As a result, for a given population of He atoms, a more evenly-dispersed set of He nanobubbles is expected at a GB, as compared to the bulk. After this saturation stage, new He atoms would arrive directly to the reconstructed GB region, re-initiating the bubble growth process. Further investigation is required to determine the bubble nucleation and growth process in other types of GBs to determine the generality of this bubble growth mode.

7. Discussion

The key defining characteristic of the AMD methods is their ability to extend the time scale of atomistic simulations to times further, in some cases much further, than conventional MD, while retaining near-full fidelity with the underlying atomistic interactions. This enables simulations that are simply impossible otherwise, probing behavior that is otherwise opaque. Many of the examples discussed in this overview exemplify this behavior, as we have elucidated atomic-scale mechanisms that would not be found if limited to MD time scales.

Just as importantly, other methods for accelerating time scales, such as Adaptive Kinetic Monte Carlo (AKMC) [53,54], would be challenging, if not impossible, to apply to some of the scenarios discussed here. In ParRep, the definition of a state can be made very generally, and this was particularly valuable when simulating bubbles, where the dynamics of the fast He atoms could be ignored. Methods such as AKMC and, to be clear, TAD, which require explicitly finding saddle points for all events, would not be applicable to systems with such fast moving degrees of freedom. Thus, while those methods are certainly powerful and provided new insight into some of the problems described here, there are limitations in using those methods in simulations involving gas bubbles.

For example, in scenarios in which the He gas is still in the W matrix, TAD proved effective at describing the kinetics of the resulting interstitial clusters. These TAD simulations revealed complex behavior versus cluster size that would be difficult to guess a priori. However, once gas bubbles nucleate, methods such as TAD would simply be impossible to apply. Using ParRep, where the states could be defined based on the W subsystem, the types of simulations can be extended significantly. In particular, ParRep enabled long-time simulations of bubbles in multiple environments with varying types of boundary conditions, providing new insight into how bubbles grow.

The most important consequence of extending the time scale of these simulations is that new mechanisms start to dominate. In most of the examples discussed here, there are multiple atomic-scale mechanisms with different rates that compete to drive the evolution of the material. If the time scale is not extended, the external drive applied to the system, in this case, the introduction of He, has to be so fast that other processes simply cannot occur on the time scale of the simulation. For example, if He is introduced too fast into a growing bubble, the emitted W interstitials are simply frozen in place, and the bubble grows qualitatively differently than if those interstitials can migrate and interact with the surrounding microstructure. Similarly, if the bubbles are grown more naturally, via the arrival of He from the matrix, the rate of He introduction changes whether the He reaches the central bubbles or not. In the case of the migration of He-V clusters, there is no external drive, but extending the time scale of the simulation allowed for both Frenkel pair formation and, critically, annihilation. If that annihilation had not been observed, neither would the net migration of the cluster. Thus, we see repeatedly that the observed behavior strongly depends on which mechanisms are active over the course of the simulation.

It should be noted that AMD methods are not a panacea for all simulations at the atomic scale. There are cases where AMD approaches will not provide any significant benefit over conventional MD. For example, the dynamics of the He within the bubbles cannot be accelerated as their dynamics are simply so fast. In the cases described here, this was not a critical limitation as the W dynamics dictate the system evolution. However, if all of the key behavior was related to the dynamics of the fast moving system, the AMD methods would provide only marginal benefit. As a general rule, what is already fast cannot be significantly accelerated.

Another limitation involves the system size. As the system size increases, so does the computational cost. Depending on the nature of the scaling, the AMD methods scale worse than MD, the worst situation being when the overall rate of events increases with system size. Thus, the simulations presented here are limited in how deep the bubbles are below the surface, a direct consequence of the need for larger system sizes and the issue with scaling. There are developments to overcome these limitations, and significant progress has been made; however, this is one of the ongoing challenges in applying these methods.

However, in spite of these limitations, the insights gained from these simulations extend our knowledge of how gases behave in metals, not only for fusion-relevant conditions, but more generally. They provide direct atomic-level information about key mechanisms that can be directly used in higher level models and have been shown to significantly change predictions at the meso-scale. In conjunction with conventional MD simulations, atomic scale simulations using AMD methods can provide high fidelity information about the atomistic mechanisms that drive material evolution in harsh conditions where kinetics are all-important.

8. Conclusions

We have reviewed some recent results concerning the modeling and simulation of helium effects in tungsten in the context of nuclear fusion power, with special focus on the application of accelerated molecular dynamics methods. We have considered the implantation process of helium atoms in the tungsten matrix, the subsequent diffusion of helium clusters, the process of nucleation and growth of helium bubbles, the interaction between helium clusters and bubbles, and the role of tungsten defects (interstitials, dislocations, and grain boundaries). A manifest outcome of this review corresponds to the importance of taking into account an extended range of time scales in order to include the dominant mechanisms determining the evolution of the surface microstructure in tungsten as a plasma-facing material.

Funding: This research was funded by the U.S. DOE, Office of Science, Office of Fusion Energy Sciences, and Office of Advanced Scientific Computing Research through the Scientific Discovery through Advanced Computing (SciDAC) project on Plasma-Surface Interactions. A.F.V. was supported by the U.S. DOE, Office of Basic Energy Sciences, Materials Sciences and Engineering Division. This research used resources of the National Energy Research Scientific Computing Center, which is supported by the Office of Science of the U.S. DOE under Contract No. DE-AC02-05CH11231, and the resources of the Oak Ridge Leadership Computing Facility at Oak Ridge National Laboratory, which is supported by the Office of Science of the U.S. DOE under Contract No. DE-AC05-00OR22725. Los Alamos National Laboratory is operated by Los Alamos National Security, LLC, for the National Nuclear Security Administration of the U.S. DOE, under Contract No. DE-AC52-O6NA25396.

Conflicts of Interest: The authors declare no conflict of interest.

References

1. Valles, G.; Martin-Bragado, I.; Nordlund, K.; Lasa, A.; Björkas, C.; Safi, E.; Perlado, J.M.; Rivera, A. Temperature dependence of underdense nanostructure formation in tungsten under helium irradiation. *J. Nucl. Mater.* **2017**, *490*, 108–114. [CrossRef]

2. Lhuillier, P.E.; Belhabib, T.; Desgardin, P.; Courtois, B.; Sauvage, T.; Barthe, M.F.; Thomann, A.L.; Brault, P.; Tessier, Y. Helium retention and early stages of helium-vacancy complexes formation in low energy helium-implanted tungsten. *J. Nucl. Mater.* **2013**, *433*, 305–313. [CrossRef]

3. Wilson, W.D.; Bisson, C.L.; Baskes, M.I. Self-trapping of helium in metals. *Phys. Rev. B* **1981**, *24*, 5616. [CrossRef]

4. Baštecká, J. Interaction of dislocation loop with free surface. *Cechoslovackij Fiziceskij Zurnal B* **1964**, *14*, 430–442.

5. Groves, P.P.; Bacon, D.J. The dislocation loop near a free surface. *Philos. Mag.* **1970**, *22*, 83–91. [CrossRef]

6. Ohr, S.M. Elastic fields of a dislocation loop near a stress-free surface. *J. Appl. Phys.* **1978**, *49*, 4953–4955. [CrossRef]

7. Borovikov, V.; Voter, A.F.; Tang, X.Z. Reflection and implantation of low energy helium with tungsten surfaces. *J. Nucl. Mater.* **2014**, *447*, 254–270. [CrossRef]

8. Li, M.; Wang, J.; Hou, Q. Molecular dynamics studies of temperature effects on low energy helium bombardments on tungsten surfaces. *J. Nucl. Mater.* **2012**, *423*, 22–27. [CrossRef]

9. Ackland, G.J.; Thetford, R. An improved N-body semi-empirical model for body-centred cubic transition metals. *Philos. Mag. A* **1987**, *56*, 15–30. [CrossRef]

10. Juslin, N.; Wirth, B.D. Interatomic potentials for simulation of He bubble formation in W. *J. Nucl. Mater.* **2013**, *432*, 61–66. [CrossRef]

11. Beck, D.E. A new interatomic potential function for helium. *Mol. Phys.* **1968**, *15*, 311–315. [CrossRef]

12. Morishita, K.; Sugano, R.; Wirth, B.D.; Díaz de la Rubia, T. Thermal stability of helium-vacancy clusters in iron. *Nucl. Instrum. Meth. B* **2003**, *202*, 76–81. [CrossRef]

13. Nastasi, M.; Mayer, J.; Hirvonen, J.K. *Ion-Solid Interactions*; Cambridge University Press: Cambridge, UK, 1996.

14. Hammond, K.D.; Wirth, B.D. Crystal orientation effects on helium ion depth distributions and adatom formation processes in plasma-facing tungsten. *J. Appl. Phys.* **2014**, *116*, 143301. [CrossRef]

15. Ding, Y.; Ma, C.; Li, M.; Hou, Q. Molecular dynamics study on the interactions between helium projectiles and helium bubbles pre-existing in tungsten surfaces. *Nucl. Instrum. Meth. B* **2016**, *368*, 50–59. [CrossRef]

16. Li, M.; Hou, Q.; Cui, J.; Wang, J. A molecular dynamics study of helium bombardments on tungsten nanoparticles. *Nucl. Instrum. Meth. B* **2018**, *425*, 43–49. [CrossRef]

17. Becquart, C.S.; Domain, C. Solute–point defect interactions in bcc systems: Focus on first principles modelling in W and RPV steels. *Curr. Opin. Solid State Mater. Sci.* **2012**, *16*, 115–125. [CrossRef]

18. van Veen, A.; Caspers, L.M.; Kornelsen, E.V.; Fastenau, R.; van Gorkum, A.; Warnaar, A. Vacancy creation by helium trapping at substitutional krypton in tungsten. *Phys. Status Solidi A* **1977**, *40*, 235–246. [CrossRef]

19. Sørensen, M.R.; Voter, A.F. Temperature-accelerated dynamics for simulation of infrequent events. *J. Chem. Phys.* **2000**, *112*, 9599–9606. [CrossRef]

20. Kim, J.; Straub, J.E.; Keyes, T. Statistical-Temperature Monte Carlo and Molecular Dynamics Algorithms. *Phys. Rev. Lett.* **2006**, *97*, 50601. [CrossRef]

21. Hansmann, U.H.E.; Okamoto, Y.; Eisenmenger, F. Molecular dynamics, Langevin and hybrid Monte Carlo simulations in a multicanonical ensemble. *Chem. Phys. Lett.* **1996**, *259*, 321–330. [CrossRef]

22. Junghans, C.; Perez, D.; Vogel, T. Molecular Dynamics in the Multicanonical Ensemble: Equivalence of Wang–Landau Sampling, Statistical Temperature Molecular Dynamics, and Metadynamics. *J. Chem. Theory Comput.* **2014**, *10*, 1843–1847. [CrossRef] [PubMed]
23. Perez, D.; Vogel, T.; Uberuaga, B.P. Diffusion and transformation kinetics of small helium clusters in bulk tungsten. *Phys. Rev. B* **2014**, *90*, 014102. [CrossRef]
24. Boisse, J.; De Backer, A.; Domain, C.; Becquart, C.S. Modeling of the self trapping of helium and the trap mutation in tungsten using DFT and empirical potentials based on DFT. *J. Mater. Res.* **2014**, *29*, 2374–2386. [CrossRef]
25. Henkelman, G.; Uberuaga, B.P.; Jónsson, H. A climbing image nudged elastic band method for finding saddle points and minimum energy paths. *J. Chem. Phys.* **2000**, *113*, 9901–9904. [CrossRef]
26. Becquart, C.S.; Domain, C. Migration Energy of He in W Revisited by Ab Initio Calculations. *Phys. Rev. Lett.* **2006**, *97*, 196402. [CrossRef] [PubMed]
27. Hu, L.; Hammond, K.D.; Wirth, B.D.; Maroudas, D. Interactions of mobile helium clusters with surfaces and grain boundaries of plasma-exposed tungsten. *J. Appl. Phys.* **2014**, *115*, 173512. [CrossRef]
28. Hu, L.; Hammond, K.D.; Wirth, B.D.; Maroudas, D. Dynamics of small mobile helium clusters near tungsten surfaces. *Surf. Sci.* **2014**, *626*, L21–L25. [CrossRef]
29. Wan, C.; Yu, S.; Ju, X. Energetics of small helium clusters near tungsten surface by ab initio calculations. *J. Nucl. Mater.* **2018**, *499*, 539–545. [CrossRef]
30. Maroudas, D.; Blondel, S.; Hu, L.; Hammond, K.D.; Wirth, B.D. Helium segregation on surfaces of plasma-exposed tungsten. *J. Phys. Condens. Matter* **2016**, *28*, 064004. [CrossRef]
31. Blondel, S.; Bernholdt, D.E.; Hammond, K.D.; Wirth, B.D. Continuum-scale modeling of helium bubble bursting under plasma-exposed tungsten surfaces. *Nucl. Fusion* **2018**, *58*, 126034. [CrossRef]
32. Martin-Bragado, I.; Rivera, A.; Valles, G.; Gomez-Selles, J.L.; Caturla, M.J. MMonCa: An Object Kinetic Monte Carlo simulator for damage irradiation evolution and defect diffusion. *Comput. Phys. Commun.* **2013**, *184*, 2703–2710. [CrossRef]
33. Valles, G.; González, C.; Martin-Bragado, I.; Iglesias, R.; Perlado, J.M.; Rivera, A. The influence of high grain boundary density on helium retention in tungsten. *J. Nucl. Mater.* **2015**, *457*, 80–87. [CrossRef]
34. Perez, D.; Sandoval, L.; Blondel, S.; Wirth, B.D.; Uberuaga, B.P.; Voter, A.F. The mobility of small vacancy/helium complexes in tungsten and its impact on retention in fusion-relevant conditions. *Sci. Rep.* **2017**, *7*, 2522. [CrossRef] [PubMed]
35. González, C.; Iglesias, R. Migration mechanisms of helium in copper and tungsten. *J. Mater. Sci.* **2014**, *49*, 8127–8139. [CrossRef]
36. Sefta, F.; Hammond, K.D.; Juslin, N.; Wirth, B.D. Tungsten surface evolution by helium bubble nucleation, growth and rupture. *Nucl. Fusion* **2013**, *53*, 073015. [CrossRef]
37. Sandoval, L.; Perez, D.; Uberuaga, B.P.; Voter, A.F. Competing kinetics and He bubble morphology in W. *Phys. Rev. Lett.* **2015**, *114*, 105502. [CrossRef] [PubMed]
38. Gao, N.; Yang, L.; Gao, F.; Kurtz, R.J.; West, D.; Zhang, S. Long-time atomistic dynamics through a new self-adaptive accelerated molecular dynamics method. *J. Phys. Condens. Matter* **2017**, *29*, 145201. [CrossRef]
39. Yang, L.; Bergstrom, J.; Wirth, B.D. First-principles study of stability of helium-vacancy complexes below tungsten surfaces. *J. Appl. Phys.* **2018**, *123*, 205108. [CrossRef]
40. De Temmerman, G.; Bystrov, K.; Doerner, R.P.; Marot, L.; Wright, G.M.; Woller, K.B.; Whyte, D.G. Helium effects on tungsten under fusion-relevant plasma loading conditions. *J. Nucl. Mater.* **2013**, *438*, S78–S83. [CrossRef]
41. Sandoval, L.; Perez, D.; Uberuaga, B.P.; Voter, A.F. Growth Rate Effects on the Formation of Dislocation Loops Around Deep Helium Bubbles in Tungsten. *Fusion Sci. Technol.* **2017**, *71*, 1–6. [CrossRef]
42. Stukowski, A. Visualization and analysis of atomistic simulation data with OVITO–the Open Visualization Tool. *Model. Simul. Mater. Sci. Eng.* **2010**, *18*, 015012. [CrossRef]
43. Voter, A.F. Parallel replica method for dynamics of infrequent events. *Phys. Rev. B* **1998**, *57*, R13985–R13988. [CrossRef]
44. Xie, H.; Gao, N.; Xu, K.; Lu, G.H.; Yu, T.; Yin, F. A new loop-punching mechanism for helium bubble growth in tungsten. *Acta Mater.* **2017**, *141*, 10–17. [CrossRef]
45. Wang, J.; Niu, L.L.; Shu, X.; Zhang, Y. Energetics and kinetics unveiled on helium cluster growth in tungsten. *Nucl. Fusion* **2015**, *55*, 092003. [CrossRef]

46. Cui, J.; Wu, Z.; Hou, Q. Estimation of the lifetime of small helium bubbles near tungsten surfaces—A methodological study. *Nucl. Instrum. Meth. B* **2016**, *383*, 136–142. [CrossRef]
47. Cui, M.; Gao, N.; Wang, D.; Gao, X.; Wang, Z. Helium bubble growth under strain fields in tungsten investigated by atomic method. *Nucl. Instrum. Meth. B* **2019**, in press. [CrossRef]
48. Sandoval, L.; Perez, D.; Uberuaga, B.P.; Voter, A.F. Formation of helium-bubble networks in tungsten. *Acta Mater.* **2018**, *159*, 46–50. [CrossRef]
49. Perez, D.; Sandoval, L.; Uberuaga, B.P.; Voter, A.F. The thermodynamic and kinetic interactions of He interstitial clusters with bubbles in W. *J. Appl. Phys.* **2016**, *119*, 203301. [CrossRef]
50. Liu, X.Y.; Uberuaga, B.P.; Perez, D.; Voter, A.F. New helium bubble growth mode at a symmetric grain-boundary in tungsten: accelerated molecular dynamics study. *Mater. Res. Lett.* **2018**, *6*, 522–530. [CrossRef]
51. Henkelman, G.; Jónsson, H. Improved tangent estimate in the nudged elastic band method for finding minimum energy paths and saddle points. *J. Chem. Phys.* **2000**, *113*, 9978–9985. [CrossRef]
52. Yang, Z.; Hu, L.; Maroudas, D.; Hammond, K.D. Helium segregation and transport behavior near $\langle 100 \rangle$ and $\langle 110 \rangle$ symmetric tilt grain boundaries in tungsten. *J. Appl. Phys.* **2018**, *123*, 225104. [CrossRef]
53. Henkelman, G.; Jónsson, H. Long time scale kinetic Monte Carlo simulations without lattice approximation and predefined event table. *J. Chem. Phys.* **2001**, *115*, 9657–9666. [CrossRef]
54. Xu, L.; Henkelman, G. Adaptive kinetic Monte Carlo for first-principles accelerated dynamics. *J. Chem. Phys.* **2008**, *129*, 114104. [CrossRef]

Article

Simulation Study of Helium Effect on the Microstructure of Nanocrystalline Body-Centered Cubic Iron

Chunping Xu [1,*,†] and Wenjun Wang [2,†]

1   College of Nuclear Equipment and Nuclear Engineering, Yantai University, Yantai 264005, China
2   College of Engineering, Shanxi Agriculture University, Taigu 030801, China; wwjbuoy03@163.com
*   Correspondence: xuchunping@ytu.edu.cn
†   These authors contributed equally to this work.

Received: 20 November 2018; Accepted: 25 December 2018; Published: 28 December 2018

Abstract: Helium (He) effect on the microstructure of nanocrystalline body-centered cubic iron (BCC-Fe) was studied through Molecular Dynamics (MD) simulation and simulated X-ray Diffraction (XRD). The crack generation and the change of lattice constant were investigated under a uniaxial tensile strain at room temperature to explore the roles of He concentration and distribution played in the degradation of mechanical properties. The simulation results show that the expansion of the lattice constant decreases and the swelling rate increases while the He in the BCC region diffuses into the grain boundary (GB) region. The mechanical property of nanocrystalline BCC-Fe shows He concentration and distribution dependence, and the existence of He in GB is found to benefit the generation and growth of cracks and to affect the strength of GB during loading. It is observed that the reduction of tensile stress contributed by GB He is more obvious than that contributed by grain interior He.

Keywords: molecular dynamics; Helium effects; crack formation; nanocrystalline BCC iron; mechanical properties

---

1. Introduction

Both displacement damage and production of foreign elements can be introduced in the structural materials under nuclear radiation environment. Point defects and foreign elements generated during neutron irradiation evolve into interstitial clusters, bubbles and voids that lead to severe degradation of mechanical properties, such as hardening, embrittlement, and fracture, void swelling and irradiation creep [1]. Due to extremely low solubility of He in any metal, He as a foreign element in nuclear materials aggregates to form clusters and bubbles near the grain boundary (GB) [2,3] which eventually link up and cause intergranular failure [1,4]. It has been established in theory that the strength of GB is reduced by He or He clusters [5–7]. Studies have shown that GB acts as a neutral and unsaturable sink to radiation defects [8,9], or a source, emitting interstitials to annihilate vacancies [10]. Thus, the effects of GB and He on the nuclear material with nanocrystalline structure are widely studied to either understand the mechanisms of radiation effects or solve some practical problems such as the reduction of He embrittlement, void swelling and so on.

Improved radiation tolerance in nanostructure alloy with a high area fraction of GBhas been reported in many experimental studies. A significant grain size effect on radiation tolerance was observed in nanocrystalline body-centered cubic Mo and Fe under He ion irradiation [11,12]. The density and size of defects (dislocation loops, He bubbles) showed a grain size dependence under irradiation [11,12]. The He embrittlement might have been reduced due to the high number density of nanoclusters which acted as effective sinks to radiation defects in nanostructured ferritic alloys (NFS) [13,14]. The behavior of He on microstructure evolution in nuclear materials with GB has been investigated by Molecular Dynamic (MD) simulation and theoretical methods. The study on the nucleation and growth of He clusters in α-Fe GB [15] indicates that the interstitial He atoms tend to combine with nearby monovacancies or He_nV clusters to form larger He_nV clusters. The accumulation of He atoms and the evolution of GB structure are affected by local He concentration, temperature, the original GB structure, or their combination in α-Fe [16]. It is demonstrated that the GB tensile strength is reduced by the He segregation which breaks (substitution) or weakens (interstitial) the surrounding interfacial Fe-Fe bonds [5]. The impact of GBs on the clustering properties of He and the possible effect of He on GB decohesion were investigated in references [6,17]. Additionally, the deformation mechanisms in nano-twinned and nanocrystalline materials have been carried out by MD simulations [17–20], and intergranular crack are reported in those studies.

These studies have provided valuable insights into the He effect on the mechanism of radiation tolerance in nuclear material with GB, but there are few reports about the He effects on the microstructure of nanocrystalline body-centered cubic iron (BCC-Fe). In this work, MD simulations were employed to study the He effects on generation and growth of cracks in nanocrystalline BCC-Fe. The evolution on microstructure of nanocrystalline BCC-Fe under a uniaxial tensile strain condition was investigated through simulated X-ray Diffraction (XRD) patterns. Based on the simulation results, the relationship between the change of microstructure and the degradation of mechanical property during loading is discussed and analyzed.

2. Simulation Method

The model of nanocrystalline BCC-Fe is constructed for simulation using the Voronoi tessellation method [21] with random Euler angles assigned to each grain (see Figure 1). The average grain size of the model with volume of $(80 \times a)^3 Å^3$ is 8.28 ± 0.32 nm. The lattice constant of the model, a, equals 2.8553Å. The size of grain is represented by the average diameter of grain volume. The simulation is performed using the LAMMPS [22] and the interatomic EAM potential for BCC-Fe developed by Mendelev and his co-worker [23]. Any overlapping atoms that are separated by less than 1.7Å are deleted from the nanocrystalline model and then the final structure with periodic boundary conditions is relaxed using conjugate gradient minimization with an energy tolerance of 10^{-5} eV/Å. There are 40 grains in the resulting structure with 1,001,493 atoms.

Figure 1. Initial model of nanocrystalline BCC-Fe, the atom was colored according its grain ID (identification).

In the picture of pure BCC-Fe model with minimized nanocrystalline structure (Figure 2a), it shows that the potential energy of the iron atom in GB region is slightly larger than that in grain interior region according to the color map. The percentage of iron atoms in the GB region with non-BCC structure, about 18.3%, is calculated by the adaptive common neighbor analysis (ACNA) algorithm. Models of nanocrystalline BCC-Fe with homogeneous and intergranular distribution of He are considered in the present work, because an He atom is introduced randomly in the nuclear structure material under radiation environment, and it tends to form clusters at GBs based on studies from both experiment [11] and theoretical simulation [6]. To clarify the failure of mechanical property contributed by the He distributed at the BCC region, the model doped with He in BCC region is also built for comparison. Substituted He are introduced in the system owing to the fast migration of interstitial He, which are known to strongly interact with vacancies [3,24,25]. According to the distribution and concentration of He, three groups of cubic models are built based on the minimized model (see Figure 2a). Then these nanocrystalline models with He is subjected to the energy minimization again at 0 K.

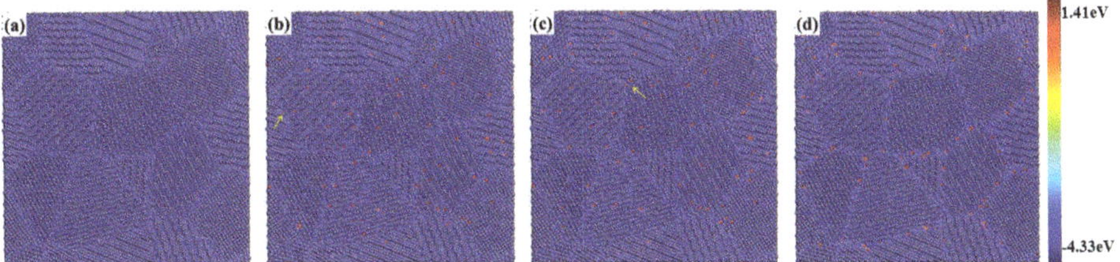

Figure 2. (**a**) Nanocrystalline model after energy minimization; (**b**) The model with BCC He (1%); (**c**) The model with uniform He (1%); (**d**) The model with GB He (1%). The atom in model (**a–d**) was colored according to potential energy (see color map). The red ball in (**b–d**) represents He atom.

The distribution of He in three groups of models is as follows: the grain-interior distribution of He is represented by "BCC He"; the homogeneous distribution of He is represented by "uniform He"; and the intergranular distribution of He is represented by "GB He". The fraction of He distributed in every group of model is 0.5%, 1% and 3%. The picture of the minimized model with the BCC He, the uniform He and the GB He (1%) are shown in Figure 2b–d, respectively. The GB He atom is observed in the model with BCC He or uniform He after the energy minimization (see Figure 2b,c, marked by yellow arrow). In addition, several He clusters are observed in Figure 2b–d since He atoms strongly aggregate into clusters rather than distribute homogeneously [6]. The simulation results about the He effects on the microstructure of polycrystalline BCC-Fe after energy minimization is detailed in Section 3.1.

In order to study the mechanical property of nanocrystalline structure, these models are annealed at 300 K for 10 ps with a Nose/Hoover isobaric-isothermal (NPT), and then subject to a uniaxial tensile load with a strain rate of 10^8 s^{-1}. The simulation step is set at 0.001 ps. The potential of He and Fe-He in references [26,27] are employed in the simulation work. The deformation of models during the process of simulation is visualized by Ovito [28,29]. The evolution on the microstructure of the present nanocrystalline models during loading is described in Section 3.2, and the calculation results is discussed and analyzed in Sections 4.1 and 4.2.

3. Results

3.1. He Effects on Microstructure of Nanocrystalline BCC-Fe after Energy Minimization

It is observed in Figure 2b–d that the potential energy of iron atom is changed owing to the introduction of He in nanocrystalline model, so the relative position of single iron atom in the system is modified. In order to investigate the He effect on the microstructure of nanocrystalline after the energy minimization, both the modification of atomic volume distribution and the expansion of lattice constant induced by He are studied. The swelling rate contributed by the expansion of atomic volume and lattice constant are calculated respectively. The atomic volume is represented by the volume of Voronoi cell for lattice site. The lattice constant of present models are obtained through the simulated XRD pattern.

The volume distribution of He and Fe are plotted in Figure 3. Figure 3a shows that the peak volume of Fe in the model doped with BCC He, uniform He and GB He moves right with the increase of He concentration. That is to say the average of atomic volume of Fe is expanded owing to the introduction of He. The small peak marked by a red arrow represents the atomic volume of iron atom affected by nearby He in the grain interior region. It is illustrated that the average distance between two perfect lattice sites which are taken up by iron atoms become further if one of them or its neighbor is replaced by a He.

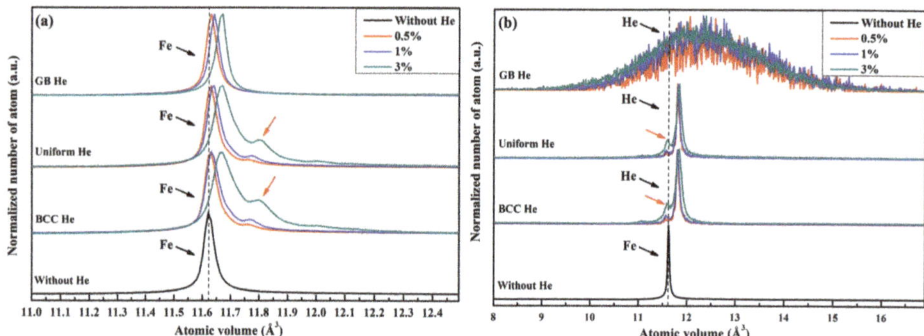

Figure 3. Volume distribution of Fe (**a**) and He (**b**) in polycrystalline model with and without He.

It is observed in Figure 3b that the average volume of He in the model with uniform He or BCC He is lower than that in the model with GB He. The volume distribution of He in the model with GB He covers a wider volume range compared with that in a model with uniform He. The small peak in volume distribution curves for models with BCC He and uniform He were marked by red arrow in Figure 3b, and these peaks represent the atomic volume of He which formed into He_2 or He_3 cluster in the grain interior region after the energy minimization. Figure 3b shows that the average volume of Voronoi cell for lattice site taken up by Fe get larger while these lattice site are occupied by He.

The existence of He, which occupied BCC lattice site in the models, leads to the distortion or disorder of the surrounding lattice site. In order to study the expansion and distortion of the lattice site of Fe, the XRD patterns of present models are simulated by LAMMPS. According to Bragg formula, the lattice constant of present models are obtained by XRD patterns. The increase of lattice constant (Figure 4b) was observed as a left shift of the {110} peak in XRD patterns (Figure 4a) when the He concentration increases. Figure 4b shows that the expansion of lattice constant caused by the given concentration of He decreases with the decreasing fraction of BCC He.

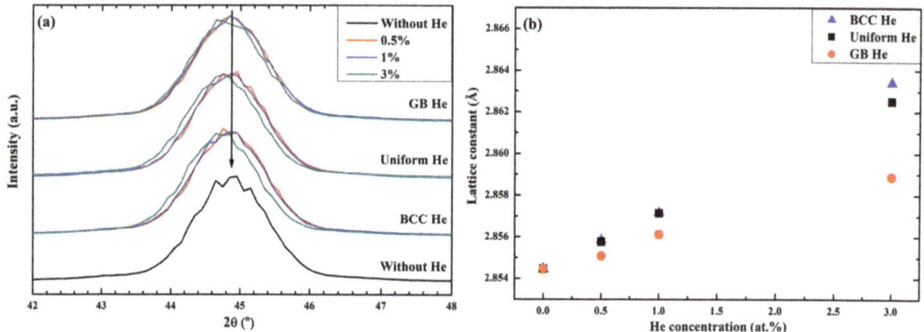

Figure 4. (**a**) Simulated XRD patterns and (**b**) lattice constant of polycrystalline model doped with He.

The expansion of atomic volume and lattice constant means that the swelling is unavoidable for the present models doped with He, thus, the swelling rate is calculated after the energy minimization. The total swelling rates contributed by the expansion of the atomic volume are plotted in Figure 5a. The total height of the bar in Figure 5a represents the swelling rate, defined as the percentage change on the average of atomic volume due to the introduction of He in the model. It is observed in Figure 5a that the swelling rate has clearly He concentration and distribution dependence. The total swelling rate of the model increases with the increasing of He concentration, and this value induced by BCC He or uniform He is lower than that induced by GB He. The swelling rate contributed by the atomic volume expansion of He has the same tendency as the total swelling rate. Thus, the expansion of model with given He concentration will continue to increase if the He which is distributed at the BCC region diffuses into the GB region. The swelling rate contributed by the change of lattice constant is shown in Figure 5b as a comparison. It is observed in Figure 5b that the swelling rate obtained by the expansion of lattice constant is slightly lower in the model with GB He than in models with uniform He or BCC He while the total He concentration remains unchanged.

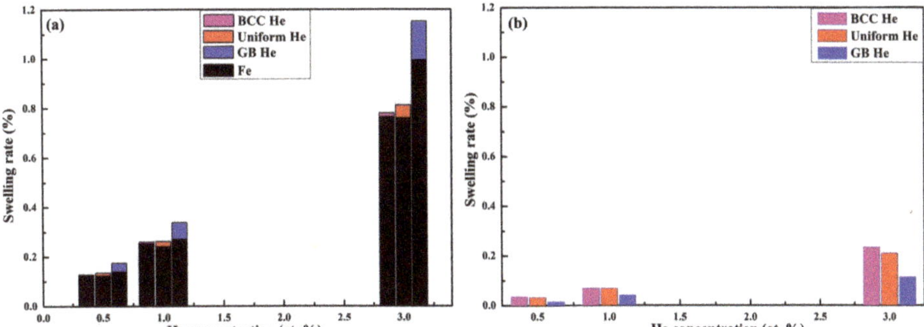

Figure 5. The swelling rate of nanocrystalline models are plotted as a function of He concentration. (a) The swelling rate is calculated by the expansion of atomic volume. The swelling rate contributed by Fe is represented by black bar. The swelling contributed by the expansion of He was colored according to the distribution of He in the models; (b) The swelling rate is contributed by the expansion of lattice site, the color of the bar represent the distribution of He.

It is clearly observed in Figure 5 that the swelling rate obtained by XRD patterns is lower than that obtained by the expansion of atomic volume. The reason is that the peak position of XRD is determined by the lattice site of BCC Fe atom rather than by the disordered Fe atom which lost the crystal structure. That is to say, the distortion of BCC lattice site (such as disordered Fe) is more severe in the models doped with GB He than in the models doped with BCC He or uniform He. Thus, it is observed that the swelling rate contributed by the expansion of lattice constant is smaller in the former than in the latter. This implies that the volume expansion of present models contributed by both doped He and disordered Fe is larger than that contributed by the expansion of lattice constant. It is deduced that the total swelling rate of present models is contributed by doped He, disordered Fe and BCC Fe. Additionally, the GB structure of nanocrystalline is changed more severely in the model with GB He than in the models with uniform He or BCC He, since the disordered Fe and the GB He is wholly located at the GB region of the former.

3.2. He Effect on the Generation and Growth of Cracks in Nanocrystalline BCC-Fe during Loading

The effect of He on the mechanical property of nanocrystalline BCC-Fe is investigated by LAMMPS. Figure 6a shows the curve of normal stress along the loading direction as a function of strain for models with different He concentration and distribution. It is observed that the peak value of normal stress decreases with increasing He concentration in the model with BCC He, and this value is further reduced if some or all of the BCC He diffuses into the GB region. It is shown in Figure 6a that the tensile strength of nanocrystalline models is also induced by BCC He (see dashed line and blue triangle), because the microstructure is changed in the BCC region (see Figures 3 and 4). It is observed in Figure 6b that the tensile strength of the model with 3% uniform He is close to that of the model with 1% GB He, that is to say, the effect of additional 2.4% uniform He or 0.4% GB He on the model with 0.6% GB He is almost same. It is implied that the reduction of tensile strength is less in the model with a larger fraction of He distributed at BCC region while the He concentration remains unchanged. Moreover, the absence of strain hardening in the present models with He is observed because the normal strength of the present nanocrystalline BCC-Fe is mainly affected by grain size [20,30].

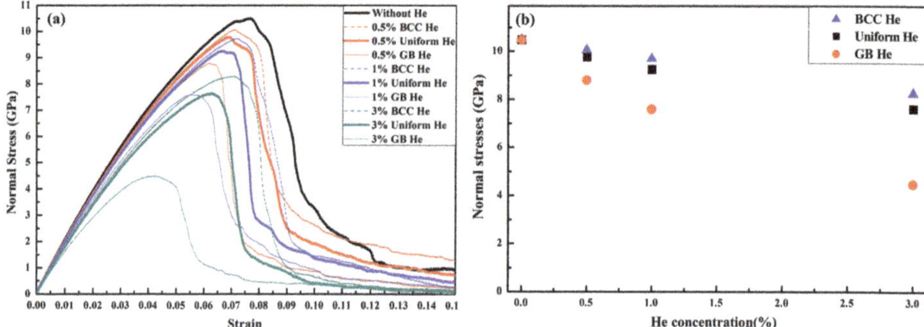

Figure 6. (**a**) Normal stress-strain curves for moles with different He concentration and distribution under uniaxial tensile strain; (**b**) The maximum normal stresses of models with BCC He, uniform He and GB He.

The deformation processes in the present models are visualized and the crack is observed in the GB region. Figure 7a–d show one of the crack appeared at 6.5% strain in the cross section of nanocrystalline model without He and with BCC He, the uniform He and GB He, respectively. Figure 7 illustrates that the crack is formed in the intergranular region during loading [17,31]. The structure of lattice site for atom in models is analyzed with the help of ACNA algorithm. It is observed that the atom of Fe with the structure of FCC or HCP formed into clusters along the GB during the deformation of the model, and those phase transitions are studied in Ref. [19]. The fraction of FCC (face-centered cubic) and HCP (hexagonal close-packed) Fe reaches the maximum at the peak of normal stress and then decreases with the increase of deformation. The existence of He is observed inside the crack shown in Figure 6c,d.

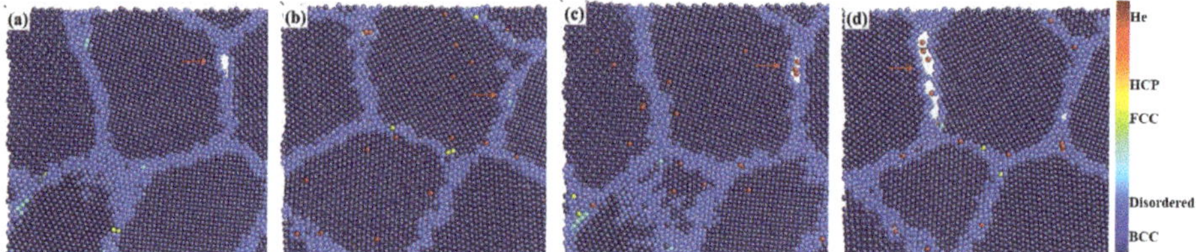

Figure 7. Cracks formed in the model (**a**) without He; (**b**) with BCC He (0.5%); (**c**) with uniform He (0.5%) and (**d**) with GB He (0.5%). The structures of Fe atom analyzed with the help of ACNA algorithm are colored according to the color map, and the red ball represents He atom.

Basically, the crack is formed due to the aggregation of vacancies in the GB region during the deformation of present models, and then the Fe with atomic volume greater than 16 Å^3 (this is the biggest atomic volume in the model without He before loading) is selected to track the generation and growth of crack. The evolution process of the He- and Fe-clusters' dependence on strain is shown in Figure 8 where the single He and selected Fe atoms are filtered out for clarity. The red arrow in Figure 8 labels the crack shown in Figure 7, and the viewpoint and size of the picture panels are same. If the size of cluster formed from the selected Fe atoms meets a certain scales during loading, it means the crack is opened. It is observed in Figure 8 that the neighboring clusters of He and selected

Fe labeled by the oval is gradually linking up and forming bigger cluster or crack [1,4] with increasing deformation of the model. The size of the crack or clusters of selected Fe is less in the model without He than in the model with He. The crack generation sources (and the crack sizes) are more (and larger) in the model with GB He than in the model with BCC He or uniform He for the given He concentration. This shows that He atoms introduced in nanocrystalline BCC-Fe can facilitate the crack formation and growth, consistent with previous results by others [2,6,17].

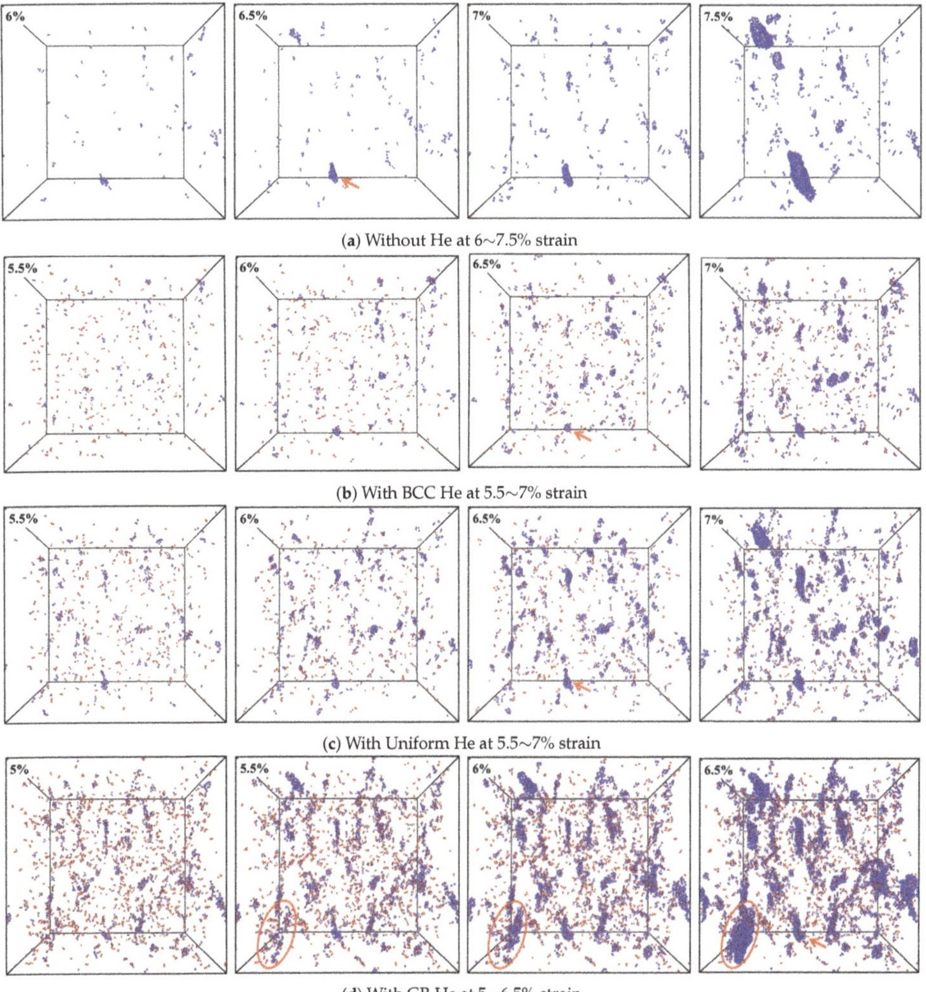

(a) Without He at 6~7.5% strain

(b) With BCC He at 5.5~7% strain

(c) With Uniform He at 5.5~7% strain

(d) With GB He at 5~6.5% strain

Figure 8. The evolution process of crack in model (**a**) without He at 6~7.5% strain; (**b**) with 0.5% BCC He at 5.5~7% strain; (**c**) with 0.5% uniform He at 5.5~7% strain; (**d**) with 0.5% GB He at 5~6.5% strain.

Figure 9 shows the size of the biggest cluster which is formed from the selected Fe and He during the deformation of models where the magenta arrow represents the strain at which the normal stress reached the maximum value. It is observed in Figure 9 that there is a cluster formed by the selected Fe atom in the model with 3% GB He before the tensile stress is loaded. Suppose that there is about 20% of the doped He in the GB region for the model with uniform He. It is shown in Figures 8 and 9 that the size of the cluster grows bigger and the number of selected Fe in the cluster gets larger with increasing fraction of GB He in the deformation of models. These results indicate that the failure of tensile property is promoted by GB He which trends to form He cluster and benefit the crack generation. Additionally, the ratio of He to vacancy also can affect the generation and growth of crack [17].

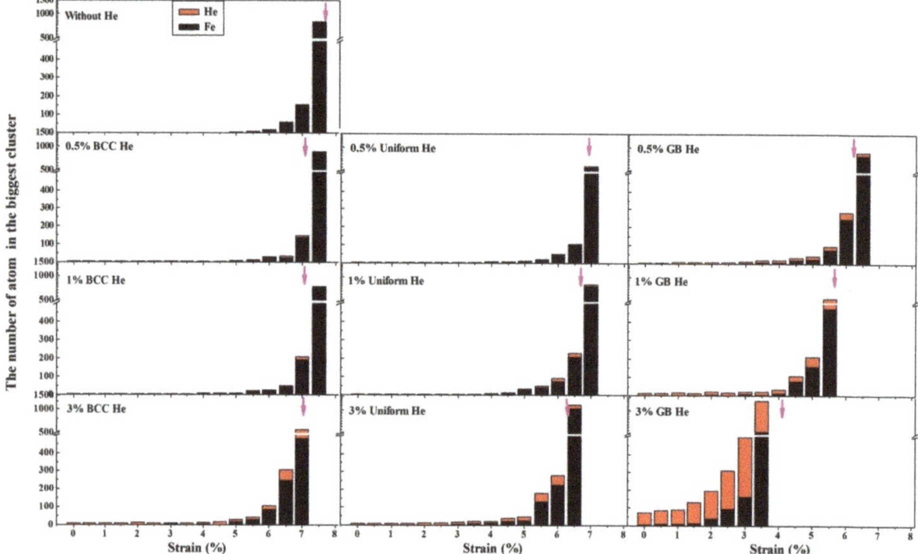

Figure 9. The size of the largest cluster at different He concentration during the deformation of models. The height of bar represents the size of the cluster, consisting of He (red) and Fe (black).

4. Discussion and Analysis

4.1. The Simulated XRD Patterns of Nanocrystalline Model during Loading

The failure of tensile strength shown in Figures 6 and 7 was induced by the change of nanocrystalline structure (shown in Figures 3 and 4) which was modified by the introduction of He. It was indicated in Figures 8 and 9 that the degradation of mechanical property was mainly affected by the GB He. In order to further interpret the effect of GB He, uniform He and BCC He on the mechanical property of nanocrystalline BCC-Fe under a uniaxial tensile load, the connection between the strain–stress curves and the corresponding change of microstructure are discussed with the help of the simulated XRD patterns.

Figure 10a shows a set of simulated XRD patterns during the deformation of model without He. As the increase of deformation, the {110} peak in the model shifts to left till reaching the maximum movement at ~8% strain, and then moves back gradually. The lattice constant of nanocrystalline model are obtained by a set of the simulated XRD patterns and plotted as a function of strain in Figure 10b. It is observed in Figures 6a and 10b that the variation of lattice constant is same as that of normal stress with the increase of deformation. So there are some connections between the generation of crack and the expansion of lattice constant.

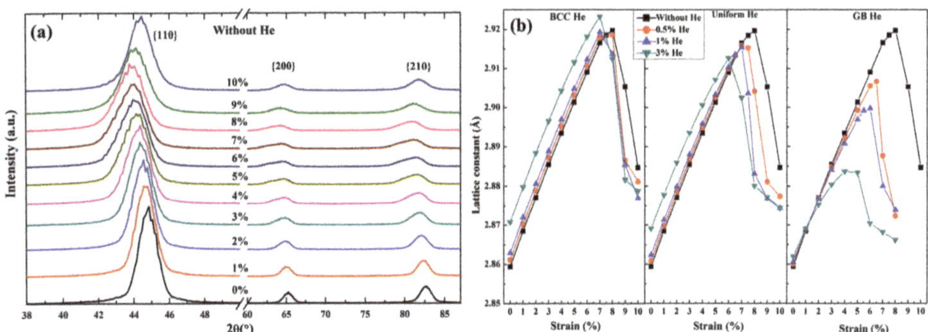

Figure 10. (**a**) Simulated XRD patterns of pure nanocrystalline BCC-Fe at the strain range from 0% to 10%. (**b**) The lattice constant of models under the tensile stresses are plotted as a function of strain.

The degradation of mechanical properties without the effect of He is analyzed firstly. It is observed in Figure 10 (see XRD patterns and the black square) and Figure 6a (see black solid line) that the more variation of the lattice constant is obtained by XRD patterns, the bigger tensile stress is loaded the deformation of the model. The variation of lattice constant is caused by the deformation of single grain whose shape is changed in turn during the loading process of nanocrystalline models. As is shown in Figure 10b (see the black square) and Figure 6a (see black solid line), the decrease of average lattice constant of single grains begins at the maximum loading. However, the crack is generated before the maximum loading is reached (see Figure 7), so there would be a recovery of lattice constant in the grain around this crack. That is to say, the dominant role is acted by the increase of lattice constant before the maximum loading and by the decrease of that after the maximum loading. Therefore the recovery of lattice constant indicates the generation of crack, which starts to appear at GB region while the ultimate strength of the GB is reached. Generally, the smaller ultimate strength of GB means the lower tensile stress needed for the generation of crack.

The role of He played in nanocrystalline model is discussed secondly. It is observed in Figure 10b that the maximum lattice constant is decrease with the increasing concentration of GB He or uniform He, and is changed slightly in the models with BCC He. It is clearly indicated that the maximum lattice constant is variated with the distribution and concentration of He doped in the models. The uniform He have a combination effect of GB He and BCC He on the microstructure of present models during loading. So the role of BCC He and GB He on the expansion of lattice constant and the strength of GB are analyzed as follows respectively.

1. BCC He. It is not difficult to understand that the number of BCC Fe atom per unit volume is reduced in grain interior region due to the BCC He is doped into the nanocrystalline models. Suppose that the lattice is a bunch of springs, which can be broken because of the existence of substitution He. The length of this bunch of springs may need to extend more to balance the given tensile stress if some of the springs are broken before loading. So the maximum lattice constant of model with the concentration of BCC He at 0.5% or 1% is changed slightly comparing that of the model without He. It is also observed that the maximum lattice constant is lower in the model without He than in the model with 3% BCC He (Figure 10), although the tensile strength of the former is stronger than that of the latter (Figure 6b). It is implied that the expansion of lattice constant is unsaturated when the ultimate strength of GB is reached in the model without He. Then the maximum lattice constant during loading is determined by the number of BCC Fe atom per unit volume if there is only BCC He in the model. The maximum of tensile stress in the models with BCC He is reduced because the GB strength is weakened by He or He clusters, which trapped by GB region during energy minimization (see Figure 2) or deformation of the models.
2. GB He. The maximum lattice constant of the models with GB He and the ultimate strength of GB decrease with the increasing concentration of GB He, because the ultimate strength of GB

is weakened by GB He and the expansion lattice constant is limited correspondingly. Thus, it is deduced in Figure 10b that the failure of mechanical property is caused due to the ultimate strength GB is reached. It implied that the expansion of lattice constant is governed by the strength of GB while the GB He is introduced in the nanocrystalline models.

According to the above analyses, it is concluded that (1) the tensile strength of nanocrystalline model is mainly determined by the GB strength, which is weakened by He distributed at GB region; (2) the number of BCC Fe atom per unit volume is reduced by He distributed at grain interior region.

4.2. The Volume Expansion of Nanocrystalline Model during Loading

The atomic volume of Fe defined as the volume of Voronoi cell for lattice site is changed with the separation of atoms in the models. The work must be done against the atomic interactions to change the distance between atoms in the system. So there is a link between the work done by tensile stress and the swelling of atomic volume of Fe during the deformation of models. In the following discussion, the expansion of atomic volume is considered as a process of energy absorption in the system, or as the result of work done by the stress loaded on the Fe atom. For example, the expansion of atomic volume (see Figure 3a) contributed by Fe is more, the potential energy is larger in the system. The expansion of atomic volume (Figure 3a) means that some certain stresses have been loaded on the models with He before loading. That is to say, the introduction of He in the nanocrystalline Fe is equivalent to doing work on this model. Based on the above description, the swelling and the average of atomic volume are calculated respectively to clearly describe the effect of GB He and BCC He on the present models during loading.

The swelling rate of present models during loading shown in Figure 11a is contributed by the expansion of atomic volume of Fe. Figure 11a shows that the swelling rate as the function of strain follows the same trend as the tensile stress while the distribution of He is unchanged, and the maximum swelling is reached when the maximum normal stress is loaded on the nanocrystalline models. The average atomic volume of Fe in the models is divided into two parts, GB region and BCC region. The variation of atomic volume as a function of strain in those regions is marked by red arrow as is shown in Figure 11b. It is observed in Figure 11b that the average atomic volume of Fe is lower in BCC region than in GB region for the model without He. It is obvious that the crack has generated in GB region after the maximum volume of Fe is reached in BCC region or GB region. It is illustrated in Figure 11b that the GB region is a weaker area when compared with BCC region due to the ultimate strength of GB is reached earlier than that of BCC region during the loading process. So the tensile strength of nanocrystalline BCC-Fe doped without He is controlled by the GB strength. This point is agreed with the conclusion in Section 4.1

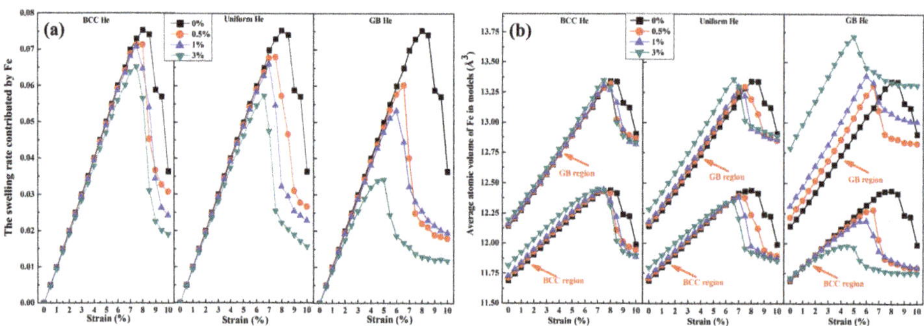

Figure 11. (**a**) Swelling rate contributed by Fe, and (**b**) average atomic volume of Fe in BCC and GB region. Atomic volume of Fe which surround the crack is set as 16 Å^3, in order to deduct the space took up by the crack.

It is observed in Figure 11b that the difference value of the maximum atomic volume between the GB region and the BCC region get larger with the increase fraction of He distributed in GB region when the He concentration is unchanged. It is implied that the concentration of stress in GB region is enhanced by the He distributed at GB region during the deformation of these models. So the failure of mechanic property of the present models can be concluded as following: (1) the crack generation is not started until the ultimate volume of single Fe is reached for the model without He; (2) the average of atomic volume of GB Fe is affected by GB He, and that of BCC Fe is affected by BCC He; (3) The maximum swelling rate of the atomic volume during loading is mainly limited by the expansion of atomic volume of GB Fe. From the perspective in this section, it is easy to find that the change of lattice constant and the atomic volume during loading contribute to understand the degradation of mechanical property in the nanocrystalline models without He or with He.

Based on the analysis above and the previous studies[5–7,17], we conclude that the introduction of He in GB region is a fatal factor for degradation of tensile property in nanocrystalline BCC-Fe comparing with that in BCC region [4]. On condition that the He concentration in models remains unchanged, then the more fraction of He is distributed in the GB region, the more severe degradation of mechanical property is occurred. The mechanical property of nanocrystalline BCC-Fe will be improved if the weaker one of GB and BCC region is strengthened, and the degradation of tensile property will be reduced if the radiation damage on the weaker one is avoided or decreased.

5. Conclusions

In summary, He effects on generation and growth of cracks in nanocrystalline BCC-Fe are studied in the present work. The modification of microstructure caused by the He distributed in the nanocrystalline is analyzed by the change of simulated XRD patterns and the atomic volume. For the nanocrystalline BCC Fe with the average grain size of 8.28 ± 0.32 nm, the following conclusions are obtained:

1. The swelling of the model with He is contributed by doped He, disordered Fe and BCC Fe; the He distributed uniformly in nanocrystalline structure causes more expansion of lattice constant than that distributed at grain boundary (GB) structure; the total swelling rate induced by uniform He is lower than that induced by GB He.
2. The tensile strength of nanocrystalline BCC-Fe is reduced after the introduction of He; the strain–stress curve shows He concentration and distribution dependence; the crack generation and growth are observed in the GB region during loading.
3. The tensile strength is affected by the strength of GB which is mainly weakened by He distributed at GB region; the change of lattice constant and the atomic volume during loading should contribute to understand the degradation of mechanical property in the nanocrystalline models.
4. The mechanical property of nanocrystalline BCC-Fe under an irradiation environment would be improved by enhancing the GB strength or reducing He concentration in GB region.

Author Contributions: Conceptualization, C.X.; Investigation, C.X. and W.W.; Methodology, C.X. and W.W.; Software, C.X. and W.W.; Visualization, C.X.; Writing—original draft, C.X.; Writing—review and editing, C.X. and W.W.

Funding: This research received no external funding.

Acknowledgments: This work was supported by laboratory of computation science at Yantai University.

Conflicts of Interest: The authors declare no conflict of interest.

Abbreviations

The following abbreviations are used in this manuscript:

BBC   Body-centered cubic
XRD   X-ray Diffraction
GB   Grain boundary
ACNA   Adaptive common neighbor analysis

References

1. Olander, D.R. *Fundamental Aspects of Nuclear Reactor Fuel Elements*; U.S. Department of Commerce: Washington, DC, USA, 1976. [CrossRef]
2. Kramer, D.; Brager, H.R.; Rhodes, C.G.; Pard, A.G. Helium embrittlement in type 304 stainless steel. *J. Nucl. Mater.* **1968**, *25*, 121–131. [CrossRef]
3. Fu, C.-C.; Willaime, F. Ab initio study of helium in α-Fe: Dissolution, migration, and clustering with vacancies. *Phys. Rev. B* **2005**, *72*, 064117. [CrossRef]
4. Han, W.-Z.; Ding, M.-S.; Shan, Z.-W. Cracking behavior of helium-irradiated small-volume copper. *Scr. Mater.* **2008**, *147*, 1359–6462. [CrossRef]
5. Zhang, L.; Shu, X.; Jin, S.; Zhang, Y.; Lu, G.H. First-principles study of He effects in a bcc Fe grain boundary: site preference, segregation and theoretical tensile strength. *J. Phys. Condens. Matter* **2010**, *22*, 375401. [CrossRef] [PubMed]
6. Zhang, L.; Fu, C.-C.; Hayward, E.; Lu, G.-H. Properties of He clustering in α-Fe grain boundaries. *J. Nucl. Mater.* **2015**, *459*, 247–258. [CrossRef]
7. Suzudo, T.; Yamaguchi, M.; Tsuru, T. Atomistic modeling of He embrittlement at grain boundaries of α-Fe: A common feature over different grain boundaries. *Model. Simul. Mater. Sci. Eng.* **2013**, *21*, 085013. [CrossRef]
8. Singh, B.N. Effect of grain size on void formation during high-energy electron irradiation of austenitic stainless steel. *Philos. Mag.* **1974**, *29*, 25–42. [CrossRef]
9. Chen, D.; Wang, J.; Chen, T.; Shao, L. Defect annihilation at grain boundaries in alpha-Fe. *Sci. Rep.* **2013**, *2*, 1450. [CrossRef]
10. Bai, X.M.; Voter, A.F.; Hoagland, R.G.; Nastasi, M.; Uberuaga, B.P. Efficient annealing of radiation damage near grain boundaries via interstitial emission. *Science* **2010**, *327*, 1631–1634. [CrossRef]
11. Yu, K.Y.; Liu, Y.; Sun, C.; Wang, H.; Shao, L.; Fu, E.G.; Zhang, X. Radiation damage in helium ion irradiated nanocrystalline Fe. *J. Nucl. Mater.* **2012**, *425*, 140–146. [CrossRef]
12. Cheng, G.M.; Xu, W.Z.; Wang, Y.Q.; Misra, A.; Zhu, Y.T. Grain size effect on radiation tolerance of nanocrystalline Mo. *Scr. Mater.* **2016**, *123*, 90–94. [CrossRef]
13. Odette, G.R.; Alinger, M.J.; Wirth, B.D. Recent Developments in Irradiation-Resistant Steels. *Annu. Rev. Mater. Res.* **2008**, *38*, 471–503. [CrossRef]
14. Edmondson, P.D.; Parish, C.M.; Zhang, Y.; Hallén, A.; Miller, M.K. Helium entrapment in a nanostructured ferritic alloy. *Scr. Mater.* **2011**, *65*, 731–734. [CrossRef]
15. Tschopp, M.A.; Gao, F.; Solanki, K.N. He-V cluster nucleation and growth in α-Fe grain boundaries. *Acta Mater.* **2017**, *124*, 544–555. [CrossRef]
16. Yang, L.; Gao, F.; Kurtz, R.J.; Zu, X.T. Atomistic simulations of helium clustering and grain boundary reconstruction in alpha-iron. *Acta Mater.* **2015**, *82*, 275–286. [CrossRef]
17. Terentyev, D.; He, X. Effect of Cr precipitates and He bubbles on the strength of <110> tilt grain boundaries in BCC Fe: An atomistic study. *Comput. Mater. Sci.* **2011**, *50*, 925–933. [CrossRef]
18. Godon, A.; Creus, J.; Cohendoz, S.; Conforto, E.; Feaugas, X.; Girault, P.; Savall, C. Effects of grain orientation on the Hall–Petch relationship in electrodeposited nickel with nanocrystalline grains. *Scr. Mater.* **2010**, *62*, 403–406. [CrossRef]
19. Gunkelmann, N.; Bringa, E.M.; Kang, K.; Ackland, G.J.; Ruestes, C.J. Polycrystalline iron under compression: Plasticity and phase transitions. *Phys. Rev. B* **2012**, *86*, 144111. [CrossRef]
20. Hahn, E.N.; Meyers, M.A. Grain-size dependent mechanical behavior of nanocrystalline metals. *Mater. Sci. Eng. A* **2015**, *646*, 101–134. [CrossRef]

21. Finney, J.L. A procedure for the construction of Voronoi polyhedr. *J. Comput. Phys.* **1979**, *32*, 137–143. [CrossRef]
22. Available online: http://lammps.sandia.gov/index.html (accessed on 17 November 2016)
23. Mendelev, M.I.; Han, S.; Srolovitz, D.J.; Ackland, G.J.; Sun, D.Y.; Asta, M. Development of new interatomic potentials appropriate for crystalline and liquid iron. *Philos. Mag.* **2003**, *83*, 3977–3994. [CrossRef]
24. Fu, C.-C.; Willaime, F. Interaction between helium and self-defects in α-iron from first principles. *J. Nucl. Mater.* **2007**, *367*, 244–250. [CrossRef]
25. Morishita, K.; Sugano, R.; Wirth, B.D.; Diaz de la Rubia, T. Thermal stability of helium-vacancy clusters in iron. *Nucl. Instrum. Meth. B* **2003**, *202*, 76–81. [CrossRef]
26. Aziz, R.A.; Janzen, A.R.; Moldover, M.R. Ab initio calculations for helium: A standard for transport property measurements. *Phys. Rev. Lett.* **1995**, *74*, 1586–1589. [CrossRef] [PubMed]
27. Gao, F.; Deng, H.; Heinisch, H.L.; Kurtz, R.J. A new Fe-He interatomic potential based on ab initio calculations in α-Fe. *J. Nucl. Mater.* **2011**, *418*, 115–120. [CrossRef]
28. Stukowski, A. Visualization and analysis of atomistic simulation data with OVITO–the Open Visualization Tool. *Model. Simul. Mater. Sci. Eng.* **2010**, *18*, 015012. [CrossRef]
29. Zhao, X.; Lu, C.; Tieu, A.K.; Zhan, L.; Pei, L.; Huang, M. Deformation mechanisms and slip-twin interactions in nanotwinned body-centered cubic iron by molecular dynamics simulations. *Comput. Mater. Sci.* **2018**, *147*, 34–48. [CrossRef]
30. Kiener, D.; Hosemann, P.; Maloy, S.A.; Minor, A.M. In situ nanocompression testing of irradiated copper. *Nat. Mater.* **2011**, *10*, 142–149. [CrossRef]
31. Jeon, J.B.; Lee, B.-J.; Chang, Y.W. Molecular dynamics simulation study of the effect of grain size on the deformation behavior of nanocrystalline body-centered cubic iron. *Scr. Mater.* **2011**, *64*, 494–497. [CrossRef]

 *materials*

Review

Radiation-Induced Helium Bubbles in Metals

Shi-Hao Li, Jing-Ting Li and Wei-Zhong Han *

Center for Advancing Materials Performance from the Nanoscale (CAMP-Nano), State Key Laboratory for Mechanical Behavior of Materials, Xi'an Jiaotong University, Xi'an 710049, China; lsh4007025@gmail.com (S.-H.L.); lijingting235@sina.com (J.-T.L.)
* Correspondence: wzhanxjtu@mail.xjtu.edu.cn

Received: 6 March 2019; Accepted: 26 March 2019; Published: 28 March 2019

Abstract: Helium (He) bubbles are typical radiation defects in structural materials in nuclear reactors after high dose energetic particle irradiation. In the past decades, extensive studies have been conducted to explore the dynamic evolution of He bubbles under various conditions and to investigate He-induced hardening and embrittlement. In this review, we summarize the current understanding of the behavior of He bubbles in metals; overview the mechanisms of He bubble nucleation, growth, and coarsening; introduce the latest methods of He control by using interfaces in nanocrystalline metals and metallic multilayers; analyze the effects of He bubbles on strength and ductility of metals; and point out some remaining questions related to He bubbles that are crucial for design of advanced radiation-tolerant materials.

Keywords: helium bubbles; bubble evolution; interfaces; radiation hardening; helium embrittlement

---

1. Introduction

Nuclear energy is playing an increasingly important role because fossil fuels are gradually running out. Nuclear energy currently provides about 13% of electrical power worldwide [1]. However, materials deployed in fission, spallation and fusion systems suffer from intense high-energy neutron radiation [1]. Continuous collision cascades produce massive vacancy clusters and interstitials in nuclear component materials, which promotes formation of various defects, such as dislocation loops [2,3], voids [4,5], stacking fault tetrahedral (SFT) [6,7] and so on. In addition, (n, α) reactions produce abundant helium (He) atoms in materials. As He has extremely low solubility in metals, it tends to accumulate and precipitate into nanoscale He bubbles in nuclear structure materials [8–14]. Figure 1 shows typical examples of transmission electron microscope (TEM) images of He bubbles formed in Al [10], tungsten [11] and Zr [12–14] after He$^+$ ion irradiation.

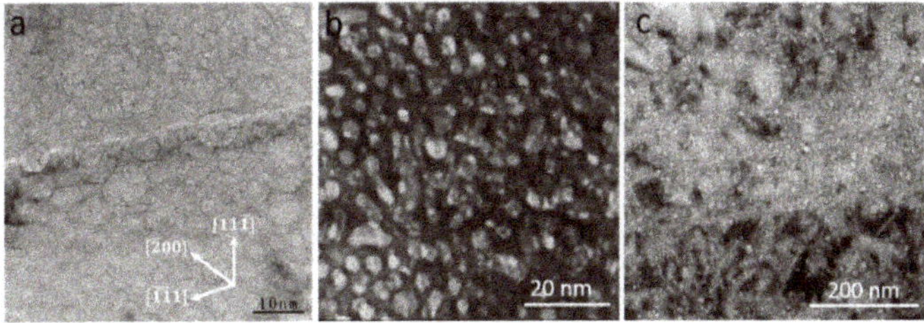

Figure 1. Dense He bubbles after He irradiation in: (**a**) Al [10]; (**b**) tungsten [11]; and (**c**) Zr [12]. Reprinted with permission from [10]; Copyright 2015 Elsevier; Reprinted with permission from [11]; Copyright 2000 Elsevier; Reprinted with permission from [12]; Copyright 2017 Elsevier.

A comprehensive understanding on the dynamic evolution of He bubbles and their effects on mechanical properties of nuclear structure materials over the lifetime remains as one of the key issues in nuclear industry. He bubbles are found to drastically deteriorate mechanical properties of metals, manifested as swelling [15,16], blistering [17] and embrittlement at high temperature [18,19]. Among these degraded properties, He-induced embrittlement has attracted particular attention since it causes catastrophic fracture in metals, particularly at high temperature [18,19]. It has been well demonstrated that even extremely low overall He concentration can lead to He embrittlement via formation of He bubbles along grain boundaries (GBs) [18,19]. In view of this, several decades of investigations have been conducted to unveil He behaviors and underlying mechanisms for He-induced degradation in metals [8–24]. Both experiments and atomic simulations are adopted to investigate He bubble formation, dynamic evolution and their effects on mechanical properties of metals.

In this review, we briefly summarize previous studies on He bubbles in metals. Section 2 reviews research on He bubble nucleation, growth and coarsening. Section 3 summarizes He bubble behaviors in nanocrystalline metals and metallic multilayers and emphasizes the important role of interfaces. Section 4 overviews the He-induced hardening in single-phase metals and metallic multilayers. Finally, we summarize the main results and discuss the critical questions remained.

2. Helium Bubble Nucleation, Growth and Coarsening

As mentioned above, due to their extremely low solubility, He atoms tend to agglomerate into He bubbles in metals. A comprehensive understanding on bubble nucleation, growth and coarsening is crucial to evaluate the role of bubbles in metals. The formation of He bubbles can be divided into homogeneous or heterogeneous nucleation. In this section, we focus on the mechanism of homogeneous bubble nucleation. In general, bubble nucleation, growth and coarsening are controlled by the concurrent operation (including diffusion, clustering, dissociation and recombination, etc.) of He atoms, vacancies and interstitials. However, limited spatial resolution and temporary resolution of various instruments hinder the direct real-time atomic-scale observation of bubble formation, which prevents unveiling the mechanisms underlying bubble nucleation, growth and coarsening. Studies on this issue using computer simulation methods [25–38] have been conducted in past decades. Because steels are widely used as nuclear reactor components and tungsten is regarded as the main candidate material for plasma facing materials (PFM) in future fusion reactors, profuse studies are focused on behaviors of He atoms, vacancies and interstitials in Fe and tungsten [25–38]. Generally, body-centered-cubic (BCC) metals have attracted more attention due to their better radiation resistance than face-centered-cubic (FCC) metals.

2.1. He–V Clusters

Rimmer and Cottrell [39] investigated He behaviors in metals and demonstrated that He accumulation can be ascribed to two reasons: first, the interstitial He has low migration energy in metals; and, second, the substitutional He is easily trapped at vacancies because of their high binding energy. Several studies [29,34,37,40–42] point out that interstitial He is energetically favorable to occupy the tetrahedral interstitial site (TIS) with low formation energy in BCC metals. Becquart [41] evaluated the formation energies of interstitial He in tungsten using ab initial calculations and found that the formation energies of interstitial He at TIS is 6.18 eV, which is slightly lower than that in octahedral interstitial sites (6.40 eV). Other studies [42–45] propose a similar trend that TISs are preferable He traps. Notably, due to the low migration energy [27,31], interstitial He in tetrahedral sites and small He clusters (mainly He_2) tend to diffuse easily in metals. These isolated He and small He clusters are easily trapped by vacancies and form a sphere-like configuration of He–V clusters with low mobility [25–44]. Consequently, He–V clusters deliver high binding energy, performing as sinks for interstitial He and small He clusters and giving rise to He bubble nucleation in metals [43–47]. Figure 2a schematically illustrates an isolated He–V cluster containing four He atoms in Fe, which can be regarded as embryo of He bubble [27]. Fu et al. [48] showed that small He–V clusters (up to four helium atoms) will lead to bubble nucleation in initial vacancy-free lattices. The binding energy for additional He atoms to combine with He–V cluster in Fe is plotted in Figure 2b [43–47]. The binding energy of an interstitial He to He–V clusters is initially high, then it decreases sharply with increasing He atoms, and keeps almost constant with a value of about 1.3 eV with further increasing He atoms. It should be noted that the binding energy reaches a local peak value with six He atoms in He–V cluster, and an additional He induces significant reduction in binding energy. He–V clusters containing six He deliver compact octahedral shape with He located at corners and the vacancy at the center [44]. An extra He will trigger the kick-out of Fe around the cluster, leading to reduced binding energy [44]. The kick-out mechanism is discussed below. Finally, as shown in Figure 2c, the matrix in molecular dynamics (MD) simulations is quenched to 0 K, followed by annealing at 800 K for 1.2 ns, and then a few He–V clusters are formed in Fe containing 685 appm He [27].

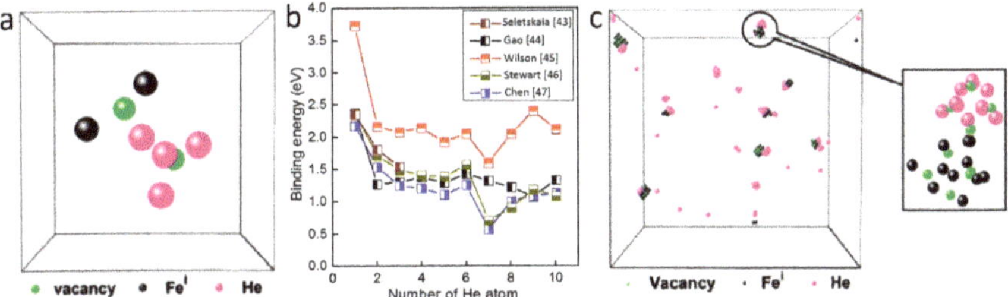

Figure 2. (**a**) Schematic illustration of He–V cluster in Fe [27]; (**b**) the dependence of binding energy of interstitial He to He–V cluster on the number of He atom in He–V cluster [43–47]; and (**c**) He clusters and other defect distribution in Fe containing 125 He atoms. The inserted image is the detailed arrangements of a He-interstitial cluster [27]. Reprinted with permission from [27]; Copyright 2013 Elsevier.

2.2. Kick-Out Mechanism

With gradual accumulation of He in He–V clusters, the pressure induced by He–V clusters on surrounding lattice will increase significantly, and then the matrix atoms surrounding He–V clusters will be pushed out due to the high pressure, producing self-interstitial atoms (SIA) and vacancies (SIA-V pairs). That is how He–V clusters lead to kick-out of matrix atom and formation of Frenkel pairs. By investigating the dependence of the binding energy of He atoms, vacancies and interstitials on He density (defined as He/V ratio) in He–V clusters using MD simulations, Morishita et al. [25] described the kick-out mechanism quantitatively in Fe. For He–V clusters with He density ranging 0–6, the binding energy of He atoms and SIAs to He–V clusters decreases with increasing He density, while the binding energy of vacancies increases with increasing of He density. For regime with He density >6, however, the binding energy of He atoms, interstitials and vacancies turns into a contrary trend in comparison with low He density regime [25]. This conspicuous transition should be ascribed to the kick-out of matrix Fe atom induced by high He density in He–V clusters. When He density is >6 in Fe, He atoms exhibit close-packed configurations and exert pressure large enough to produce Frenkel pairs. The kick-out mechanism creates additional vacancies and interstitials, lowering He density in He–V clusters. As shown in Figure 3 [25], the binding energy of He atoms to He–V clusters with low He density is quite high, which indicates that He–V clusters with low He density are strong sinks for He atoms. Therefore, the kick-out mechanism significantly enhances He bubble nucleation and growth.

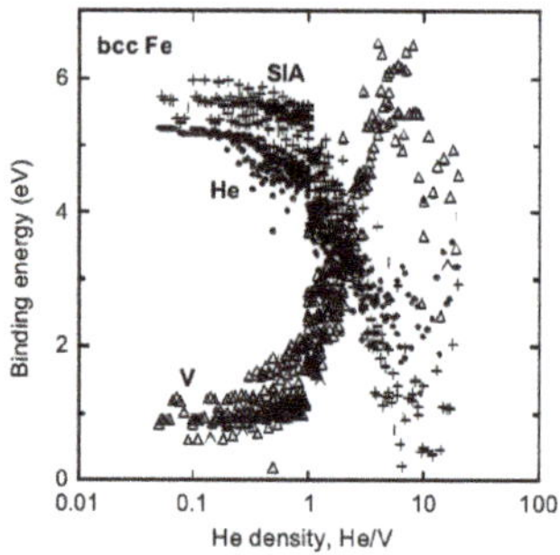

Figure 3. The variation of binding energies of a vacancy, an interstitial He atom and a SIA to He–V cluster with different He density in Fe. Reprinted with permission from [25]; Copyright 2003 Elsevier.

Sandoval et al. [33] investigated He bubble nucleation and growth accompanied by kick-out mechanism in detail. Figure 4 schematically illustrates the process of He bubble nucleation and growth accompanied by formation of Frenkel pairs in tungsten. The gray, red and blue spheres represents He atoms, interstitials and vacancies, respectively. In addition, green spheres indicate the deformed tungsten lattice around interstitials. At the very beginning, a cluster containing 8 He atoms is deliberately placed at the center of a tungsten sphere containing 23,538 tungsten atoms. He atoms are implanted at a constant rate (1×10^{19} He/s) at 1000 K, which is slow enough to allow the inserted He clusters to diffuse in tungsten matrix and interact with the pre-existing central He bubble [33]. As shown in Figure 4a, the inserted He atoms attach to the pre-existing He cluster, leading to the formation of Frenkel pair and creating a He_9V cluster. Another Frenkel pair is created with further inserting He atoms in tungsten, as shown in Figure 4b. Notably, two interstitials diffuse around He–V cluster in a coordinated way, staying in two adjacent <111> rows, which is consistent with interstitials arrangement in Fe [27]. Gao et al. [31] proposed that, before forming crowdion along <111> direction, interstitials may create stable configurations with the formation of a <110> dumbbell in Fe, which is consistent with Xiao's work [37]. In addition, a transition from <100> interstitial cluster to <111> interstitial cluster is identified in tungsten [29,49]. This process is energetically favorable with an energy release of 1.12 eV, indicating that the <111> SIAs cluster is more stable. However, similar orientation transition is not observed in Figure 4. With further He insertion, more Frenkel pairs are created around the central He–V cluster, as shown in Figure 4c,d. Notably, the orientation of SIAs at this stage is kept constant during the following simulation timescales. In addition, the interaction of inserted He atoms with interstitial configurations can be identified clearly. These inserted He atoms can be trapped by interstitials. At 1000 K, these new He atoms attached to interstitials are mobile and can diffuse along interstitial configurations or jump to the He–V cluster. With further increasing of He concentration, new He cluster containing three He atoms is formed near the central He–V cluster, as shown in Figure 4e. A new He cluster can further trigger the formation of Frenkel pairs and act as an embryo for new He bubbles, as shown in Figure 4f,g. These He–V clusters further trap the subsequently inserted He, producing more Frenkel pairs and forming He bubbles in tungsten matrix. Finally, SIAs detach from the initial He bubbles and three helium bubbles are nucleated in the matrix.

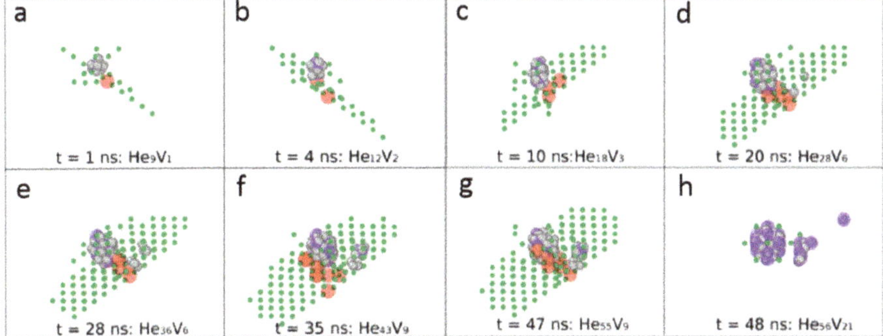

Figure 4. The evolution process of a bubble network at 1000 K with an insertion rate of 1×10^9 He/s in tungsten. (**a**) Formation of the first Frenkel pair and He_9V_1 cluster; (**b**) Another Frenkel pair is formed with further inserted of He atoms with two interstitials staying in two adjacent <111> rows; (**c,d**) More Frenkel pairs are created around the central He–V cluster; (**e**) New He cluster containing three He atoms is formed; (**f,g**) He clusters promote Frenkel pairs formation. (**h**) He bubbles in the matrix. Gray, red and blue dots represent He atoms, tungsten interstitials and vacancies, respectively. Reprinted with permission from [33]; Copyright 2018 Elsevier.

The number of He atoms in He clusters needed to create Frenkel pairs has been evaluated using MD simulations and first principle calculations under various conditions. The minimum number of He atoms needed to create Frenkel pairs in Fe is slightly different, which may be attributed to various simulation conditions. For instance, when the Fe matrix is fully dense and without vacancy, three He atoms are believed to be enough to push out matrix Fe atoms [31]. If there are pre-existing vacancies in the simulation box, six He atoms are needed to create Frenkel pairs [25]. The pre-existing vacancies may release the pressure of He clusters, and more He atoms are required to create Frenkel pairs.

2.3. Dislocation Loop Punching Mechanism

In Section 2.2, we summarize studies on kick-out mechanism that is commonly observed in He-containing metals. These SIAs pushed out by high pressure He bubbles deliver limited mobility and form crowdion structure around He bubbles. The crowdion structure is observed to be temporarily trapped by He bubbles once formed [31–34,37,48,49]. Notably, the crowdion structure usually exhibits an energetically favorable orientation along <111> direction in BCC metals [27,29,31,37,49]. With further increasing He concentration, SIAs accumulate continuously around He bubbles, giving rise to the formation and emission of dislocation loops. This well-known growing mechanism of He bubble, named as dislocation loop punching, is widely identified in various metals [27,29,49–56].

Xie and coworkers [49] performed MD simulations to investigate the loop-punching mechanism for He bubble growth in tungsten at 300 K. Figure 5 illustrates the loop-punching process in He bubble with an initial radius of 0.15 nm (containing 1 He atom). From 0.02 to 0.11 ns, He bubble grows homogeneously without emitting defects. With increasing time, He accumulates continuously in He bubble and SIAs are pushed out due to increasing pressure. These SIAs form crowdion configurations along <100> direction, as labeled in Figure 5c,d. In BCC metals, the <100> orientation crowdion configuration is not stable due to its slightly high energy [5,29,49]. At 0.32 ns, the <100> orientation crowdion configuration reorientates to a more stable <111> orientation, thus releasing the high energy of the configuration [49]. A transition from <100> direction to <111> direction of crowdion configuration is also identified in tungsten recently [29]. The <111> crowdion configuration further absorbs subsequent SIAs and evolves into a prismatic loop at 0.37 ns [49]. Finally, the well-developed prismatic loop dissociates from the He bubble and slips away, leaving a larger He bubble (about 0.55 nm) behind. In addition to the conventional loop-punching mechanism, Xie et al. [49] also proposed a new loop-punching mechanism for He bubble with larger initial radius in tungsten. They demonstrated that, for larger He bubbles, the prismatic loop is formed after two screw components cross-slip on the opposite directions. Finally, a dislocation network formed via dislocation interactions surrounds He bubbles (details are shown in Reference [49]).

The kick-out of matrix atoms plays a role of precursor for dislocation loop punching [27,29,49–55]. SIAs are pushed out first and rearrange their orientations to form dislocation loops. Generally, bubble growth is accompanied by large dislocation loop formation and punching [49–55]. It is reported that pushing out SIAs and reorientation of SIA cluster can release the high pressure of He bubbles [49]. Dislocation loop punching can only be triggered above certain critical He pressure and also releases He bubble pressure, but less work is conducted on the effect of large loops on He bubbles.

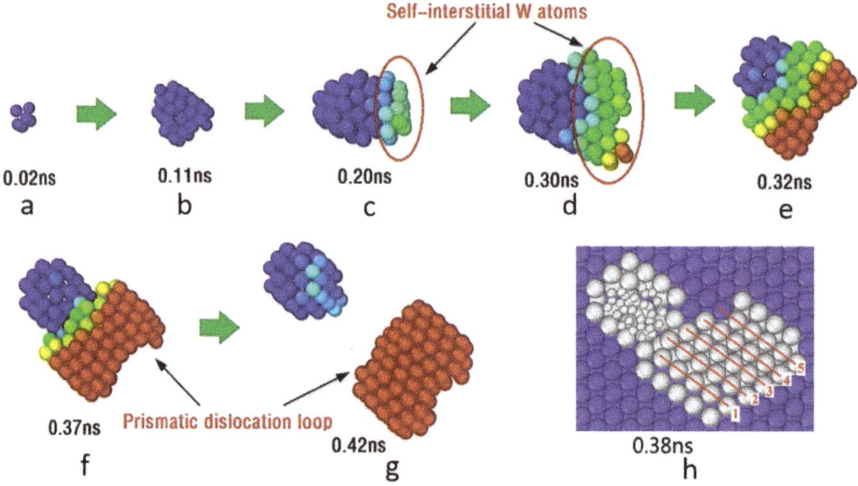

Figure 5. The evolution of He bubble with an initial radius of 0.15 nm. The color from blue to red qualitatively designates the distance from the present atomic site to the center of the He bubble. (**a,b**) Homogeneous bubble nucleation and growth; (**c,d**) self-interstitial tungsten atoms emit from matrix and then attach to the bubble surface; (**e–g**) self-interstitial tungsten atoms evolve into a prismatic dislocation loop and then dissociate from the bubble; and (**h**) the atomic structure of the self-interstitial atom cluster. Reprinted with permission from [49]; Copyright 2017 Elsevier.

2.4. Bubble Coarsening

Bubble coarsening is commonly observed when metals are annealed at a certain temperature that is higher than the temperature at which bubble nucleation and growth proceed [56–59]. With constant He dose, bubble coarsening inevitably leads to increased average bubble size and reduced bubble number density [56–59]. Marochov et al. [57] conducted thermal treatment on nickel (Ni) after they implanted He at a dose of 1×10^{17} ions/cm^3. Figure 6 shows He bubble in irradiated Ni annealed at 750 °C for 2, 12, 20 and 100 h, respectively [57]. By comparing bubble morphologies in Figure 6, it is evident that the average bubble size increases obviously with increasing of annealing time, indicating significant bubble coarsening under annealing [57].

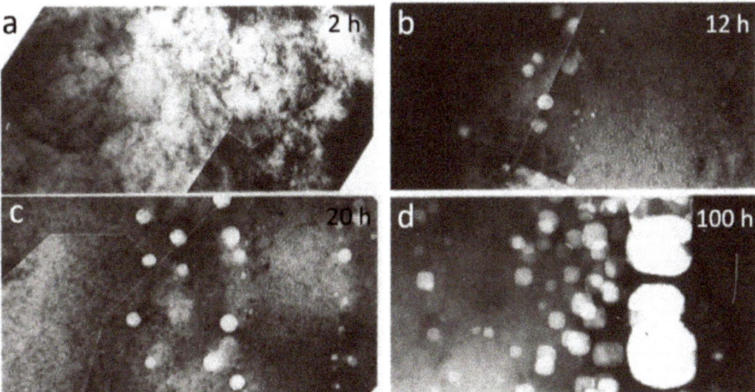

Figure 6. He bubbles in Ni implanted with 1×10^{17} ions cm^{-2} at 500 keV, annealed at 750 °C for: (**a**) 2 h; (**b**) 12 h; (**c**) 20 h; and (**d**) 100 h. He bubbles size increases with increasing annealing time. Reprinted with permission from [57]; Copyright 1987 Elsevier.

Generally, two typical mechanisms, bubble migration and coalescence (BMC) and Ostwald ripening (OR), are proposed for bubble coarsening under annealing [56–61]. BMC mechanism mainly depends on the rearrangement of bubble surface through diffusion of internal surface atom [58,59]. Marochov et al. [57] found that small bubbles are preserved for a long time in irradiated Ni under annealing. This seems to be counterintuitive as small bubbles are believed to deliver relatively high mobility. This phenomenon can be rationalized with bubble pressure. Small bubbles usually exhibit high equilibrium pressure due to the high He/V ratio in bubbles [28,57]. The high equilibrium pressure significantly suppresses the diffusion of internal surface atom, which reduces the mobility of small bubbles [57]. More vacancies are required for small bubbles before they coarsen through BMC. In addition, large bubbles develop energetically favorable facets and the mobility is controlled by ledge nucleation [57]. OR mechanism is driven by different equilibrium pressure of bubbles with different sizes [60,61]. This process is controlled by thermally activated dissociation of He–V from one bubble and recombination of He/V to another bubble [56,60,61]. Generally, He–V tends to dissolute from small bubbles and then recombines with large bubbles, leading to shrinkage or even disappearance of small bubbles and coarsening of large bubbles. Obviously, the activation energy for OR mechanism is determined by the energy for dissociation of He–V from bubbles, which is much higher than that for MC mechanism [56]. Therefore, temperature is believed to play a key role in determining the bubble coarsening mechanism under annealing.

Other studies demonstrated that microstructural characteristics (i.e., alloying elements, precipitates, defects etc.) may exert non-negligible effects on bubble behaviors, leading to different bubble coarsening mechanisms. Ono et al. [62] observed bubble migration and coalescence in He-irradiated Fe under 750 °C annealing treatment, while bubbles in Fe-9Cr show much lower mobility. In addition, Roldán et al. [63] conducted post-irradiation annealing experiments on He-irradiated two reduced activation ferritic martensitic steels (EU-ODS EUROFER and EUROFER97). Both steels contain complex microstructures that serve as barriers for bubbles. Large He bubbles are formed in EU-ODS EUROFER, while high-density small He bubbles are observed in EUROFER97. They proposed that different He bubble morphologies in samples should be attributed to different microstructures. Neither OR nor BMC fits perfectly with bubble coarsening under annealing treatment in their samples. These studies indicate that microstructures also significantly influence bubble coarsening mechanism.

3. Helium Bubble in Nanocrystalline Metals and Metallic Multilayers

In the preceding section, we overview He bubbles nucleation, growth and coarsening in the interior of grains. Interfaces, mainly GBs and heterophase interfaces (bi-metal interfaces), have attracted tremendous interests in the past decades [64–112]. Two reasons are responsible for the tremendous research interest of interfaces. Firstly, He bubbles at GBs are most harmful due to their tendency of coalescence under stress or at elevated temperature. In addition, a transition from bubble to void is widely identified in metals containing bubbles [64–67]. At certain temperature, bubbles larger than a critical size evolve into voids by absorbing vacancies quickly [64–67]. For GB He bubbles, this transition is particularly deleterious and always lead to high temperature helium embrittlement in metals [67]. More details are discussed in Section 5. Secondly, GBs and heterophase interfaces are sinks for radiation-induced defects, which are usually used to design radiation tolerant materials [64,68,69]. Serving as efficient sinks for radiation defects, interfaces can mitigate and recover various defects produced in metals after high displacement damage [68,69]. Singh [70] found that austenitic stainless steel (SS) exhibits enhanced radiation tolerance with reducing grain sizes as GBs serve as sinks for radiation defects. Similarly, Zhang et al. [23] reported that He bubble size and density all decrease with increasing interface density in He irradiated Cu/V multilayers. Using three atomistic simulation methods, Bai et al. [68] proposed a novel "loading-unloading" mechanism to rationalize the strong sink effect of GBs in Cu. They recognized that interstitials are loaded into GBs under irradiation and then are emitted to annihilate vacancies in bulk [68]. Ackland [69] pointed out that the "loading-unloading" mechanism is general and may also work at heterophase interfaces. Therefore, GBs and heterophase interfaces are introduced into metals deliberately to enhance radiation tolerance [64]. Generally, nanocrystalline (NC) metals have higher radiation resistance than their coarse-grained (CG) counterparts, and nanolaminates show enhanced radiation tolerance than their bulk single-phase counterparts. In this section, we introduce the influence of interface on He bubbles agglomeration in metals. A fundamental understanding on bubble–interface interactions is crucial for managing He bubbles, mitigating He-induced degradation of interface and designing of radiation-tolerant metals.

3.1. Helium Bubbles in Nanocrystalline Metals

GBs, as common defects in polycrystalline metals, have sharply different atomistic structures from that of interior of grains. Sun et al. [71] performed in situ Kr irradiation on both CG and NC Ni and found that high-density GBs in NC Ni significantly reduce density and size of irradiation induced defects. Similarly, some other investigations [72,73] confirm that NC metals deliver enhanced radiation resistance compared with CG counterparts. Both experiments [74–76] and simulations [77–79] report that He atoms tend to segregate to GBs and form He clusters/bubbles along GBs. Due to the sink effect of GB [79], this process is activated under irradiation and post-annealing at various temperatures [74–79]. A vivid example for He precipitation along GBs is shown in austenitic steel (Figure 7) [74]. Notably, He bubbles formed along GB are larger than bubbles located in the interior of grains. As marked by arrows in Figure 7, bubble-denuded zone (BDZ), with a width of tens of nanometers, is observed on both sides of GB [74]. BDZs, similar to void-denuded zones in Cu [80,81], indicate that GB may absorbs adjacent He and vacancies and produces poor-He zones on both sides. Similarly, approximately 10 nm-wide BDZs are identified in He-implanted ferritic alloy [75]. Taking Fe as model material, Kurtz et al. [79] evaluated the trapping efficiency of He at substitutional and interstitial sites in and near GBs. They found that interstitial He atom was strongly binding to GB core with an energy of 0.5–2.7 eV [79]. Surprisingly, even at 0 K, GBs still serve as activated He sinks and deliver a He capture radius ranges from 0.3 to 0.7 nm [79]. Considering the high mobility of He atoms in metals [27,31,64], the effective capture radius of GBs may be 1–2 orders of magnitude higher than that of calculated here. This indicates that the effective capture radius may reach several nanometers, even tens of nanometers. This simulation rationalizes the formation of BDZs in He irradiated austenitic steel [74] and ferritic alloys [75]. Series of complex factors, including intrinsic factors (matrix, GB misorientations [79,80], GB energies [80], etc.) and extrinsic factors (temperature [82], He concentration [64], etc.), are supposed to exert a non-negligible effect on BDZs formation and evolution.

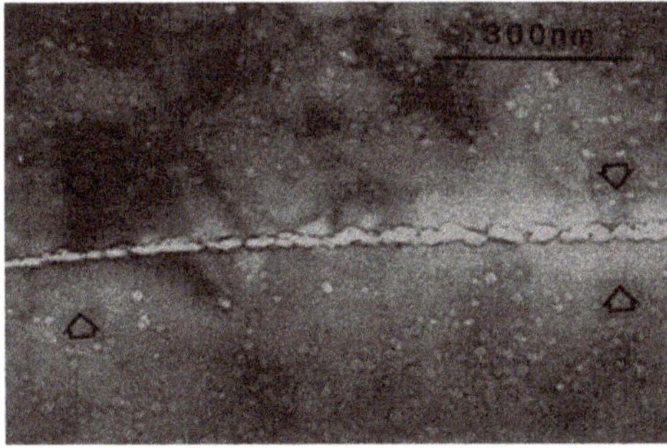

Figure 7. Large He bubbles along GB in He irradiated austenitic steel. As indicated by arrows, BDZs with width of tens of nanometers distribute symmetrically on both sides of GB. Reprinted with permission from [74]; Copyright 1983 Informa UK Limited.

Many studies [82–85] are performed to unveil the dependence of radiation tolerance on grain size and explore the underlying mechanisms on both CG metals and NC metals. Due to their huge disparity in the ratio of GB area to total volume, CG metals and NC metals irradiated by He at the same condition deliver quite different bubble morphologies and distributions [82–85]. For He irradiation, the enhanced radiation tolerance in NC metals is usually manifested as small and low-density bubbles [82–85]. A representative example for He precipitation in NC Mo and CG Mo is shown in Figure 8 [83]. In NC Mo with an average grain size of 44 nm, small and sparse bubbles are formed after He irradiation (Figure 8a). The average bubble size is about 0.6 nm. The CG Mo are decorated with dense bubbles with an average size of about 1.2 nm. This finding is also confirmed in NC Fe and CG Fe. Yu et al. [84] proposed that smaller and lower-density bubbles in NC metals should be attributed to the depletion of vacancies from grain interior induced by high-density GBs. As mentioned above, Bai et al. [68] reported that GBs can capture interstitials under irradiation, and then play a role of interstitial emitter and fire interstitials back into the lattice to recombine with vacancies that stay within a few nanometers to GBs (recombination area). NC metals possess a rather high ratio of GB area to total volume, which means many traps for interstitials are produced in collision cascade. More importantly, with reducing grain size, the ratio of width of recombination area to grain size will generally reach 1, which indicates that the overall vacancies in the interior of grains have a great chance to recombine with interstitials emitted by GBs, leading to the low concentration of vacancies [68,84]. The depletion of vacancies gives rise to smaller and lower-density of He bubbles in NC grains.

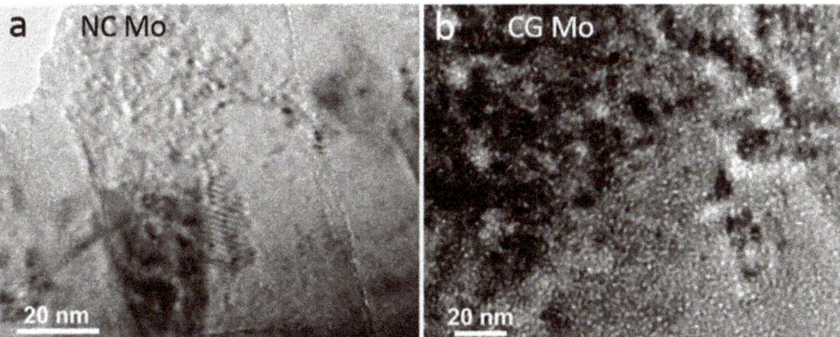

Figure 8. He bubbles induced by He irradiation in: NC Mo (**a**); and CG Mo (**b**) [83]. Small He bubbles with an average size of 0.6 nm are observed in NC Mo, while large He bubbles with an average size of 1.2 nm are identified in CG Mo. Reprinted with permission from [83]; Copyright 2016 Elsevier.

Bullough et al. [86] evaluated the dependence of sink strength on grain size. Using the cellular model, they concluded that the sink strength for GB can be describe as

$$k_{\text{gb}}^2 = 15/R^2 \tag{1}$$

where k_{gb}^2 is the GB sink strength and R is grain size. The cellular model focuses on a single spherical isolated grain without identified surrounding medium and uses the average point defect concentration within the grain. When the embedding model is used, the sink strength of GB can be calculated as

$$k_{\text{gb}}^2 = 14.4/R^2 \tag{2}$$

where k_{gb}^2 and R have the same meaning as in Equation (1). It is evident that sink strength of GB predicted using both models delivers similar values. This result unveils the trend that the sink strength increases as the grain size decreases. However, it should be noted that there are some limits for these two formulas, although they predict the dependence of sink strength on grain size [87]. Firstly, according to these formulas, the sink strength can become larger and larger as the grain size decreases. This seems to be unreasonable [87]. Secondly, these formulas ignore the dependence of sink strength on GB characters. For instance, high angle GBs (HAGBs) usually possess higher sink strength compared to low angle GBs (LAGBs), as HAGBs deliver lower formation energy for interstitials and vacancies [79,80,87]. Thirdly, the sink strength of GBs is reduced after absorbing tremendous point defects [87]. In addition, grain growth in NC metals is also inevitable under irradiation, especially at high temperature [113]. Considering these issues listed above, an alterable factor that determined by GB energy (γ) and GB orientation (θ) should be involved. The sink strength of GB can be calculated using a modified formula [87]

$$k_{\text{gb}}^2 = 15f(\theta,\gamma)/R^2 \tag{3}$$

Generally, the $f(\theta,\gamma)$ ranges from 0 to 1. Further effort is needed to investigate the dependence of $f(\theta,\gamma)$ on GB energy (γ) and GB orientation (θ), which is out of the scope of this review.

3.2. Helium Bubbles in Metallic Multilayers

Tremendous studies [74–85] have demonstrated that NC metals deliver dramatically enhanced radiation tolerance compared with their CG counterparts. However, NC metals are not widely adopted as candidate structural materials in nuclear reactors that are subjected to a high level of irradiation because of their thermal instability [113]. Generally, for NC metals containing grains with size of about 10 nm, an arresting grain growth at room temperature or even below can be identified if the equilibrium melting temperature is lower than about 873 K [113]. The temperature triggering grain growth rises with increasing grain size and melting temperature, but is still not high enough compared with that of real service environment.

Interface engineering is regarded as an important method to achieve high radiation resistance in metals [88–112]. In fact, interface engineering shares the same fundamental principle with NC metals. That is, in short, increasing the ratio of interface area to total volume enhances the radiation tolerance. As interfaces serve as efficient defect sinks, recombination of irradiation-induced point defects, mainly interstitials and vacancies, is promoted with increasing interface density, hence giving rise to enhanced radiation tolerance [88–112]. Radiation resistant multilayers consisting of alternative nanoscale layers are designed by introducing high-density heterophase interfaces and exhibit much better radiation tolerance than either of its single-phase components [94,107,108]. The evolution of radiation defects, including dislocation loops, He bubbles, voids etc., are dramatically suppressed by the high-density of heterophase interfaces [88–112]. Yu et al. [111] provided in situ observation of the phenomenon that heterophase interfaces capture and annihilate Kr irradiation-induced defect clusters in immiscible Ag/Ni multilayers.

Design of multilayers containing high-density heterophase interfaces to alleviate He irradiation damage is based on the theory that interfaces can trap He and vacancies efficiently and provide abundant nucleation sites for He bubbles [64]. Studies on He-irradiated multilayers demonstrated that heterophase interfaces play a critical role in improving radiation tolerance. Li et al. [93] characterized He bubble size and distribution in Cu and 5 nm Cu/Nb multilayers after 7 at.% He irradiation. They found that the average bubble radius in Cu/Nb multilayers is about 0.5 nm, which is much smaller than bubbles in Cu (1.35 nm). In addition, the bubble volume fraction in Cu/Nb multilayers is only about 1.74%, while the bubble volume fraction in Cu reaches 5.77%. Similarly, compared to Ag, He bubbles in 6 nm V/Ag multilayers are smaller in both size and volume friction after He irradiation [105]. VDZs [80,81] and BDZs [108] are also identified in multilayers, suggesting the critical role of heterophase interface in suppressing irradiation defects. In past decades, there are many successful examples in the design of multilayers containing high-density heterophase interfaces to achieve outstanding radiation tolerance and mechanical properties. These studies are mainly dedicated to radiation behaviors in a series of multilayers, including Cu/Nb [80,88–97,112], Cu/V [98–100], Cu/Mo [101], Cu/Co [102],Cu/W [103], Fe/W [104], V/Ag [105,106], Al/Nb [107], Cu/Ag [108,109] and Ag/Ni [110,111]. Among them, semi-coherent immiscible FCC/BCC heterophase interfaces [32,88–97,103,105–107] have been investigated both experimentally and theoretically, and this review focuses on this system.

Different laboratory methods have been developed to manufacture multilayers with various layer thickness, such as magnetron sputtering [92,93] and accumulated roll bonding (ARB) [80]. By tuning manufacture parameters, multilayers with various layer thickness can be obtained for experimental investigations. Figure 9 schematically illustrates the manufactured Cu/Nb multilayers by using ARB method [64,112]. The original bulk Cu and Nb plates are stacked together and then processed by continuous rolling, and repeated cutting and stacking; as a result, Cu/Nb multilayers with layer thickness ranging from millimeters to nanometers can be produced. Figure 9 [64,112] shows typical layered microstructures formed in ARB Cu/Nb multilayers. The Cu/Nb interfaces are planar and sharp. The controllable layer thickness makes it easy to tune the density of heterophase interface, which provides an opportunity to evaluate the dependence of radiation tolerance on the density of heterophase interface. Notably, novel interfaces are found in multilayers processed using ARB. In ARB Cu/Nb multilayers, $\{112\}_{FCC} | | \{112\}_{BCC}$ and $\{110\}_{FCC} | | \{112\}_{BCC}$ are two of the dominant interface orientations [64,112]. Contrary to ARB, FCC/BCC multilayers synthesized using magnetron sputtering usually have similar interface orientation relationship (except for V/Ag multilayers [90,105,106]). For instance, all heterophase interfaces in Cu/Nb magnetron sputtered multilayers have the Kurdjumov–Sachs orientation relationship with interface plane of $\{111\}_{FCC} | | \{110\}_{BCC}$ [89–97].

Figure 9. Multilayers manufactured using ARB with layer thickness ranging from millimeters to nanometers. Reprinted with permission from [64]; Copyright 2015 Elsevier; Reprinted with permission from [112]; Copyright 2013 Cambridge University Press.

As mentioned above, He has extremely low solubility in metals. He atoms, once introduced into metals, are supposed to precipitate and form bubbles quickly. Surprisingly, Demkowicz et al. [90] found that no He bubbles were identified in region with low but not zero He concentration via TEM in magnetron sputtered Cu/Nb multilayers after He^3 irradiation [90]. In Figure 10, the upper image shows Cu/Nb multilayers after He^3 irradiation, while the lower figure describes the evolution of He^3 concentration along depth. The He^3 concentration is measured by nuclear reaction analysis (NRA), instead of estimation by using the Stopping and Range of Ions in Matter (SRIM) [114], as NRA accurately reflects the true He^3 depth profile [90]. He bubbles are only observed at high He concentration region, while no He bubbles are detected in low He concentration region. Based on this sharp comparison, it is inferred that small He clusters, which are too small to be detected by TEM, are captured by Cu/Nb heterophase interfaces in low He concentration region. With increasing He concentration, these small clusters evolve into visible bubbles, as shown in the middle part of Figure 10. This comparison indicates that there exists a critical He concentration to form visible He bubbles (under TEM) along interfaces.

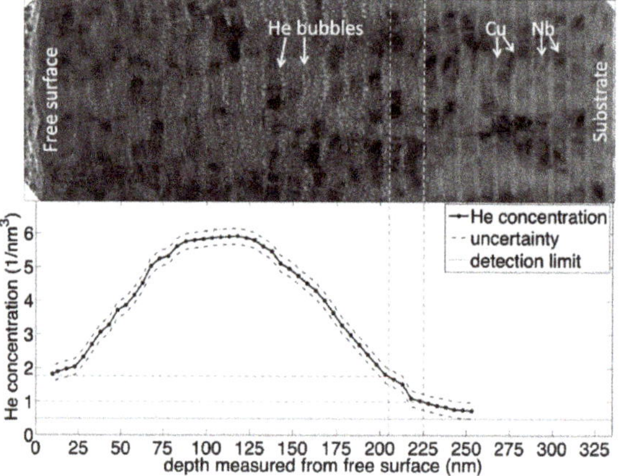

Figure 10. TEM image of He^3-implanted Cu–Nb multilayers with layer thickness of 5.6 nm and He concentration profile measured using NRA. For region below a critical depth around 215 nm, no He bubble can be observed. Reprinted with permission from [90]; Copyright 2010 AIP.

To determine the critical He concentration to form visible He bubbles in Cu/Nb multilayers, Demkowicz et al. [90] performed He^3 irradiation on several multilayers with different layer thicknesses. These irradiated multilayers are characterized using TEM with under-focus mode to detect He bubbles. He^3 concentration as a function of depth is evaluated using NRA. When converted to number of He atoms per unit of interface area (rather than sample volume), the critical He concentration for formation of visible He bubbles is about 8.5 atoms/nm^2 in Cu/Nb multilayer, which is irrelevant to layer thickness [90]. Analogous studies find that the critical He concentrations to form visible He bubbles in Cu/V multilayers [98–100] and Cu/Mo [101] multilayers are 1.9 atoms/nm^2 and 3 atoms/nm^2, respectively. The sharply different critical He concentrations in various systems are ascribed to different areal densities of misfit dislocation intersections (MDIs) in various heterophase interfaces [90,98–101].

Atomistic modeling revealed that $\{111\}_{FCC}$ | | $\{110\}_{BCC}$ interfaces contain two sets of parallel misfit dislocations [95], forming a series of MDIs on interfaces. As demonstrated using atomistic modeling in Reference [95], these MDIs are preferred He bubble nucleation sites. Figure 11a shows the effect of creating a vacancy near a misfit dislocation in Cu plane. After annealing for 10 ps at 300 K followed by energy minimization, the vacancy occupies the MDI with a low formation energy of −0.13 eV. This configuration indicates that interface decorated by constitutional vacancy is energetically favorable to the vacancy-free interface. Constitutional vacancies were continuously added to the heterophase interface until no negative formation energy sites remain. The final Cu/Nb interface containing about 5 at.% constitutional vacancies gathered near MDIs has about 25 mJ/m^2 lower energy than a perfect interface. That is, about 2.5 constitutional vacancies gather near one MDI (Figure 11c). Similarly, a Cu/V interface with 2.5 constitutional vacancies on single MDI (Figure 11d) and an energy reduction of 3.4 mJ/m^2 is formed finally. It should be noted that, due to different lattice parameters, the areal density of MDIs on Cu/Nb interface is over a factor of five larger than that on Cu/V interface [90,98–100]. This results in the atomic percentage of constitutional vacancies (about 0.8 at.%) on Cu/V interface being much smaller than that on Cu/Nb interface (about 5 at.%) [90]. Due to the high binding energy, He atoms are easily trapped by constitutional vacancies locating on MDIs and form He–V clusters. These He–V clusters can be regarded as He bubble embryo and may develop into He bubbles. Hence, MDIs are preferred He bubble nucleation sites. Notably, the amount of He atoms trapped by a MDI should be constant before forming a visible He bubble. The areal density of MDIs is five times higher for Cu/Nb interface than that for Cu/V interface, which corresponds very well with the fact that the critical He concentration to form visible He bubbles on Cu/Nb interface (about 8.5 atoms/nm^2) is also about five times higher than that on Cu/V interface (about 1.9 atoms/nm^2). This attests that different areal densities of MDIs are responsible for various critical He concentration in different multilayers.

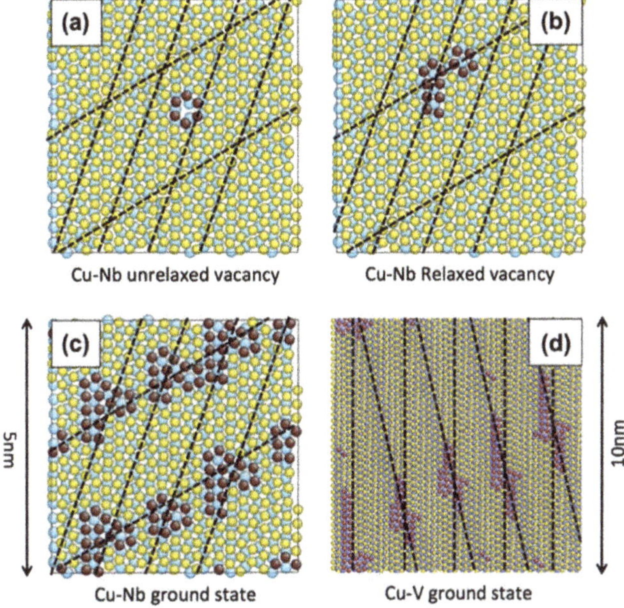

Figure 11. Unrelaxed (**a**) and relaxed (**b**) vacancies on the Cu side of a Cu–Nb interface; (**c**) about 5 at.% constitutional vacancies locates in the ground state Cu–Nb interface; and (**d**) about 0.8 at.% constitutional vacancies locates in the ground state Cu–V interface. Vacancies tend to gather around MDIs. Reprinted with permission from [95]; Copyright 2012 Elsevier.

MDIs in heterophase interfaces are efficient sinks for He. This deduction is further proved in bi-crystal Au film, as shown in Figure 12 [115]. He implantation is conducted on both single-crystal and bi-crystal Au foil containing a twist boundary. Dense He bubbles with various bubble size are homogeneously distributed in single-crystal Au foil, as shown in Figure 12b. Interestingly, He bubbles are observed to array periodically on dislocation intersections in bi-crystal Au. The twist boundary comprises a square grid of screw dislocations [115]. All dislocation intersections are occupied by

He bubbles, while no bubbles are identified in matrix. Similar bubble distribution on dislocation intersections is further demonstrated using Metropolis Monte Carlo simulation (Figure 12c) [115]. From the comparison between single-crystal Au and bi-crystal Au, it is evident that dislocation intersections are preferred He bubble nucleation sites. Similarly, MDIs in heterophase interfaces decorated with vacancies can trap He efficiently [95].

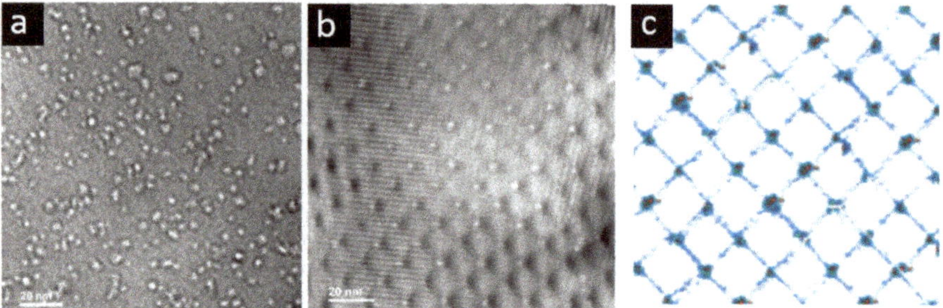

Figure 12. (**a**) Dense He bubbles in single-crystal Au after He irradiation; (**b**) He bubbles distribute on MDIs in a (001) twist grain boundary with a misorientation angle about 1°; and (**c**) Metropolis Monte Carlo simulation shows dislocation network in blue and He bubble at MDIs in red. Reprinted with permission from [115]; Copyright 2012 Elsevier.

Figure 13 [95] plots the areal density of MDIs and the critical He concentration to form visible bubbles as a function of the lattice parameters ratio. The areal density of MDIs in FCC/BCC heterophase interfaces can be evaluated using O-lattice theory, which is described in details in Ref. [116]. The areal density of MDIs in Ag/V, Cu/V, Cu/Mo and Cu/Nb interfaces calculated adopting O-lattice theory are plotted in Figure 13 [95]. Generally, the MDI areal densities vary with different lattice parameter ratio without a clear trend for ratio ranging from 0.7 to 0.95. However, for lattice parameter ratio ranges from about 0.82 to 0.95, the areal density of MDIs calculated using O-lattice theory increases monotonically with increasing lattice parameter ratio (Figure 13). The Cu/Nb interfaces have the largest lattice parameter ratio and highest areal density of MDIs. Similarly, the measured critical He concentration also increases monotonically with increasing lattice parameter ratio, as shown in Figure 13. The dependence of critical concentrations on lattice parameter ratios matches well with that of the MDI areal densities. This further attests that various critical He concentrations should be ascribed to different areal densities of MDIs in FCC/BCC multilayers. Based on the areal densities of MDIs and the critical concentrations, it can be determined that about 25 He atoms can be stored on each MDI without forming a visible He bubble in Cu/Nb, Cu/V and Cu/Mo multilayers [90]. According to Figure 13, it seems that Ag/V multilayers contain extremely low-density of MDIs. This should be ascribed to the fact that neighboring Ag and V layers in magnetron sputtered Ag/V multilayers deliver a variety of orientation relations, including Kurdjumov–Sachs, Nishiyama–Wasserman, Bain, and Pitsch orientation relations [105]. Moreover, each successive layer in Ag/V multilayers is polycrystalline. These heterogeneous structures in Ag/V multilayers do not mean that Ag/V multilayers have lower radiation tolerance compared to other FCC/BCC multilayers. In fact, Ag/V multilayers exhibit similar He-irradiation tolerance to Cu/Nb multilayers [90,105].

Heterophase interfaces in semi-coherent FCC/BCC multilayers show unprecedented He storage capacity, which is far beyond the equilibrium solubility of He in metals. The phenomenon that one MDI can sustains 25 He atoms without forming visible bubbles is not unexpected and attracted tremendous interests [88–111]. Recent atomistic modeling work focused on He atom behaviors at MDIs rationalized why one MDI can sustain so many He atoms without forming visible bubbles. A He cluster evolution process, briefly described as "platelet to bubble" transition, is proposed [96] to explain the unprecedented He storage capacity of interfaces.

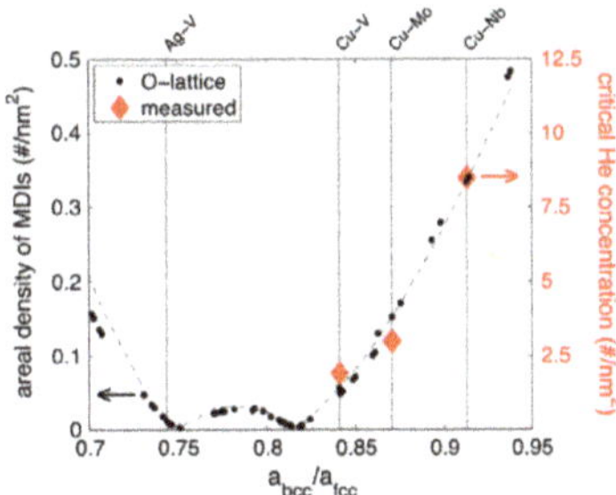

Figure 13. Black dots show the dependence of areal densities of MDIs calculated using O-lattice theory on lattice parameter ratios in FCC/BCC interfaces with identical interface orientation relationship. Red diamonds represent the critical He concentration to form visible He bubbles with different lattice parameter ratios. Reprinted with permission from [95]; Copyright 2012 Elsevier.

He bubbles in crystalline metals tend to evolve into stable spherical bubbles due to high isotropic equilibrium pressure. However, low concentration He may form stable He platelets at interfaces [96]. As shown in Figure 14a–c, He platelet is formed by wetting high-energy interface regions and extends along the interface with increasing He concentration in Cu/Nb multilayers. The shape and location of the He platelet is determined by three surface energies. These surface energies are the Cu/He surface energy $\gamma_{Cu/He}$, the Nb/He surface energy $\gamma_{Nb/He}$ and the Cu/Nb interface energy $\gamma_{Cu/Nb}$. An excess interface wetting energy is then described as [96],

$$W = \gamma_{Cu/Nb} + \gamma_{Cu/He} - \gamma_{Cu/Nb} \tag{4}$$

when $W > 0$, the He platelet is expected to wet the interface at a contact angel that depends on the surface energies to minimize the total energy. The surface energies $\gamma_{Cu/He}$ and $\gamma_{Nb/He}$ are calculated to be 1.93 and 2.40 J/m^2, respectively. The average interface energy $\gamma_{Cu/Nb}$ is 0.54 J/m^2 [96]. The excess interface wetting energy should always be positive, indicating that, regardless of size, He cluster will always wet the interface in the form of He platelet. However, this hypothesis is valid for He platelet containing fewer than 20 He atoms. This is generally consistent with the former result of 25 He atoms [90]. With further increasing He concentration, the He platelet grows through adding a new layer of He and keeps its area along the interface constant, as shown in Figure 14d. This indicates that a transition from platelet to bubble happens. Further increasing He concentration promotes bubble growth, as shown in Figure 14e.

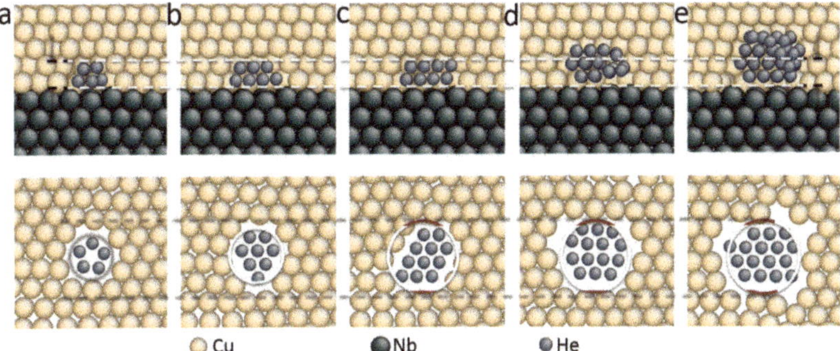

Figure 14. The evolution process of He cluster along Cu/Nb interface: (**a–c**) He platelet is formed by wetting high-energy interface regions and extends along the interface; and (**d,e**) He platelet transforms into He bubble by adding a new layer of He and keeps its area along the interface constant. Reprinted with permission from [96]; Copyright 2013 NLM.

Due to the existence of MDIs in semi-coherent Cu/Nb interfaces, $\gamma_{Cu/Nb}$ is not homogeneous across the interface and varies with positions, as shown in Figure 15 [96]. The high energy areas deriving from MDIs in Figure 15 have an interface energy of about 0.8 J/m^2, while the low energy areas have an interface energy of about 0.4 J/m^2, leading to an average interface energy of about 0.54 J/m^2 [90]. The heterogeneity of interface energy across the interface is responsible for the platelet to bubble transition [96]. For high interface energy regions, the excess interface wetting energy W is always positive, giving rise to formation of He platelet by wetting the interface. With increasing of He concentration, the He platelet extends by further wetting the interface [96]. However, once the He platelet is large enough and exceeds the original high interface energy region, W will be negative, which inhibits further wetting. As a result, the growth of He platelet proceeds into Cu due to the lower vacancy formation energy in Cu. Finally, the platelet transforms into bubble and visible bubbles are detected along interfaces [96].

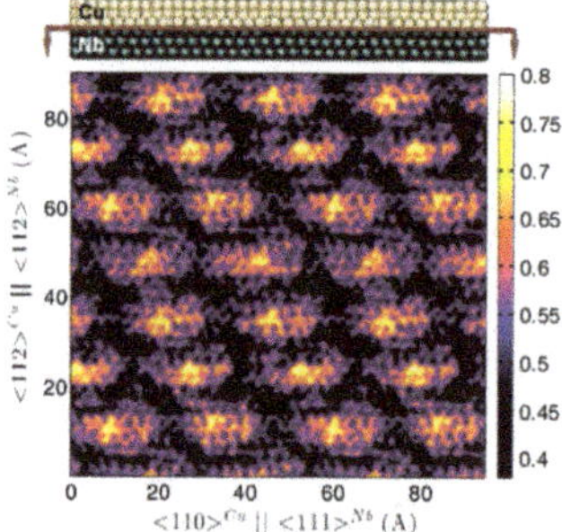

Figure 15. Location dependence of Cu/Nb interface energy. Areas occupied by MDIs deliver the highest interface energy. Reprinted with permission from [96]; Copyright 2013 NLM.

The above model describes He precipitation along semi-coherent FCC/BCC interfaces innovatively and reasonably. MDIs along interfaces play a critical role in tuning radiation defects and contribute to the high radiation resistance in NC metals and multilayers. MDIs deliver outstanding He storage capacity besides enhancing interstitials and vacancies recombination. An important conclusion derived from above model is that design of interfaces with high-density MDIs is a potential way to achieve enhanced radiation tolerance in materials. These insights provide a practical way to predict He behaviors along interfaces, such as formation of He platelets and bubbles. In addition, the He concentration that a certain multilayer may sustain before forming visible He bubbles along interfaces can be calculated using MDI density. This would surely facilitate the prediction of lift-time of various multilayers exposed to He irradiation. More importantly, these insights guide the design of multilayers and provide new avenues for manufacturing radiation-tolerant materials. However, to apply the interface engineering strategy in real application, a new method that can be scale-up to fabricate interface-dominated metals in industry is still needed.

4. Helium Radiation Hardening

In this part, we overview irradiation hardening induced by He bubbles. Most investigations on the strengthening effect of He bubbles are conducted using direct He implantation to introduce a high density of He bubbles within a short range of time [88–111]. However, He implantation usually forms He bubbles within a shallow region due to the limitation of ion energy. The depth of He-bubbled region generally ranges from hundreds of nanometers to several micrometers [88–101], which depends on the energies of incident He ions. Hence, nano/micro-mechanical tests are widely adopted to evaluate He bubble induced hardening in these sample. Nanoindentation [63,116,117], and in situ mechanical testing in SEM [118–121] and TEM [122–127] are widely performed to investigate the effect of He bubbles on the mechanical properties of various metals. We mainly focus on three parts next: He irradiation hardening in single-phase metals, He irradiation hardening in metallic multilayers and theoretical modeling of He irradiation hardening.

4.1. Helium Radiation Hardening in Single-Phase Metals

We start the discussion with He irradiation hardening in single-phase metals. Several studies are performed on He irradiated steel [63], Ni [120], Cu [118,122–126] and Al–4Cu alloy [127] to evaluate He radiation hardening and interactions of He bubble with dislocations/deformation twin [63,120–127]. Using nanoindentation teats, Roldan et al. [63] found that He-irradiated steels after annealing demonstrate much higher hardness compared with as-received samples. We [122] demonstrated that He bubbles with an equilibrium internal pressure slightly less than 1 GPa serve as shearable obstacles and internal dislocation sources in single-crystal Cu pillars. As shown in Figure 16a, pillars containing He bubbles deliver higher flow stress and more stable deformation compared with the deformation of fully dense Cu pillars. He bubbles reduce dislocation velocity and mean free path, increase shear stress, and promote the storage of dislocations, giving rise to enhanced yield strength, flow stress and strain hardening in Cu pillars. This is verified by in situ nanomechanical measurement and videos of dislocation–He bubbles interaction in both fully dense and He-bubbled Cu pillars [122]. Sharp slip steps can be identified on surface of deformed fully dense Cu pillar (Figure 16b), while He-bubbled pillars form smooth front surface after large strain of compression (Figure 16c). More importantly, we found that He bubbles can enhance ductility in small-volume Cu pillars [122], which is contrary to the widely accepted He-induced ductility lost in metals [18,19]. The reasons for enhanced ductility are discussed in next section. Similarly, another related study on He irradiated Cu [126] indicates that He bubbles performed as obstacles for both dislocations and twins, leading to enhanced yield strength and flow stress of Cu pillars. In addition, the authors of [120,126] reported a dose-dependent He irradiation hardening in Ni and Cu. The flow stress in both Cu and Ni pillars increases significantly with increasing of He concentration [120,126]. In tensile tests, the ultimate tensile strengths (UTSs) for fully dense single-crystal Ni, and Ni after He implantation with a fluence of 2×10^{17} ions cm^{-2} and 3.8×10^{17} ions cm^{-2} are 241, 384 and 503 MPa, respectively. The dose-dependent He irradiation hardening should be common in various monolithic metals, but there may exist a critical He concentration over which the He irradiation hardening keeps constant or even decreases with further increasing of He concentration, as demonstrated in He implanted pure Ag [106]. Below the critical He concentration, the spacing of He bubbles reduces continuously with increasing He concentration due to bubble growth and new bubble nucleation. Once the critical He concentration is reached, bubble coalescence, bubble coarsening or bubble-to-void transition may take place, leading to a complex mechanical response. The detailed bubble evolution and their mechanical effects at high He concentration need to be further studied experimentally and theoretically.

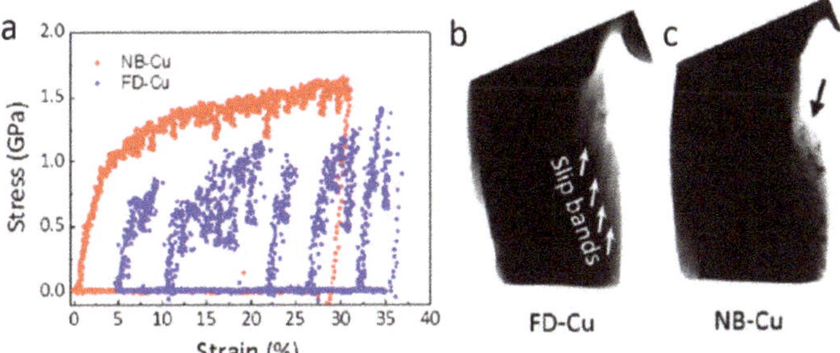

Figure 16. Compression tests on FD-Cu (unirradiated) and NB-Cu (irradiated) pillars: (**a**) NB-Cu containing nanoscale He bubbles deliver stable deformability and higher flow strength; (**b**) sharp slip bands are identified in FD-Cu pillar; and (**c**) smooth deformation profile is identified in NB-Cu pillar. Reprinted with permission from [122]; Copyright 2016 American Chemical Society.

4.2. Helium Radiation Hardening in Metallic Multilayers

In this section, we review He radiation hardening in multilayers. Nanoindentation tests on Cu/V multilayers with various layer thickness reveal a size-dependent He bubbled-induced hardening [100]. For layer thickness ranging from 200 to 2.5 nm, the magnitude of He bubble-induced hardening decreases with reducing layer thickness, and the hardening effect is negligible for layer thickness of 2.5 nm or less. Notably, this size-dependent irradiation hardening is also identified in other multilayers, such as V/Ag [106], Ag/Ni [111], Fe/W [104], Cu/Cu–Zr [117], etc. (Figure 17a). Many studies indicate that the size-dependent irradiation hardening is valid for layer thickness h below a critical value h_c. This trend disappears when h gradually increases to the critical thickness [106]. The size-dependent irradiation hardening in multilayers can be rationalized as follows. Firstly, radiation damage decreases with reducing layer thickness in multilayers [88–94], as discussed in Section 3. TEM images of He irradiated 50 and 2.5 nm Cu/V multilayers (Figure 17b,c) confirm that 2.5 nm Cu/V multilayers exhibit much better radiation tolerance, as manifested by the formation of low-density of He bubbles and negligible irradiation hardening. Secondly, the size-dependent He irradiation hardening is closely related to the deformation mechanism transition in nanolaminates with reducing layer thickness [128–132]. The deformation mechanisms in multilayers can be classified into three regions with the variation of layer thickness [128–132]. (1) For layer thickness ranging from millimeters to sub-micrometers, dislocation pile-up (Hall–Petch model) in single layer is the dominant deformation mechanism. (2) For layer thickness ranging from hundreds to tens of nanometers, confined layer slip (CLS) is the dominant mode. (3) For layer thickness of several nanometers, dislocation transmitting interface is the main deformation mechanism [128–132]. He irradiation hardening in multilayers with various layer thickness can be well explained based on the three dominant deformation models discussed above.

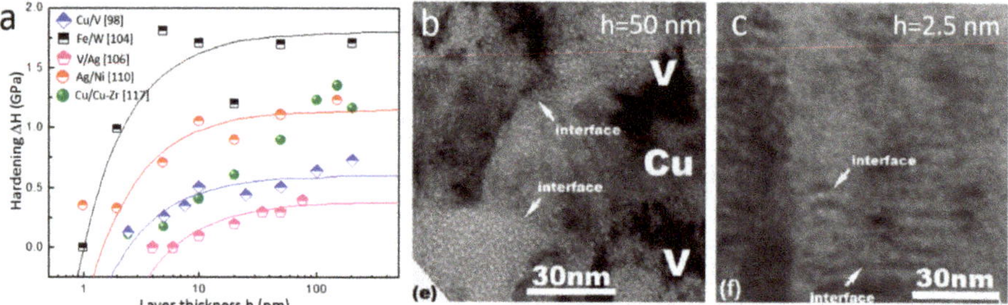

Figure 17. (**a**) Size dependent radiation hardening in various multilayers [98,104,106,110,117]; (**b**) TEM images of He irradiated Cu/V with a layer thickness of 50 nm [100]; and (**c**) He irradiated 2.5 nm Cu/V multilayers [100]. Reprinted with permission from [100]; Copyright 2009 Elsevier.

In the cases of dislocation pile-up and CLS, dislocations are expected to cut through bubbles to carry plasticity. Wei et al. [106] proposed that the force of interaction between bubbles and dislocations is confined to a small segment of dislocations. Thus, the total resistant force produced by bubbles can be calculated as the sum of individual interaction force (Figure 18a). The irradiation-induced hardening can be determined by the equilibrium between the line force and sum of resistant force [106]

$$\Delta\tau = \frac{n\tau_i bl}{bh'} \tag{5}$$

where n is the number of bubbles along dislocation, τ_i is the average shear strength of bubble, b is the Burgers vector, l is bubble spacing, and h' is layer thickness parallel to glide plane. Parameter n in Equation (5) can be substituted using $n = (h' - l)/l$, and then Equation (5) can be rewritten as

$$\Delta\tau = \tau_i \left(1 - \frac{l}{\sqrt{2}h} \right) \tag{6}$$

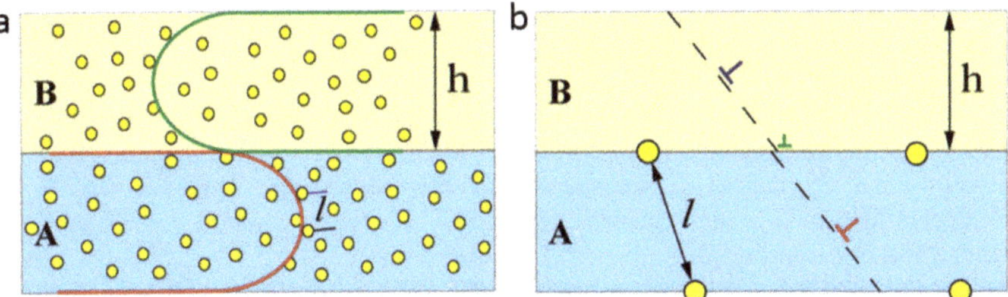

Figure 18. Schematic illustration of dislocation-He bubble interactions in multilayers with various layer thicknesses: (**a**) for large layer thickness, CLS is the dominant deformation mechanism and He bubbles serve as obstacles for dislocations; and (**b**) for small layer thickness, dislocation crossing is the main deformation mechanism and He bubbles on interfaces have a less significant effect on dislocation behavior. Reprinted with permission from [106]; Copyright 2011 Elsevier.

For multilayers with layer thickness h ranging from millimeters to sub-micrometers, irradiation hardening derives from bubble–dislocation interactions in single layers. In this region, h is far larger than the average bubble spacing l. In addition, because h is larger than the critical layer thickness h_c, the irradiation hardening is about τ_i [106].

For multilayers with layer thickness h ranging from hundreds to tens of nanometers, CLS is the dominant deformation mechanism and bubbles distributing inside layers serve as obstacles for dislocation motion. According to Equation (6), irradiation hardening $\Delta\tau$ decreases with reducing layer thickness h, manifested as size-dependent irradiation hardening (Figure 18a).

For multilayers with layer thickness h of several nanometers (<5 nm), dislocation transmitting interface serve as the dominant mechanism. For nanolaminates, He bubbles tend to array on interfaces and only some bubbles stay inside layers (Figure 18b). Because Equations (5) and (6) are established based on the situation that bubble–dislocation interactions occur in a single layer (dislocation pile-up and CLS dominate), they are not applicable to calculate He radiation hardening in extremely thin multilayers [64,93]. Li et al. [93] reported that 5 nm and 2.5 nm multilayers after implant of 7 at% He demonstrate similar irradiation hardening as that of after 1 at. % He implantation. This implies that the interface He bubbles/platelet produce a similar hardening effects and only give rise to modest hardening in multilayers containing extremely thin layers [93,96]. The detailed mechanism of interaction between dislocation and He platelet/bubble still needs further study.

4.3. Modeling of Helium Radiation Hardening

A widely used model to calculate He irradiation hardening is the Friedel–Kroupa–Hirsch (FKH) model [126,128,133,134]. He irradiation hardening in monolithic metals and metallic multilayers containing thick layers ($h > 5$ nm) can be evaluated as [133,134]

$$\Delta\sigma = \frac{1}{8}M\mu b d N^{2/3} \tag{7}$$

where M is the Taylor factor, μ is the shear modulus, b is the Burgers vector, d is the average bubble diameter, and N is the average bubble density. This model [133] was initially presented to describe elastic interactions of dislocations and dislocation loops, but it works well in evaluating hardening induced by weak obstacles for dislocation motion, such as He bubbles [126]. Generally, the FKH model is applicable to obstacles with a barrier strength lower than 0.25 [134]. As He bubbles deliver a barrier strength of about 0.2 [135], this low barrier strength makes it reasonable to calculate He irradiation hardening using the FKH model.

A simplified Orowan model is also proposed to evaluate radiation-induced hardening in metals [93,136–138]. The model can be expressed as follow.

$$\Delta\tau = \mu b / Kl \tag{8}$$

where l is the average obstacle spacing, K is a factor related to obstacle strength. Obviously, strong obstacles usually deliver small K [136,137]. Osetsky et al. [136,137] demonstrated that K is about 1.8–10 for impenetrable defects depending on obstacle size, and is 2–5 for voids. He bubbles with $l = 7.9$ nm have $K = 5.5$ in Ag/V multilayers. Generally, He bubbles have larger K value compared to voids due to relatively low obstacle strength.

He bubbles with low equilibrium pressure usually serve as shearable obstacles for dislocation motion [123]. For shearable bubbles, dislocations are forced to bow out and then cut bubbles to further carry plasticity, as identified in He irradiated Cu [122]. The resolved shear stress required for a dislocation to cut two adjacent He bubbles is calculated using the modified Orowan equation [106,139].

$$\tau = \frac{\mu b}{2\pi l} \ln\left(\frac{l}{r}\right) (\cos \varphi_c)^{1/2} \tag{9}$$

where μ, b and l have the same meaning as in Equations (8) and (9). r is the radius of He bubbles and φ_c is defined as half of the critical angle of a bow-out dislocation. φ_c can be regarded as a reference to evaluate dislocations and obstacles interactions. $\varphi_c = 0°$ corresponds to strong obstacles and Equation (9) deduces to Orowan formula. Very weak obstacles deliver φ_c of about 90°. Wei et al. [106] found that $\varphi_c = 76°$ for 0.8 nm He bubbles in V/Ag multilayers. Similarly, we calculated that φ_c is about 60° for 4 nm He bubbles in Al–4Cu alloy [127]. These relatively large φ_c indicate that shearable He bubbles are weak obstacles for dislocation motion.

5. Effect of Helium Bubbles on Ductility of Metals

5.1. High Temperature Helium Embrittlement in Polycrystalline Metals

As one form of well-known radiation damage, high temperature He embrittlement in metals has attracted tremendous interests in past decades [140–163]. A very low overall helium concentration could significantly degenerate the mechanical properties of metals, leading to catastrophic fracture of nuclear reactor components [140–163]. Kramer et al. [142] performed tensile tests on 304 SS after He implantation. They reported that an extremely low He concentration, about 1×10^{-7} atom fraction of He above 650 °C and 3×10^{-5} atom fraction of He above 540 °C resulting in severe ductility loss in 304 SS. A representative work to evaluate He effects on metals was conducted on oxygen free high purity Cu as well [140]. It is found that only about 2 appm He concentration can exert a profound effect on microstructure of Cu [140]. As shown in Figure 19a, a series of intergranular cavities with various size are formed along GB, which originate from coalescence of He bubbles formed along GBs. Specimens containing He introduced by tritium decay also show significant ductility loss (Figure 19b). In addition, the fracture surface (Figure 19c) has an intergranular fracture morphology, presumably due to radiation-induced cavities along GBs (Figure 19a).

Investigations performed in the past decades have gained valuable insights into high temperature He embrittlement in metals. However, the evolution process of He bubbles under stress that leads to final embrittlement is rather complicated. Stress enhanced bubble growth along GBs is widely accepted as one of dominant failure mechanisms underlying high temperature He embrittlement [141–151], but there still leaves many opening questions. Creep tests on samples after low temperature He implantation or during He implantation (named "in beam" [141]) are regarded as the most instructive way to investigate high temperature He embrittlement, as "in beam" creep tests provide an opportunity to monitor the complete bubble evolution processes under stress. Here, we briefly discuss creep tests [141] performed on 316 SS and DIN 1.4970 SS after room temperature He implantation or during He implantation [141]. Figure 20a plots the dependence of creep rupture time on creep stress in 316 SS at 1023 K. Under the same creep stress, He-irradiated 316 SS delivers much shorter creep rupture time than their counterparts without radiation [141]. Among irradiated 316 SS, the "in beam" samples show the shortest creep rupture time, indicating that high-temperature He irradiation may significantly reduce the service life of metals. In addition, all irradiated samples exhibit brittle, intergranular fracture, while samples without radiation show transgranular cup-cone fracture mixed with ductile tearing [141]. Figure 20b shows typical TEM image of 316 SS "in beam" tests after 25 hours under creep stress of 50 MPa at 1023 K. The corresponding He concentration is 2500 appm. He bubbles with mean diameter of about 100 nm distributed along GB, which is perpendicular to the applied stress. These creep tests further demonstrate that stress-enhanced bubble growth along GBs is responsible for high temperature He embrittlement [141–151].

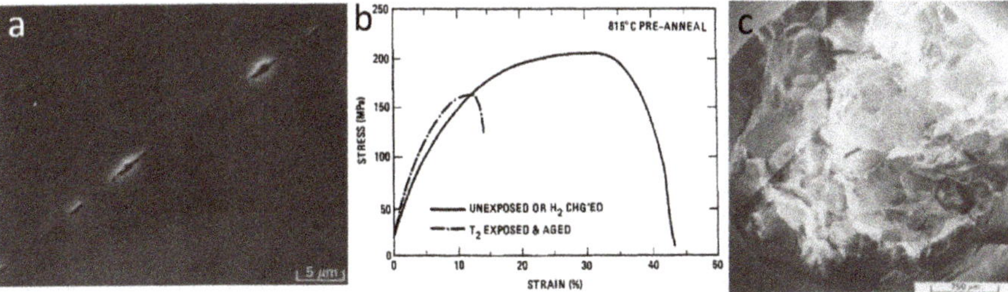

Figure 19. (**a**) Intergranular cavities along GB in Cu after T_2 exposure; (**b**) tensile behavior of unexposed and T_2 charged and aged Cu; and (**c**) intergranular fracture surface morphology of Cu after tritium exposure treatment. Reprinted with permission from [140]; Copyright 1986 Elsevier.

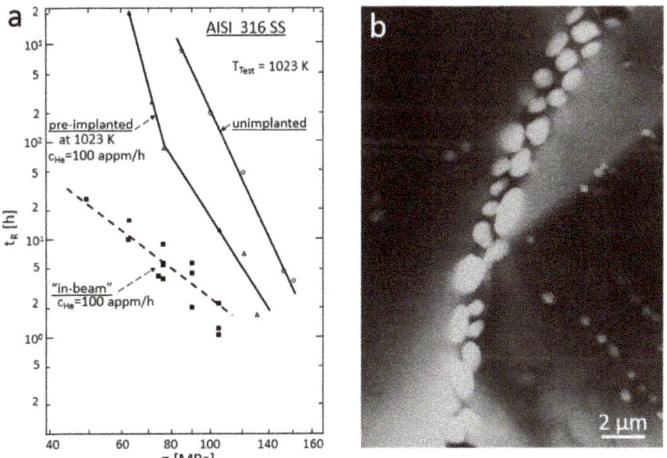

Figure 20. (**a**) The dependence of creep rupture time on creep stress of 316 SS at 1023 K with different irradiation parameters; and (**b**) large He bubbles distributed along GB, which is perpendicular to the applied stress after creep test. Reprinted with permission from [141]; Copyright 1983 Elsevier.

Next, we focus on He bubble evolution at high temperature during "in beam" creep tests. As schematically illustrated in Figure 21a, high-temperature He embrittlement consists of a sequence of processes including bubble nucleation, stable gas-driven bubble growth, unstable stress-driven void growth and crack nucleation by cavity coalescence [144–151,158]. "In beam" creep tests are accompanied by a constant He irradiation rate ($\dot{c}_{He}$). At the very beginning, He forms He–V clusters and develops into He bubbles gradually along GBs by absorbing additional He atoms and vacancies. With further increasing He concentration, the bubble concentration along GBs, c_{Bubble}^{GB}, will increase, as shown in nucleation part in Figure 21a. c_{Bubble}^{GB} reaches a peak value with further increasing of He concentration [151,158]. Thus, bubble nucleation is suppressed and bubble growth is dominant [150]. c_{Bubble}^{GB} is supposed to reduce due to bubble growth and coarsening (Figure 21a). Before reaching the critical radius c_r, He bubbles grow in a stable gas-driven growth mode by absorbing He and vacancies at an equilibrium pressure $p = 2\gamma/r$, where γ is surface energy of matrix and r is radius of He bubbles [56,144–151,158]. However, irradiation or stress-induced effective vacancy supersaturation will enhance the transition from gas-driven bubble growth to unstable stress-driven void growth at an equilibrium pressure $p < 2\gamma/r$ [56,146–148,158]. Once the He bubble reaches the critical radius c_r, a transition from bubble-to-void happens, and the samples will fracture in a short time once exposed to creep stress. An important conclusion on creep tests is that creep life time is dominated by gas-driven growth of bubbles along GBs [56,145–147]. This indicates that the time from bubble nucleation to bubble–void transition is regarded as a rupture criterion [145–147]. After the bubble-to-void transition, voids will grow unstably under creep stress. Finally, voids coalescence leads to crack nucleation and final catastrophic intergranular fracture.

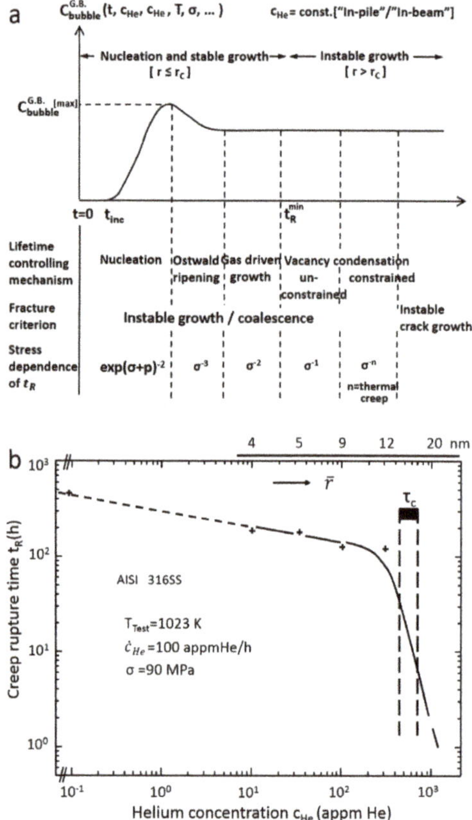

Figure 21. (**a**) The bubble evolution process along GB in creep test [158]; and (**b**) the creep rupture time decreases with increasing He concentration. The creep rupture time decreases sharply when He bubble size reaches the critical size [18]. Reprinted with permission from [158]; Copyright 1983 Informa UK Limited; Reprinted with permission from [18]; Copyright 1985 Elsevier.

Based on the discussion above, it is obvious that the critical radius c_r plays as an important role in determining the creep life. For GBs containing He bubbles under creep stress, the critical size for bubble-to-void transition can be calculated as [145]

$$c_r = 4\gamma/3\sigma \tag{10}$$

where γ is surface energy of matrix and σ is the creep stress. When the creep stress is removed, the critical size for bubble-to-void transition is [56,145]

$$c_{r0} = c_r/\sqrt{3} = 4\gamma/3\sqrt{3}\sigma \tag{11}$$

Below this critical size, He bubbles remain stable and high-temperature He embrittlement will be suppressed. Two practical approaches are proposed to mitigate high-temperature He embrittlement in metals [64]: (1) maximize the critical size for bubble-to-void transition by reducing irradiation-induced vacancies; and (2) maximize bubble nucleation sites to increase the number of stable He bubble. In fact, NC metals [70–73], multilayers [80,88–101] and oxide dispersion-strengthened (ODS) steels [164,165] are examples designed based on the above two approaches. Abundant interfaces in NC metals, multilayers and ODS steels significantly enhance the recombination of interstitials and vacancies, which reduce the production rate of irradiation-induced vacancies [68,69]. Moreover, these interfaces are preferable nucleation sites for He bubbles [97,98], which suppresses overall bubble growth and keeps He bubble below the critical size for bubble-to-void transition.

5.2. Helium Bubbles Enhance Ductility in Small-Volume Single-Crystal Metals

Contrary to polycrystalline metals, nanoscale He bubbles are demonstrated to enhance ductility in small-volume single-crystal metals [122,127] and metallic glass [166]. The underlying mechanisms are discussed as follows. Firstly, because they serve as shearable obstacles, He bubbles hinder dislocation motion and reduce dislocation mean free path, and hence promote dislocation storage, which gives rise to stable and homogeneous plasticity [122,126,127]. Secondly, He bubbles are preferable internal dislocation nucleation sources [122]. As shown in Figure 22, both experiments and simulations have confirmed this novel deformation mechanism. Figure 22a,c presents snapshots recording dislocation operation in sample containing dense He bubbles in Cu. A partial dislocation nucleates from the adjacent region of He bubble within 0.1 s (Figure 22b). This indicates that the He bubble surface and sample surface are near equal preferable dislocation nucleation sites [122]. Similarly, dislocations are also observed to nucleate and propagate homogeneously across the whole small-volume Al–4Cu sample containing He bubbles [127], which is in sharp contrast to localized dislocation operation in bubble-free sample. Subsequently, He bubbles are cut through by dislocations, as indicated by the steps and stacking faults in Figure 22b,c. Atomic simulations indicate that He bubbles are preferable dislocation nucleation sites compared to pillar surface/corner at the resolved shear stress above 1.5 GPa, which is usually the critical resolved shear stress required for dislocation nucleation in molecular dynamic simulations. Above this critical shear stress, the dislocation nucleation frequency is higher and the activation energy is lower for dislocation nucleation from He bubbles than that of dislocation nucleation from pillar surface/corner (Figure 22d,e). He bubble–dislocation interaction can further produce massive slip steps at He bubble surface, which are new internal dislocation nucleation sources, as shown in Figure 22f. Therefore, He bubbles play a combined role of shearable obstacles and internal dislocation nucleation sources, which promote dislocation storage in small-volume samples, thus enhance stable and homogeneous deformability in SC pillars. In addition, besides bubble coalescence, bubble cleavage is also identified as an alternative micro-damage mechanism in deformation of He-bubbled SC pillars [123]. This provides new insight into the failure mechanism of He-bubbled metals.

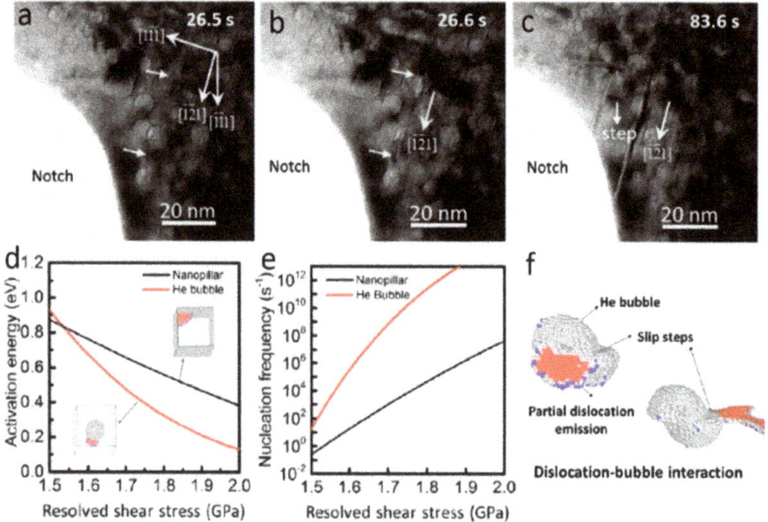

Figure 22. (**a–c**) A partial dislocation nucleates from He bubble and cuts through adjacent He bubbles, leaving a stacking fault and slip steps on bubbles. (**d,e**) A simulation demonstrates that He bubbles are preferable dislocation nucleation sites. For resolved shear stress higher than 1.5 GPa, the activation energy for dislocation nucleation from He bubbles is lower than that from pillar surface/corner, leading to higher dislocation nucleation frequency from He bubbles. (**f**) Slip steps on He bubbles also serve as dislocation nucleation sites. Reprinted with permission from [122]; Copyright 2016 American Chemical Society.

The surprising finding that He bubbles enhance ductility in single-crystal metals may provide new applications for ion beam engineering. By deliberately introducing bubbles or other defects using ion beam engineering, mechanically robust single-crystal devices can be manufactured [121]. Moreover, this finding provides a new approach to suppress high temperature He embrittlement in nuclear reactor components. Single-crystal metals may be considered as a potential candidates for nuclear reactor components, but a series of issues, such as recrystallization of deformed SC metals at elevated temperature, the difficulty of fabricating bulk single-crystal metals, etc., remain as challenges.

6. Summary and Outlook

Helium irradiation-induced damages have been studied extensively in the past decades and remain an important issue today. In this review, we briefly summarize previous studies on He bubble evolution and their effects on mechanical properties of metals. He bubble nucleation and growth are promoted with the aid of radiation produced vacancies. Matrix atoms are pushed out and form Frenkel pairs with increasing He atoms in He–V clusters. In addition, He bubbles grow accompanied by dislocation loop punching. Bubble migration and coalescence and Ostwald ripening are two main mechanisms of bubble coarsening under annealing. Interface engineering is widely adopted to improve radiation tolerance of metals. Interfaces serve as sinks for point defects (vacancies, interstitials and He), which promote vacancy–interstitial recombination and assist the stable storage of He. Dense He bubbles act as shearable obstacles under straining, and induce hardening in metals. Multilayers demonstrate size-dependent hardening due to the transition of deformation mechanism with reducing layer thickness. High temperature He embrittlement consists of four stages: bubble nucleation, stable gas-driven bubble growth, unstable stress-driven void growth and crack nucleation by cavity coalescence. Once the bubble size is larger than a critical value, a transition from bubble-to-void takes place and leads to catastrophic intergranular fracture. He bubbles can enhance the strength and ductility in small-volume metals because He bubbles play a combined role of shearable obstacles and dislocation sources.

As discussed in previous sections, significant progresses have been made in understanding of He bubbles in metals. However, there are still paramount problems left that need to be addressed in the future. In the following, we only propose some of representative issues that need special attention.

First, most nuclear reactor components serve at medium or high temperatures, and the thermal stability of interfaces in NC metals and multilayers should be considered with particular concern. Radiation-assisted grain growth in NC metals and layer pinch-off in multilayers will lead to sharp reduction in the density of interfaces, which dramatically deteriorates radiation tolerance in NC metals and multilayers. There are increasing studies on thermal stability of GBs and heterophase interfaces, but there are still very limited efficient approaches to improve the thermal stability of GBs and heterophase interfaces in metals, especially under radiation condition. Improving the thermal stability of interfaces without sacrificing the mechanical properties will give rise to unprecedented radiation tolerance in NC metals and multilayers.

Second, details of He–interface interactions are still not very clear and need further study. Experimental evidence from previous studies confirm that interfaces are efficient sinks for He, but existing studies do not provide a clear picture on details of He–interface interactions. Atomic simulations may provide an effective way to unveil He–interface interactions, but studies using atomic simulations on this issue are still limited. Moreover, in situ He irradiation inside TEM may serves as a direct and visual method to further investigate this question.

Third, microstructures of GBs and heterophase interfaces are supposed to evolve due to direct irradiation and interactions with point defects induced by collision cascade. As a result, the capacity of interfaces in trapping point defects (He, vacancies and interstitials) and the ability of promoting vacancy–interstitial recombination may alter or reduce with continuous interface–defect interactions. Thus, a method to design self-healing interface is highly desired to further improve the radiation resistance of metals.

Funding: This research was funded by the National Key Research and Development Program of China (2017YFB0702301), the National Natural Science Foundation of China (Nos. 51471128 and 51621063), the 111 Project of China (No. BP2018008) and the Innovation Project of Shaanxi Province (No. 2017KTPT-12).

Conflicts of Interest: The authors declare no conflict of interest.

References

1. Zinkle, S.J.; Was, G.S. Materials challenges in nuclear energy. *Acta Mater.* **2013**, *61*, 735–758. [CrossRef]
2. Mahajan, S.; Eyre, B.L. Formation of dislocation channels in neutron irradiated molybdenum. *Acta Mater.* **2017**, *122*, 259–265. [CrossRef]
3. Yi, X.O.; Jenkins, M.L.; Kirk, M.A.; Zhou, Z.F.; Roberts, S.G. In-situ TEM studies of 150 keV W$^+$ ion irradiated W and W-alloys: Damage production and microstructural evolution. *Acta Mater.* **2016**, *112*, 105–120. [CrossRef]
4. Chen, Y.; Yu, K.Y.; Liu, Y.; Shao, S.; Wang, H.; Kirk, M.A.; Wang, J.; Zhang, X. Damage-tolerant nanotwinned metals with nanovoids under radiation environments. *Nat. Commun.* **2015**, *6*, 7036. [CrossRef] [PubMed]
5. Atwani, O.E.; Esquivel, E.; Efe, M.; Aydogan, E.; Wang, Y.Q.; Martinez, E.; Maloy, S.A. Loop and void damage during heavy ion irradiation on nanocrystalline and coarse grained tungsten: Microstructure, effect of dpa rate, temperature, and grain size. *Acta Mater.* **2018**, *149*, 206–219. [CrossRef]
6. Kiener, D.; Hosemann, P.; Maloy, S.A.; Minor, A.M. In situ nanocompression testing of irradiated copper. *Nat. Mater.* **2011**, *10*, 608–613. [CrossRef] [PubMed]
7. Dai, Y.; Victoria, M. Defect cluster structure and tensile properties of copper single crystals irradiated with 600 MeV protons. *Mat. Res. Symp. Proc.* **1996**, *439*, 319–324. [CrossRef]
8. Tyler, S.; Goodhew, P. The growth of helium bubbles in niobium and Nb-1% Zr. *J. Nucl. Mater.* **1978**, *74*, 27–33. [CrossRef]

9.   Wei, Q.M.; Li, N.; Sun, K.; Wang, L.M. The shape of bubbles in He-implanted Cu and Au. *Scr. Mater.* **2010**, *63*, 430–433. [CrossRef]

10. Zhang, F.F.; Wang, X.; Wierschke, J.B.; Wang, L.M. Helium bubble evolution in ion irradiated Al/B_4C metal metrix composite. *Scr. Mater.* **2015**, *109*, 29–33. [CrossRef]

11. Iwakiri, H.; Yasunaga, K.; Morishita, K.; Yoshida, N. Microstructure evolution in tungsten during low-energy helium ion irradiation. *J. Nucl. Mater.* **2000**, *283–287*, 1134–1138. [CrossRef]

12. Tunes, M.A.; Harrison, R.W.; Greaves, G.; Hinks, J.A.; Donnelly, S.E. Effect of He implantation on the microstructure of zircaloy-4 studied using in situ TEM. *J. Nucl. Mater.* **2017**, *493*, 230–238. [CrossRef]

13. Shen, H.H.; Peng, S.M.; Chen, B.; Naab, F.N.; Sun, G.A.; Zhou, W.; Xiang, X.; Sun, K.; Zu, X.T. Helium bubble evolution in a Zr-Sn-Nb-Fe-Cr alloy during post-annealing: An in-situ investigation. *Mater. Charact.* **2015**, *107*, 309–316. [CrossRef]

14. Liu, S.M.; Li, S.H.; Han, W.Z. Helium bubble superlattice impacts the plasticity of α-Zr. *J. Mater. Sci. Technol.* **2019**. [CrossRef]

15. Yutani, K.; Kishimoto, H.; Kasada, R.; Kimura, A. Evaluation of helium effects on swelling behavior of oxide dispersion strengthened ferritic steels under ion irradiation. *J. Nucl. Mater.* **2007**, *367–370*, 423–427. [CrossRef]

16. Bond, G.M.; Mazey, D.J.; Lewis, M.B. Helium-bubble formation and void swelling in nimonic PE16 alloy under dual-ion (He^+, Ni^+) irradiation. *Nucl. Instrum. Methods Phys. Res. Sect. B* **1983**, *209–210*, 381–386. [CrossRef]

17. Soria, S.R.; Tolley, A.; Sanchez, E.A. The influence of microstructure on blistering and bubble formation by He ion irradiation in Al alloys. *J. Nucl. Mater.* **2015**, *467*, 357–367. [CrossRef]

18. Ullmaier, H. Helium in fusion materials: High temperature embrittlement. *J. Nucl. Mater.* **1985**, *133*, 100–104. [CrossRef]

19. Baskes, M.I. Recent advances in understanding helium embrittlement in metals. *MRS Bull.* **1986**, *11*, 14–18. [CrossRef]

20. Odette, G.R.; Alinger, M.J.; Wirth, B.D. Recent developments in irradiation-resistant steels. *Annu. Rev. Mater. Res.* **2008**, *38*, 471–503. [CrossRef]

21. Zinkle, S.J.; Busby, J.T. Structural materials for fission & fusion energy. *Mater. Today* **2009**, *12*, 12–19.

22. Ventelon, L.; Wirth, B.; Domain, C. Helium-self-interstitial atom interaction in α-iron. *J. Nucl. Mater.* **2006**, *351*, 119–132. [CrossRef]

23. Zhang, X.; Fu, E.G.; Misra, A.; Demkowicz, M.J. Interface-enabled defect reduction in He ion irradiated metallic multilayers. *JOM* **2010**, *62*, 75–78. [CrossRef]

24. Miura, T.; Fujii, K.; Fukuya, K. Micro-mechanical investigation for effects of helium on grain boundary fracture of austenitic stainless steel. *J. Nucl. Mater.* **2015**, *457*, 279–290. [CrossRef]

25. Morishita, K.; Sugano, R.; Wirth, B.D. MD and KMC modeling of the growth and shrinkage mechanisms of helium–vacancy clusters in Fe. *J. Nucl. Mater.* **2003**, *323*, 243–250. [CrossRef]

26. Morishita, K.; Sugano, R.; Wirth, B.D.; Diaz de la Rubia, T. Thermal stability of helium–vacancy clusters in iron. *Nucl. Instrum. Methods Phys. Res. Sect. B* **2003**, *202*, 76–81. [CrossRef]

27. Yang, L.; Deng, H.Q.; Gao, F.; Heinisch, H.L.; Kurtz, R.J.; Hu, S.Y.; Li, Y.L.; Zu, X.T. Atomistic studies of nucleation of He clusters and bubbles in bcc iron. *Nucl. Instrum. Methods Phys. Res. Sect. B* **2013**, *303*, 68–71. [CrossRef]

28. Henriksson, K.O.E.; Nordlund, K.; Keinonen, J. Molecular dynamics simulations of helium cluster formation in tungsten. *Nucl. Instrum. Methods Phys. Res. Sect. B* **2006**, *244*, 377–391. [CrossRef]

29. Wang, J.L.; Niu, L.L.; Shu, X.L.; Zhang, Y. Energetics and kinetics unveiled on helium cluster growth in tungsten. *Nucl. Fusion* **2015**, *55*, 092003. [CrossRef]

30. Li, X.C.; Shu, X.L.; Tao, P.; Yu, Y.; Niu, G.J.; Xu, Y.P.; Gao, F.; Luo, G.N. Molecular dynamics simulation of helium cluster diffusion and bubble formation in bulk tungsten. *J. Nucl. Mater.* **2014**, *455*, 544–548. [CrossRef]

31. Gao, N.; Victoria, M.; Chen, J.; Swygenhoven, H.V. Helium-vacancy cluster in a single bcc iron crystal lattice. *J. Phys. Condens. Matter* **2011**, *23*, 245403. [CrossRef] [PubMed]

32. Kobayashi, R.; Hattori, T.; Tamura, T.; Ogata, S. A molecular dynamics study on bubble growth in tungsten under helium irradiation. *J. Nucl. Mater.* **2015**, *463*, 1071–1074. [CrossRef]

33. Sandoval, L.; Perez, D.; Uberuaga, B.P.; Voter, A.F. Formation of helium-bubble networks in tungsten. *Acta Mater.* **2018**, *159*, 46–50. [CrossRef]

34. Guo, S.H.; Zhu, B.E.; Liu, W.C.; Pan, Z.Y.; Wang, Y.X. Pressure of stable He-vacancy complex in bcc iron: Molecular dynamics simulations. *Nucl. Instrum. Methods Phys. Res. Sect. B* **2009**, *267*, 3278–3281. [CrossRef]

35. Zhou, H.B.; Liu, Y.L.; Zhang, Y.; Jin, S.; Lu, G.H. First-principles investigation of energetics and site preference of He in a W grain boundary. *Nucl. Instrum. Methods Phys. Res. Sect. B* **2009**, *267*, 3189–3192. [CrossRef]

36. Ventelon, L.; Willaime, F.; Fu, C.C.; Heran, M.; Ginoux, I. Ab initio investigation of radiation defects in tungsten: Structure of self-interstitials and specificity of di-vacancies compared to other bcc transition metals. *J. Nucl. Mater.* **2012**, *425*, 16–21. [CrossRef]

37. Xiao, W.; Zhang, X.; Geng, W.T.; Lu, G. Helium bubble nucleation and growth in α-Fe: Insights from first–principles simulations. *J. Phys. Condens. Matter* **2014**, *26*, 255401. [CrossRef] [PubMed]

38. Derlet, P.M.; Nguyen-Manh, D.; Dudarev, S.L. Multiscale modeling of crowdion and vacancy defects in body-centered-cubic transition metals. *Phys. Rev. B Condens. Matter* **2007**, *76*, 54107. [CrossRef]

39. Rimmer, D.E.; Cottrell, A.H. The solution of inert gas atoms in metals. *Philos. Mag.* **1957**, *2*, 1345–1353. [CrossRef]

40. Samaras, M. Multiscale Modelling: The role of helium in iron. *Mater. Today* **2009**, *12*, 46–53. [CrossRef]

41. Becquart, C.S.; Domain, C. Migration energy of He in W revisited by ab initio calculations. *Phys. Rev. Lett.* **2006**, *97*, 196402. [CrossRef] [PubMed]

42. Seletskaia, T.; Osetsky, Y.; Stoller, R.E.; Stocks, G.M. Magnetic interactions influence the properties of helium defects in iron. *Phys. Rev. Lett.* **2005**, *94*, 046403. [CrossRef] [PubMed]

43. Seletskaia, T.; Osetsky, Y.N.; Stoller, R.E.; Stocks, G.M. Calculation of helium defect clustering properties in iron using a multi-scale approach. *J. Nucl. Mater.* **2006**, *351*, 109–118. [CrossRef]

44. Gao, F.; Deng, H.Q.; Heinisch, H.L.; Kurtz, R.J. A new Fe–He interatomic potential based on ab initio calculations in a-Fe. *J. Nucl. Mater.* **2011**, *418*, 115–120. [CrossRef]

45. Wilson, W.D.; Johnson, R.A. Rare gases in metals. In Proceedings of the Battelle Institute Materials Science Colloquia, Columbus, OH, USA, 17 September 1972; pp. 375–390.

46. Stewart, D.M.; Osetsky, Y.N.; Stoller, R.E.; Golubov, S.I.; Seletskaia, T.; Kamenski, P.J. Atomistic studies of helium defect properties in bcc iron: Comparison of He–Fe potentials. *Philos. Mag. A* **2010**, *90*, 935–944. [CrossRef]

47. Chen, P.H.; Lai, X.C.; Liu, K.Z.; Wang, X.L.; Bai, B.; Ao, B.Y.; Long, Y. Development of a pair potential for Fe–He by lattice inversion. *J. Nucl. Mater.* **2010**, *405*, 156–159. [CrossRef]

48. Fu, C.C.; Willaime, F. Ab initio study of helium in α-Fe: Dissolution, migration, and clustering with vacancies. *Phys. Rev. B Condens. Matter* **2005**, *72*, 4117. [CrossRef]

49. Xie, H.X.; Gao, N.; Xu, K.; Lu, G.H.; Yu, T.; Yin, F.X. A new loop-punching mechanism for helium bubble growth in tungsten. *Acta Mater.* **2017**, *141*, 10–17. [CrossRef]

50. Sandoval, L.; Perez, D.; Uberuaga, B.P.; Voter, A.F. Competing kinetics and He bubble morphology in W. *Phys. Rev. Lett.* **2015**, *114*, 105502. [CrossRef]

51. Yang, L.; Gao, F.; Kurtz, R.J.; Zu, X.T.; Peng, S.M.; Long, X.G.; Zhou, X.S. Effects of local structure on helium bubble growth in bulk and at grain boundaries of bcc iron: A molecular dynamics study. *Acta Mater.* **2015**, *97*, 86–93. [CrossRef]

52. Trinkaus, H.; Wolfer, W.G. Conditions for dislocation loop punching by helium bubbles. *J. Nucl. Mater.* **1984**, *122*, 552–557. [CrossRef]

53. Wolfer, W.G. The pressure for dislocation loop punching by a single bubble. *Philos. Mag. A* **1988**, *58*, 285–297. [CrossRef]

54. Weatherly, G.C. Loss of coherency of growing particles by the prismatic punching of dislocation loops. *Philos. Mag.* **1968**, *17*, 791–799. [CrossRef]

55. Wolfer, W.G. Dislocation loop punching in bubble arrays. *Philos. Mag. A* **1989**, *59*, 87–103. [CrossRef]

56. Trinkaus, H.; Singh, B.N. Helium accumulation in metals during irradiation—Where do we stand? *J. Nucl. Mater.* **2003**, *323*, 229–242. [CrossRef]

57. Marochov, N.; Perryman, L.J.; Goodhew, P.J. Growth of inert gas bubbles after implantation. *J. Nucl. Mater.* **1987**, *149*, 296–301. [CrossRef]

58. Goodhew, P.J.; Tyler, S.K. Helium bubble behaviour in b.c.c. metals below $0.65T_\mathrm{m}$. *Proc. R. Soc. Lond. Ser. A* **1981**, *377*, 151–184. [CrossRef]

59. Singh, B.N.; Trinkaus, H. An analysis of the bubble formation behaviour under different experimental conditions. *J. Nucl. Mater.* **1992**, *186*, 153–165. [CrossRef]

60. Markworth, A.J. On the coarsening of gas-filled pores in solids. *Metall. Trans. A* **1973**, *4*, 2651–2656. [CrossRef]
61. Greenwood, G.W.; Boltax, A. The role of fission gas re-solution during post-irradiation heat treatment. *J. Nucl. Mater.* **1962**, *5*, 234–240. [CrossRef]
62. Ono, K.; Arakawa, K.; Hojou, K. Formation and migration of helium bubbles in Fe and Fe-9Cr ferritic alloy. *J. Nucl. Mater.* **2002**, *307–311*, 1507–1512. [CrossRef]
63. Roldan, M.; Fernandez, P.; Rams, J.; Sanchez, F.J.; Gomez-Herrero, A. Nanoindentation and TEM to study the cavity fate after post-irradiation annealing of He implanted EUROFER97 and EU-ODS EUROFER. *Micromachines* **2018**, *9*, 633. [CrossRef]
64. Beyerlein, I.J.; Demkowicz, M.J.; Misra, A.; Uberuaga, B.P. Defect-interface interactions. *Prog. Mater. Sci.* **2015**, *74*, 125–210. [CrossRef]
65. Mansur, L.K.; Coghlan, W.A. Mechanisms of helium interaction with radiation effects in metals and alloys: A review. *J. Nucl. Mater.* **1983**, *119*, 1–25. [CrossRef]
66. Dai, Y.; Odette, G.R.; Yamamoto, T. The effects of helium in irradiated structural alloys. In *Comprehensive Nuclear Materials*; Elsevier Science: Amsterdam, The Netherlands, 2012; pp. 141–193.
67. Schroeder, H.; Kestehrnich, W.; Ullmaier, H. Helium effects on the creep and fatigue resistance of austenitic stainless steels at high temperatures. *Nucl. Eng. Des.* **1985**, *2*, 65–95. [CrossRef]
68. Bai, X.M.; Voter, A.F.; Hoagland, R.G.; Nastasi, M.; Uberuaga, B.P. Efficient annealing of radiation damage near grain boundaries via interstitial emission. *Science* **2010**, *327*, 1631–1634. [CrossRef] [PubMed]
69. Ackland, G. Controlling radiation damage. *Science* **2010**, *327*, 1587–1588. [CrossRef]
70. Singh, B.N. Effect of grain size on void formation during high-energy electron irradiation of austenitic stainless steel. *Philos. Mag.* **1974**, *29*, 25–42. [CrossRef]
71. Sun, C.; Song, M.; Yu, K.Y.; Chen, Y.; Kirk, M.; Li, M.; Wang, H.; Zhang, X. In situ evidence of defect cluster absorption by grain boundaries in Kr ion irradiated nanocrystalline Ni. *Metall. Trans. A* **2013**, *44*, 1966–1974. [CrossRef]
72. Rose, M.; Balogh, A.G.; Hahn, H. Instability of irradiation induced defects in nanostructured materials. *Nucl. Instrum. Methods Phys. Res. Sect. B* **1997**, *127–128*, 119–122. [CrossRef]
73. Chimi, Y.; Iwase, A.; Ishikawa, N.; Kobiyama, M.; Inami, T.; Okuda, S. Accumulation and recovery of defects in ion-irradiated nanocrystalline gold. *J. Nucl. Mater.* **2001**, *297*, 355–357. [CrossRef]
74. Lane, P.L.; Goodhew, P.J. Helium bubble nucleation at grain boundaries. *Philos. Mag. A* **1983**, *48*, 965–986. [CrossRef]
75. Li, Q.; Parish, C.M.; Powers, K.A.; Miller, M.K. Helium solubility and bubble formation in a nanostructured ferritic alloy. *J. Nucl. Mater.* **2014**, *445*, 165–174. [CrossRef]
76. Thorsen, P.A.; Bilde-Sorensen, J.B.; Singh, B.N. Influence of grain boundary structure on bubble formation behaviour in helium implanted copper. *Trans Tech Publ.* **1996**, *207–209*, 445–448. [CrossRef]
77. Xia, J.X.; Hu, W.Y.; Yang, J.Y.; Ao, B.Y.; Wang, X.L. A study of the behavior of helium atoms at Ni grain boundaries. *Phys. Status Solidi B* **2006**, *243*, 2702–2710. [CrossRef]
78. Hetherly, J.; Martinez, E.; Nastasi, M.; Caro, A. Helium bubble growth at BCC twist grain boundaries. *J. Nucl. Mater.* **2011**, *419*, 201–207. [CrossRef]
79. Kurtz, R.J.; Heinisch, H.L. The effects of grain boundary structure on binding of He in Fe. *J. Nucl. Mater.* **2004**, *329–333*, 1199–1203. [CrossRef]
80. Han, W.Z.; Demkowicz, M.J.; Mara, N.A.; Fu, E.G.; Sinha, S.; Rollett, A.D.; Wang, Y.Q.; Carpenter, J.S.; Beyerlein, I.J.; Misra, A. Design of radiation tolerant materials via interface engineering. *Adv. Mater.* **2013**, *25*, 6975–6979. [CrossRef]
81. Han, W.Z.; Demkowicz, M.J.; Fu, E.G.; Wang, Y.Q.; Misra, A. Effect of grain boundary on sink efficiency. *Acta Mater.* **2012**, *60*, 6341–6351. [CrossRef]
82. El-Atwani, O.; Nathaniel, J.E.; Leff, A.C.; Baldwin, J.K.; Hattar, K.; Taheri, M.L. Evidence of a temperature transition for denuded zone formation in nanocrystalline Fe under He irradiation. *Mater. Res. Lett.* **2016**, 1–6. [CrossRef]
83. Cheng, G.M.; Xu, W.Z.; Wang, Y.Q.; Misra, A.; Zhu, Y.T. Grain size effect on radiation tolerance of nanocrystalline Mo. *Scr. Mater.* **2016**, *123*, 90–94. [CrossRef]
84. Yu, K.Y.; Liu, Y.; Sun, C.; Wang, H.; Shao, L.; Fu, E.G.; Zhang, X. Radiation damage in helium ion irradiated nanocrystalline Fe. *J. Nucl. Mater.* **2012**, *425*, 140–146. [CrossRef]

85. El-Atwani, O.; Hattar, K.; Hinks, J.A.; Greaves, G.; Harilal, S.S.; Hassanein, A. Helium bubble formation in ultrafine and nanocrystalline tungsten under different extreme conditions. *J. Nucl. Mater.* **2015**, *458*, 216–223. [CrossRef]

86. Bullough, R.; Hayns, M.R.; Wood, M.H. Sink strengths for thin film surfaces and grain boundaries. *J. Nucl. Mater.* **1980**, *90*, 44–59. [CrossRef]

87. Zhang, X.H.; Hattar, K.; Chen, Y.X.; Shao, L.; Li, J.; Sun, C.; Yu, K.Y.; Li, N.; Taheri, M.L.; Wang, H.Y.; et al. Radiation damage in nanostructured materials. *Prog. Mater. Sci.* **2018**, *96*, 217–321. [CrossRef]

88. Han, W.Z.; Mara, N.A.; Wang, Y.Q.; Misra, A.; Demkowicz, M.J. He implantation of bulk Cu-Nb nanocomposites fabricated by accumulated roll bonding. *J. Nucl. Mater.* **2014**, *452*, 57–60. [CrossRef]

89. Mitchell, T.E.; Lu, Y.C.; Griffin, A.J., Jr.; Nastasi, M.; Kung, H. Structure and mechanical properties of copper/niobium multilayers. *J. Am. Ceram. Soc.* **1997**, *80*, 1673–1676. [CrossRef]

90. Demkowicz, M.J.; Bhattacharyya, D.; Usov, I.; Wang, Y.Q.; Nastasi, M.; Misra, A. The effect of excess atomic volume on He bubble formation at fcc–bcc interfaces. *Appl. Phys. Lett.* **2010**, *97*, 161903. [CrossRef]

91. Demkowicz, M.J.; Hoagland, R.G. Structure of Kurdjumov–Sachs interfaces in simulations of a copper–niobium bilayer. *J. Nucl. Mater.* **2008**, *372*, 45–52. [CrossRef]

92. Hattar, K.; Demkowicz, M.J.; Misra, A.; Robertson, I.M.; Hoagland, R.G. Arrest of He bubble growth in Cu–Nb multilayer nanocomposites. *Scr. Mater.* **2008**, *58*, 541–544. [CrossRef]

93. Li, N.; Nastasi, M.; Misra, A. Defect structures and hardening mechanisms in high dose helium ion implanted Cu and Cu/Nb multilayer thin films. *Int. J. Plast.* **2012**, *32–33*, 1–16. [CrossRef]

94. Misra, A.; Demkowicz, M.J.; Zhang, X.; Hoagland, R.G. The radiation damage tolerance of ultra-high strength nanolayered composites. *JOM* **2007**, *59*, 62–65. [CrossRef]

95. Demkowicz, M.J.; Misra, A.; Caro, A. The role of interface structure in controlling high helium concentrations. *Curr. Opin. Solid State Mater. Sci.* **2012**, *16*, 101–108. [CrossRef]

96. Kashinath, A.; Misra, A.; Demkowicz, M.J. Stable storage of helium in nanoscale platelets at semicoherent interfaces. *Phys. Rev. Lett.* **2013**, *110*, 086101. [CrossRef] [PubMed]

97. Demkowicz, M.J.; Hoagland, R.G.; Hirth, J.P. Interface structure and radiation damage resistance in Cu-Nb multilayer nanocomposites. *Phys. Rev. Lett.* **2008**, *100*, 136102. [CrossRef] [PubMed]

98. Fu, E.G.; Misra, A.; Wang, H.; Shao, L.; Zhang, X. Interface enabled defects reduction in helium ion irradiated Cu/V nanolayers. *J. Nucl. Mater.* **2010**, *407*, 178–188. [CrossRef]

99. Fu, E.G.; Wang, H.; Carter, J.; Shao, L.; Wang, Y.Q.; Zhang, X. Fluence-dependent radiation damage in helium (He) ion-irradiated Cu/V multilayers. *Philos. Mag.* **2013**, *93*, 883–898. [CrossRef]

100. Fu, E.G.; Carter, J.; Swadener, G.; Misra, A.; Shao, L.; Wang, H.; Zhang, X. Size dependent enhancement of helium ion irradiation tolerance in sputtered Cu/V nanolaminates. *J. Nucl. Mater.* **2009**, *385*, 629–632. [CrossRef]

101. Li, N.; Carter, J.J.; Misra, A.; Shao, L.; Wang, H.; Zhang, X. The influence of interfaces on the formation of bubbles in He-ion-irradiated Cu/Mo nanolayers. *Philos. Mag.* **2011**, *91*, 18–28. [CrossRef]

102. Chen, Y.; Liu, Y.; Fu, E.G.; Sun, C.; Yu, K.Y.; Song, M.; Li, J.; Wang, Y.Q.; Wang, H.; Zhang, X. Unusual size-dependent strengthening mechanisms in helium ion-irradiated immiscible coherent Cu/Co nanolayers. *Acta Mater.* **2015**, *84*, 393–404. [CrossRef]

103. Gao, Y.; Yang, T.F.; Xue, J.M.; Yan, S.; Zhou, S.Q.; Wang, Y.G.; Kwok, D.T.K.; Chu, P.K.; Zhang, Y.W. Radiation tolerance of Cu/W multilayered nanocomposites. *J. Nucl. Mater.* **2011**, *413*, 11–15. [CrossRef]

104. Li, N.; Fu, E.G.; Wang, H.; Carter, J.J.; Shao, L.; Maloy, S.A.; Misra, A.; Zhang, X. He ion irradiation damage in Fe/W nanolayer films. *J. Nucl. Mater.* **2009**, *389*, 233–238. [CrossRef]

105. Wei, Q.M.; Wang, Y.Q.; Nastasi, M.; Misra, A. Nucleation and growth of bubbles in He ion-implanted V/Ag multilayers. *Philos. Mag.* **2011**, *91*, 553–573. [CrossRef]

106. Wei, Q.M.; Li, N.; Mara, N.; Nastasi, M.; Misra, A. Suppression of irradiation hardening in nanoscale V/Ag multilayers. *Acta Mater.* **2011**, *59*, 6331–6340. [CrossRef]

107. Li, N.; Martin, M.S.; Anderoglu, O.; Misra, A.; Shao, L.; Wang, H.; Zhang, X. He ion irradiation damage in Al/Nb multilayers. *J. Appl. Phys.* **2009**, *105*, 123522. [CrossRef]

108. Wang, M.; Beyerlein, I.J.; Zhang, J.; Han, W.Z. Defect-interface interactions in irradiated Cu/Ag nanocomposites. *Acta Mater.* **2018**, *160*, 211–223. [CrossRef]

109. Zheng, S.J.; Shao, S.; Zhang, J.; Wang, Y.Q.; Demkowicz, M.J.; Beyerlein, I.J.; Mara, N.A. Adhesion of voids to bimetal interfaces with non-uniform energies. *Sci. Rep.* **2015**, *5*, 15428. [CrossRef]

110. Yu, K.Y.; Liu, Y.; Fu, E.G.; Wang, Y.Q.; Myers, M.T.; Wang, H.; Shao, L.; Zhang, X. Comparisons of radiation damage in He ion and proton irradiated immiscible Ag/Ni nanolayers. *J. Nucl. Mater.* **2013**, *440*, 310–318. [CrossRef]

111. Yu, K.Y.; Sun, C.; Chen, Y.; Liu, Y.; Wang, H.; Kirk, M.A.; Li, M.; Zhang, X. Superior tolerance of Ag/Ni multilayers against Kr ion irradiation: An in situ study. *Philos. Mag.* **2013**, *93*, 3547–3562. [CrossRef]

112. Beyerlein, I.J.; Mara, N.A.; Carpenter, J.S.; Nizolek, T.; Mook, W.M.; Wynn, T.A.; McCabe, R.J.; Mayeur, J.R.; Kang, K.; Zheng, S.J.; et al. Interface-driven microstructure development and ultra high strength of bulk nanostructured Cu-Nb multilayers fabricated by severe plastic deformation. *J. Mater. Res.* **2013**, *28*, 1799–1812. [CrossRef]

113. Gleiter, H. Nanocrystalline materials. *Prog. Mater. Sci.* **1989**, *33*, 223–315. [CrossRef]

114. Ziegler, J.F.; Ziegler, M.D.; Biersack, J.P. SRIM-The stopping and range of ions in matter. *Nucl. Instrum. Methods Phys. Res. Sect. B* **2008**, *268*, 1818–1823. [CrossRef]

115. Hetherly, J.; Martinez, E.; Di, Z.F.; Nastasi, M.; Caro, A. Helium bubble precipitation at dislocation networks. *Scr. Mater.* **2012**, *66*, 17–20. [CrossRef]

116. Yuryev, D.V.; Demkowicz, M.J. Computational design of solid-state interfaces using O-lattice theory: An application to mitigating helium-induced damage. *Appl. Phys. Lett.* **2014**, *105*, 221601. [CrossRef]

117. Zhang, J.Y.; Wang, Y.Q.; Liang, X.Q.; Zeng, F.L.; Liu, G.; Sun, J. Size-dependent He-irradiated tolerance and plastic deformation of crystalline/amorphous Cu/Cu–Zr nanolaminates. *Acta Mater.* **2015**, *92*, 140–151. [CrossRef]

118. Li, N.; Mara, N.A.; Wang, Y.Q.; Nastasi, M.; Misra, A. Compressive flow behavior of Cu thin films and Cu/Nb multilayers containing nanometer-scale helium bubbles. *Scr. Mater.* **2011**, *64*, 974–977. [CrossRef]

119. Li, N.; Demkowicz, M.; Mara, N.; Wang, Y.Q.; Misra, A. Hardening due to interfacial He bubbles in nanolayered composites. *Mater. Res. Lett.* **2015**, *4*, 1–8. [CrossRef]

120. Reichardt, A.; Ionescu, M.; Davis, J.; Edwards, L.; Harrison, R.P.; Hosemann, P.; Bhattacharyya, D. In situ micro tensile testing of He^{+2} ion irradiated and implanted single crystal nickel film. *Acta Mater.* **2015**, *100*, 147–154. [CrossRef]

121. Han, W.Z.; Zhang, J.; Ding, M.S.; Lv, L.; Wang, W.H.; Wu, G.H.; Shan, Z.W.; Li, J. Helium nanobubbles enhance superelasticity and retard shear localization in small-volume shape memory alloy. *Nano Lett.* **2017**, *17*, 3725–3730. [CrossRef]

122. Ding, M.S.; Du, J.P.; Wan, L.; Ogata, S.; Tian, L.; Ma, E.; Han, W.Z.; Li, J.; Shan, Z.W. Radiation-induced helium nanobubbles enhance ductility in submicron-sized single-crystalline copper. *Nano Lett.* **2016**, *16*, 4118–4124. [CrossRef]

123. Ding, M.S.; Tian, L.; Han, W.Z.; Li, J.; Ma, E.; Shan, Z.W. Nanobubble fragmentation and bubble-free-channel shear localization in helium-irradiated submicron-sized copper. *Phys. Rev. Lett.* **2016**, *117*, 215501. [CrossRef]

124. Han, W.Z.; Ding, M.S.; Shan, Z.W. Cracking behavior of helium-irradiated small-volume copper. *Scr. Mater.* **2018**, *147*, 1–5. [CrossRef]

125. Han, W.Z.; Ding, M.S.; Narayan, R.L.; Shan, Z.W. In situ study of deformation twinning and detwinning in helium irradiated small-volume copper. *Adv. Eng. Mater.* **2017**. [CrossRef]

126. Wang, Z.J.; Allen, F.I.; Shan, Z.W.; Hosemann, P. Mechanical behavior of copper containing a gas-bubble superlattice. *Acta Mater.* **2016**, *121*, 78–84. [CrossRef]

127. Li, S.H.; Zhang, J.; Han, W.Z. Helium bubbles enhance strength and ductility in small-volume Al-4Cu alloys. *Scr. Mater.* **2019**, *165*, 112–116. [CrossRef]

128. Misra, A.; Hirth, J.P.; Hoagland, R.G. Length-scale-dependent deformation mechanisms in incoherent metallic multilayered composites. *Acta Mater.* **2005**, *53*, 4817–4824. [CrossRef]

129. Li, Q.Z.; Anderson, P.M. Dislocation-based modeling of the mechanical behavior of epitaxial metallic multilayer thin films. *Acta Mater.* **2005**, *53*, 1121–1134. [CrossRef]

130. Sergueeva, A.; Mara, N.; Mukherjee, A.K. Structure and high-temperature mechanical behavior relationship in nano-scaled multilayered materials. *Mat. Res. Symp. Proc.* **2001**, *673*, 803–806. [CrossRef]

131. Misra, A.; Hoagland, R.G. Effects of elevated temperature annealing on the structure and hardness of copper/niobium nanolayered films. *J. Nucl. Mater.* **2005**, *20*, 2046–2054. [CrossRef]

132. Phillips, M.A.; Clemens, B.M.; Nix, W.D. A model for dislocation behavior during deformation of Al/Al$_3$Sc (fcc/L1$_2$) metallic multilayers. *Acta Mater.* **2003**, *51*, 3157–3170. [CrossRef]

133. Kroupa, P. The interaction between prismatic dislocation loops and straight dislocations. part I. *Philos. Mag.* **1962**, *7*, 783–801. [CrossRef]
134. Zinkle, S.J.; Matsukawa, Y. Observation and analysis of defect cluster production and interactions with dislocations. *J. Nucl. Mater.* **2004**, *329–333*, 88–96. [CrossRef]
135. Lucas, G.E. The evolution of mechanical property change in irradiated austenitic stainless steels. *J. Nucl. Mater.* **1993**, *206*, 287–305. [CrossRef]
136. Osetsky, Y.N.; Bacon, D.J. Atomic-scale mechanisms of void hardening in bcc and fcc metals. *Philos. Mag.* **2010**, *90*, 945–961. [CrossRef]
137. Osetsky, Y.N.; Bacon, D.J.; Mohles, V. Atomic modelling of strengthening mechanisms due to voids and copper precipitates in α-iron. *Philos. Mag.* **2003**, *83*, 3623–3641. [CrossRef]
138. Knapp, J.A.; Follstaedt, D.M.; Myers, S.M. Hardening by bubbles in He-implanted Ni. *J. Nucl. Mater.* **2008**, *103*, 013518. [CrossRef]
139. Kocks, U.F. The theory of an obstacle-controlled yield strength—Report after an international workshop. *Mater. Sci. Eng.* **1977**, *27*, 291–298. [CrossRef]
140. Goods, S.H. The influence of tritium exposure and helium build-in on the properties of OFHC copper. *Scr. Metall.* **1986**, *20*, 565–569. [CrossRef]
141. Schroeder, H.; Batfalsky, P. The dependence of the high temperature mechanical properties of austenitic stainless steels on implanted helium. *J. Nucl. Mater.* **1983**, *117*, 287–294. [CrossRef]
142. Kramer, D.; Brager, H.R.; Rhodes, C.G.; Pard, A.G. Helium embrittlement in type 304 stainless steel. *J. Nucl. Mater.* **1968**, *25*, 121–131. [CrossRef]
143. Fabritsiev, S.A.; Yaroshevich, V.D. Mechanism of helium embrittlement of metals and alloys. *Strength Mater.* **1986**, *18*, 1194–1202. [CrossRef]
144. Trinkaus, H. On the modeling of the high-temperature embrittlement of metals containing helium. *J. Nucl. Mater.* **1983**, *118*, 39–49. [CrossRef]
145. Trinkaus, H. Modeling of helium effects in metals: High temperature embrittlement. *J. Nucl. Mater.* **1985**, *133–134*, 105–112. [CrossRef]
146. Trinkaus, H.; Ullmaier, H. The effect of helium on the fatigue properties of structural materials. *J. Nucl. Mater.* **1988**, *155*, 148–155. [CrossRef]
147. Trinkaus, H.; Ullmaier, H. High temperature embrittlement of metals due to helium: Is the lifetime dominated by cavity growth or crack growth? *J. Nucl. Mater.* **1994**, *212–215*, 303–309. [CrossRef]
148. Singh, B.N.; Leffers, T.; Victoria, M.; Green, W.V. Mechanisms controlling high temperature embrittlement due to helium. *Radiat. Eff.* **1987**, *101*, 1–4. [CrossRef]
149. Trinkaus, H.; Ullmaier, H. A model for the high-temperature embrittlement of metals containing helium. *J. Nucl. Mater.* **1979**, *118*, 563–580. [CrossRef]
150. Bullough, R.; Harries, D.R.; Hayns, M.R. The effect of stress on the growth of gas bubbles during irradiation. *J. Nucl. Mater.* **1980**, *88*, 312–314. [CrossRef]
151. Hull, D.; Rimmer, D.E. The growth of grain-boundary voids under stress. *Philos. Mag.* **1959**, *4*, 673–687. [CrossRef]
152. Gilbert, M.R.; Dudarev, S.L.; Zheng, S.; Packer, L.W.; Sublet, J. An integrated model for materials in a fusion power plant: Transmutation, gas production, and helium embrittlement under neutron irradiation. *Nucl. Fusion* **2012**, *52*, 083019. [CrossRef]
153. Yamamoto, N.; Chuto, T.; Murase, Y.; Nagakawa, J. Correlation between embrittlement and bubble microstructure in helium-implanted materials. *J. Nucl. Mater.* **2004**, *329–333*, 993–997. [CrossRef]
154. Borodin, V.A.; Manichev, V.M.; Ryazanov, A.I. Grain boundary cavities and cracks during high temperature irradiation embrittlement. *J. Nucl. Mater.* **1992**, *191–194*, 1305–1308. [CrossRef]
155. Schroeder, H.; Dai, Y. Helium concentration dependence of embrittlement effects in DIN 1.4970, 13% cw austenitic stainless steel at 873 K. *J. Nucl. Mater.* **1992**, *191–194*, 781–785. [CrossRef]
156. Imasaki, K.; Hasegawa, A.; Nogami, S.; Satou, M. Helium effects on the tensile property of 316FR stainless steel at 650 and 750 °C. *J. Nucl. Mater.* **2011**, *417*, 1030–1033. [CrossRef]
157. Magnusson, P.; Chen, J.; Jung, P.; Sauvage, T.; Hoffelner, W.; Spätig, P. Helium embrittlement of a lamellar titanium aluminide. *J. Nucl. Mater.* **2013**, *434*, 252–258. [CrossRef]
158. Schroeder, H. High temperature embrittlement of metals by helium. *Radiat. Eff.* **1983**, *78*, 297–314. [CrossRef]

159. Schroeder, H. High temperature helium embrittlement in austenitic stainless steels—Correlations between microstructure and mechanical properties. *J. Nucl. Mater.* **1988**, *155–157*, 1032–1037. [CrossRef]
160. Zhang, T.; Vieh, C.; Wang, K.; Dai, Y. Irradiation-induced evolution of mechanical properties and microstructure of Eurofer 97. *J. Nucl. Mater.* **2014**, *450*, 48–53. [CrossRef]
161. Ullmaier, H.; Chen, J. Low temperature tensile properties of steels containing high concentrations of helium. *J. Nucl. Mater.* **2003**, *318*, 228–233. [CrossRef]
162. Wang, K.; Dai, Y.; Spatig, P. Microstructure and fracture behavior of F82H steel under different irradiation and tensile test conditions. *J. Nucl. Mater.* **2016**, *468*, 246–254. [CrossRef]
163. Li, N.; Demkowicz, M.J.; Mara, N.A. Microstructure evolution and mechanical response of nanolaminate composites irradiated with helium at elevated temperatures. *JOM* **2017**, *69*, 2206–2213. [CrossRef]
164. Kim, I.S.; Hunn, J.D.; Hashimoto, N.; Larson, D.L.; Maziasz, P.J.; Miyahara, K.; Lee, E.H. Defect and void evolution in oxide dispersion strengthened ferritic steels under 3.2 MeV Fe^+ ion irradiation with simultaneous helium injection. *J. Nucl. Mater.* **2000**, *280*, 264–274. [CrossRef]
165. Bhattacharyya, D.; Dickerson, P.; Odette, G.R.; Maloy, S.A.; Misra, A.; Nastasi, M.A. On the structure and chemistry of complex oxide nanofeatures in nanostructured ferritic alloy U14YWT. *Philos. Mag.* **2012**, *92*, 2089–2107. [CrossRef]
166. Liontas, R.; Gu, X.W.; Fu, E.G.; Wang, Y.Q.; Li, N.; Mara, N.; Greer, J.R. Effects of helium implantation on the tensile properties and microstructure of $Ni_{73}P_{27}$ metallic glass nanostructures. *Nano Lett.* **2014**, *14*, 5176–5183. [CrossRef] [PubMed]

Communication

In-Situ Helium Implantation and TEM Investigation of Radiation Tolerance to Helium Bubble Damage in Equiaxed Nanocrystalline Tungsten and Ultrafine Tungsten-TiC Alloy

Osman El Atwani [1,*], Kaan Unal [1], William Streit Cunningham [2], Saryu Fensin [1], Jonathan Hinks [3], Graeme Greaves [3] and Stuart Maloy [1]

[1] Materials Science and Technology Division, Los Alamos National Laboratory, Los Alamos, NM 87545, USA; kunal@lanl.gov (K.U.); saryuj@lanl.gov (S.F.); maloy@lanl.gov (S.M.)
[2] Department of Materials Science and Chemical Engineering, Stony Brook University, Stony Brook, NY 11790, USA; william.cunningham@stonybrook.edu
[3] School of Computing and Engineering, University of Huddersfield, Huddersfield HD1 3DH, UK; j.a.hinks@hud.ac.uk (J.H.); g.greaves@huc.ac.uk (G.G.)
*   Correspondence: oelatwan25@gmail.com

Received: 3 December 2019; Accepted: 3 February 2020; Published: 10 February 2020

Abstract: The use of ultrafine and nanocrystalline materials is a proposed pathway to mitigate irradiation damage in nuclear fusion components. Here, we examine the radiation tolerance of helium bubble formation in 85 nm (average grain size) nanocrystalline-equiaxed-grained tungsten and an ultrafine tungsten-TiC alloy under extreme low energy helium implantation at 1223 K via in-situ transmission electron microscope (TEM). Helium bubble damage evolution in terms of number density, size, and total volume contribution to grain matrices has been determined as a function of He^+ implantation fluence. The outputs were compared to previously published results on severe plastically deformed (SPD) tungsten implanted under the same conditions. Large helium bubbles were formed on the grain boundaries and helium bubble damage evolution profiles are shown to differ among the different materials with less overall damage in the nanocrystalline tungsten. Compared to previous works, the results in this work indicate that the nanocrystalline tungsten should possess a fuzz formation threshold more than one order of magnitude higher than coarse-grained tungsten.

Keywords: nanocrystalline tungsten; alloy; in-situ electron microscopy; helium bubbles; radiation tolerance

---

1. Introduction

Fusion energy applications require materials to possess superior properties able to withstand extreme environments of high thermal loads, transient heat fluxes, high fluxes of helium (He) plasma particles, and fast neutrons [1]. Furthermore, neutron irradiation can also lead to solid transmutation products and He gas formation [2]. These challenging conditions can exacerbate damage in materials facing the plasma. Tungsten is currently one of the best candidates for the divertor armor and is to be used in ITER (originally the International Thermonuclear Experimental Reactor), but has demonstrated several drawbacks (in terms of microstructural and mechanical property changes) when exposed to extreme irradiation conditions [3–5]. One of the most studied microstructural changes is the formation of fuzz: a high density of dendritic structures that can develop on the surface of tungsten at high temperatures and He plasma fluxes [3]. While the precise mechanisms of fuzz formation are still under investigation, it is understood to occur due to a high density of He bubbles forming near the tungsten surface [6,7]. Several routes have been suggested in the quest to design irradiation resistant materials capable of mitigating microstructural changes and fuzz formation in a plasma facing material. Refining the grain size to the nanocrystalline regime, and thereby increasing the density of interfaces that act as defect sinks, has been proposed as a method to mitigate these changes [8,9]. Particle reinforcement, where the particles act as defect annihilation sites, is also a suggested route to increase the irradiation resistance of the material and enhance its mechanical properties and thermal stability [10–12]. The hypothesis is that materials that are radiation tolerant to bubble formation should lead to an increased He^+ fluence threshold for fuzz formation (at fuzz formation conditions) if fuzz formation depends on the same conditions as proposed in the literature [6]. El Atwani et al. studied nanocrystalline tungsten with a 35 nm average grain size [13] and an ultrafine tungsten alloy with TiC dispersoids (W-TiC 1.1%) [14] under heavy ion irradiation (to mimic fast neutron damage). Both types of materials have shown enhanced irradiation tolerance when compared to commercially-available tungsten. Damage due to loop formation was shown to be minimal in the nanocrystalline tungsten and decreased with time (due to loop annihilation and dissolution under heavy ion cascades) in the W-TiC (1.1%). The response of these materials, however, due to low energy He^+ implantation is still unknown. The testing of severe plastically deformed (SPD) nanocrystalline and ultrafine tungsten, where both elongated ultrafine (<500 nm) and nanocrystalline (<100 nm) grains coexist under low energy He^+ implantation and high temperatures (1223 K), showed a trend in both damage and He bubble formation as a function of grain size [15]. Grains of less than 60 nm (in elongated grains, size was defined to be the distance across the middle in the narrower direction) demonstrated a high irradiation tolerance to bubble formation and grains of ~35 nm were shown to possess minimum damage.

Here, we present a detailed study investigating the radiation tolerance to bubble formation under low energy (2 keV). Implantation of He^+ at a high temperature (1223 K) (similar conditions to previous SPD work [15]) of the equiaxial nanocrystalline tungsten with an average grain-size 85 nm (referred to as NCW herein) was formed by magnetron deposition, and the tungsten-TiC (1.1%) (labelled as W-TiC (1.1%)) formed by hot iso-static pressing of tungsten powders with TiC dispersoids. The experiments were performed in-situ within the transmission electron microscope (TEM) in the MIAMI-2 system (Microscope and Ion Accelerators for Materials Investigations) at the University of Huddersfield [16]. Bubble formation, distribution, and evolution in the materials were studied by quantifying bubble density, average size, and total change in volume (in the grain matrices of the material due to bubble formation) as a function of implantation He^+ fluence. Together with the previous work in SPD tungsten, the results constitute a comparable set of bubble/damage evolution profiles and distribution for different ultrafine and nanocrystalline grades (pure and alloyed) that should assist in the understanding of nanocrystalline and ultrafine material behavior under He^+ implantation and the design of materials with higher irradiation resistance for fusion applications.

2. Materials and Methods

The TEM sample (~100 nm thickness) preparation methodology and detailed morphology of the samples prior to implantation have been described in detail previously [13,14]. The implantations were incident at 18.7° from the surface normal with fluxes of 8.8×10^{13} and 6.8×10^{13} ion.cm^{-2}.s^{-1} for the NCW and W-TiC (1.1%), respectively, to a total He^+ fluence of 3.6×10^{16} ion.cm^{-2}. Displacement damage and He distributions were found by the Kinchin-Pease model in the Stopping Range of Ions in Matter (SRIM) Monte Carlo computer code (version 2013) [17], using 70 eV as a displacement threshold [18]. He bubbles were characterized using bright-field TEM imaging at under-focused (bubbles appear bright) conditions [19]. For every He^+ fluence reported, different grains (about 7 grains in the W-TiC (1.1%) sample and 15 grains in the NCW sample) in every sample were quantified at different He^+ fluences. In every grain, several small circles (3–8 depending on the grain size) of the same area were drawn randomly and the number of He bubbles and their corresponding sizes were found. Averages were calculated from the results in all grains. A detailed illustration of the quantification process was published in the supplemental of reference [20]. Figure 1 shows a schematic diagram of the sample, implantation conditions and overlapping ion and displacement damage distributions. The projected He peak is ~12–15 nm, which is much less than the nominal thickness of the film (~100 nm). This is an important consideration when comparing and illustrating phenomena (such as bubble formation and defect annihilation/recombination) that can be affected by the proximity to free surfaces. However, surface proximity effects becomes negligible in nanocrystalline and ultrafine grains when the grain boundary to surface ratio approaches a value of 1 [21]. The grain boundary to surface ratios on the equiaxed 85 nm tungsten and the W-TiC (1.1%) were measured taking into consideration the upper and the bottom surface of the foils and assuming edge-on grain boundaries, and were found to be 3.8 and 0.3, respectively. Interfaces of TiC particles and the tungsten matrix inside the grains were not considered in the case of the W-TiC (1.1%) grade. These values are relatively high compared to fine- or coarse-grained grades. For example, a corresponding grain boundary to surface ratio for a fine-grained tungsten grade with an average grain size of ~2 μm would be ~0.04 [13]. Due to the shallow depth of implanted He in this work, no considerable variations in proximity effects are expected between the two grades investigated.

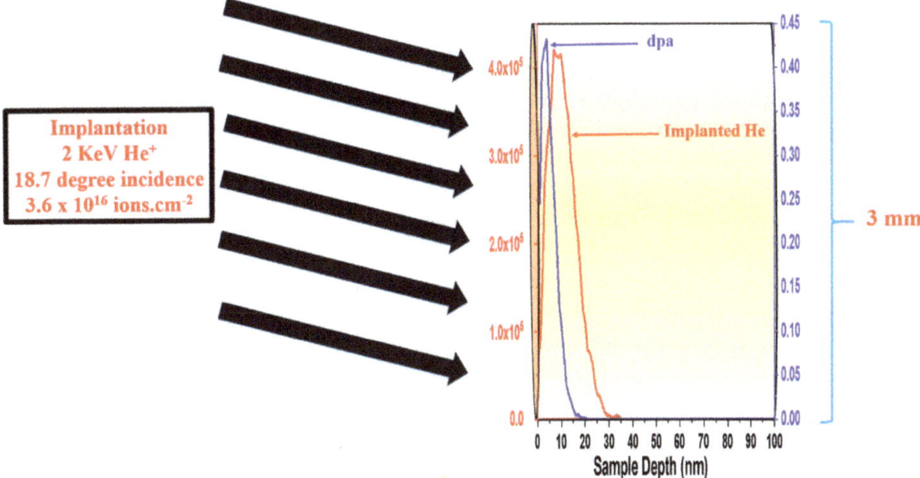

Figure 1. A schematic showing the sample shape, implanted He (red curve), and displacement per atom (dpa) damage (blue curve) distributions of 2 keV He$^+$ (as determined by SRIM). The thickness of the sample is magnified for the purpose of overlapping the implanted He and displacement damage distributions. (For interpretation of the references to color in this figure legend, the reader is referred to the web version of this article.)

3. Results and Discussion

3.1. He Bubble Formation and Growth

At the temperature studied in this work (1223 K), vacancies, interstitials, and small He–vacancy complexes are mobile and hence He bubbles are expected to form [22]. The He bubble density, average size, and the consequential change in volume in the grain matrices due to He bubble formation, as a function of He$^+$ fluence, not only reveal the total damage (as defined here) in the sample but also demonstrate the damage evolution profiles and elucidate the variations in the response of the two materials. Figures 2 and 3 show bright-field TEM images taken under Fresnel conditions (under-focused conditions where bubbles appear bright) of the NCW and W-TiC (1.1%) as a function of He$^+$ implantation fluence and indicate He bubble formation. In both cases (Figures 2b and 3b), He bubble formation was more evident at the grain boundaries than in the bulk grain matrices. As the fluence was increased, grain boundaries demonstrated larger He bubble formation than that in the grain matrices. While this demonstrates efficient He trapping by the grain boundaries, the mechanisms involved in this process can be complicated [23]. At the He$^+$ energy in this work (2 keV), the number of vacancies generated per ion is 0.5 for a 70 eV displacement threshold [18] according to SRIM calculations. Therefore, He$_n$V$_m$ complexes of n/m larger than 1 are expected to dominate especially taking the inevitable dynamic annealing of vacancies into account. Such complexes have very large migration energies and are not expected to migrate [24]. Only complexes with migration energies close to that of vacancies (1.7 eV) [25] can contribute to He bubble formation at the grain boundaries under these conditions and these are mainly complexes with a higher vacancy content [24,26]. Interstitial-He trapping by the grain boundaries (2D trapping) can also occur and then He bubbles can form through trap mutation processes as well as via vacancy migration to the grain boundary [27]. The probability of trap mutation and He bubble formation on the grain boundaries can be high at this temperature due to higher gas pressure and the decrease in self-interstitial formation energy [28]. Such mechanisms are therefore expected to contribute to the observed formation of larger He bubbles at the grain boundaries; nevertheless, understanding large He bubble growth on grain boundaries requires further coordinated and complimentary experimental and modelling work.

3.2. He Bubble Evolution as a Function of He$^+$ Fluence

He bubble damage evolution in the grain matrices, as quantified by changes in the bubble number density, area, and the total change in volume (found using $\frac{\Delta v}{v} = \frac{4}{3}\pi\, r_c^3 N_v$ where N_v is the bubble density in a 100 nm thick foils and r_c is the average radius of the bubbles) due to bubble formation in the grain matrices for both tungsten grades, are shown in Figure 4 (data adjusted for the NCW for variations in focus, in different video segments, occurring during the dynamic in-situ experiments, as described in Figure S1 and the Supplemental Materials). The He bubble density in the NCW grade increased with He$^+$ fluence and reached a plateau (at a density of ~ 0.125 nm^{-2}), while the average bubble size showed a decrease as a function of He$^+$ fluence and reached an average size of 3 nm^2. The corresponding change in volume showed an increase with He$^+$ fluence at the start of implantation, peaked at ~0.7, and then decreased to ~0.4 at the last He$^+$ fluence (3.5×10^{16} ion.cm^{-2}). In the case of W-TiC (1.1%), the number density increased first (when the bubbles became visible) to 0.025 nm^{-2} and then saturated throughout the implantation, while the average bubble area increased as a function of time (up to 8 nm^2) in a similar trend to the corresponding change in volume, which reached a value of 0.6.

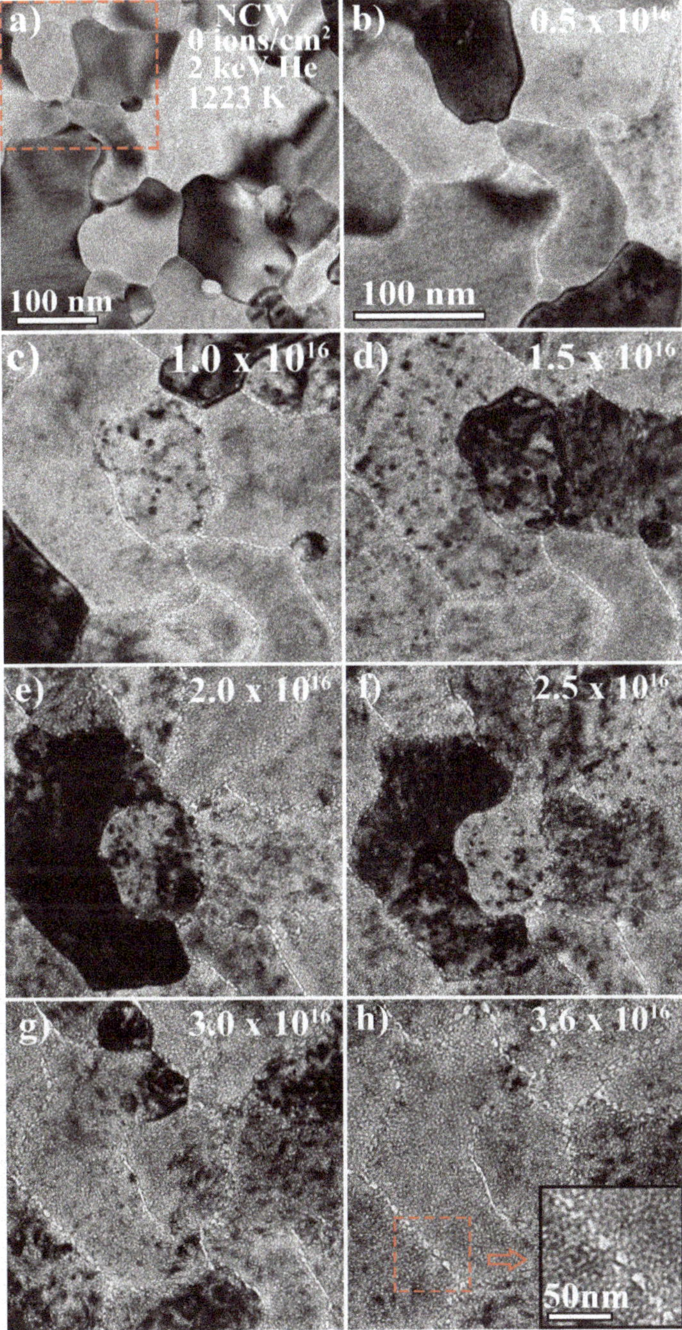

Figure 2. (**a–h**): Bright-field TEM micrographs of a small implanted region taken under Fresnel conditions (under-focused) showing He bubble formation and evolution as a function of He$^+$ fluence in the grain matrices and grain boundaries in equiaxial nanocrystalline tungsten (NCW) with an average grainsize of 85 nm implanted in-situ with 2 keV He$^+$ at 1223 K. Scale bar of (**b–h**) is the same and is shown in (**b**). Red box in (**a**) approximately represents a magnified region presented in (**b**) to (**h**).

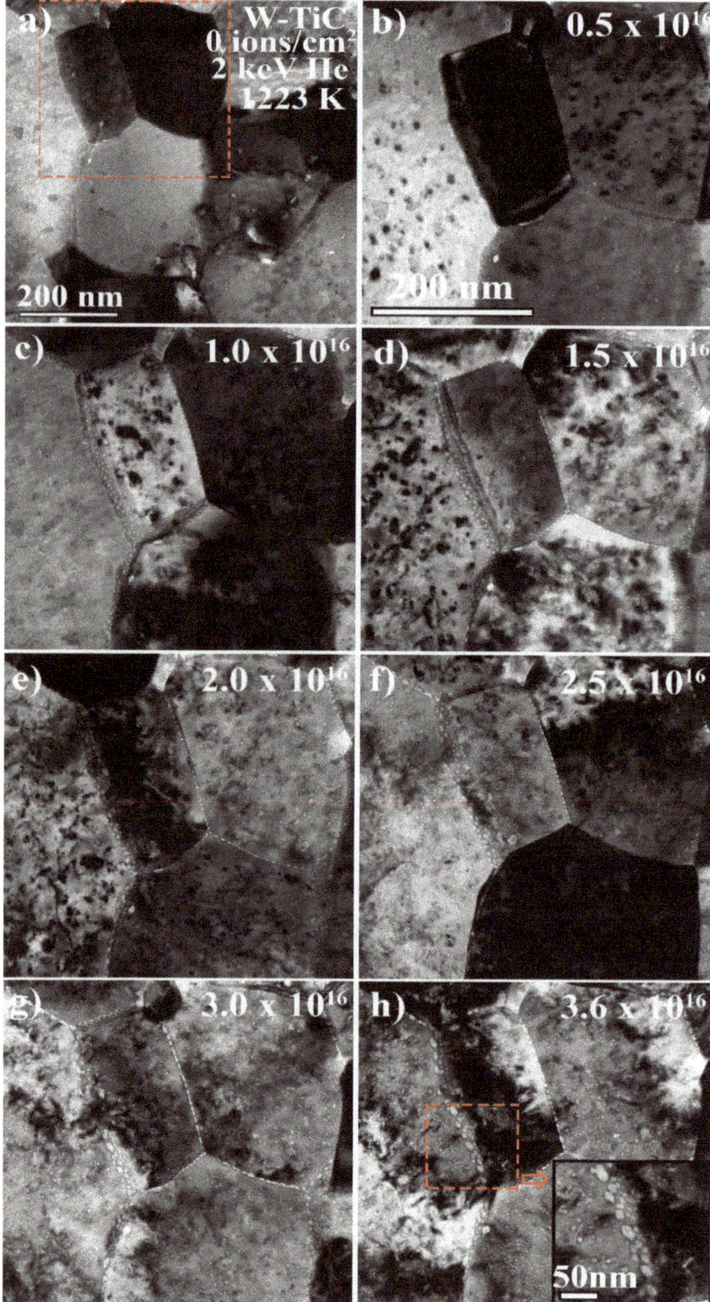

Figure 3. (**a–h**): Bright-field TEM micrographs of a small implanted region taken under Fresnel conditions (under-focused) showing He bubble formation and evolution as a function of He^+ fluence in the grain matrices and grain boundaries in W-TiC (1.1%) implanted in-situ with 2 keV He^+ at 1223 K. Scale bar of (**b–h**) is the same and is shown in (**b**). Red box in (**a**) approximately represents a magnified region presented in (**b**) to (**h**).

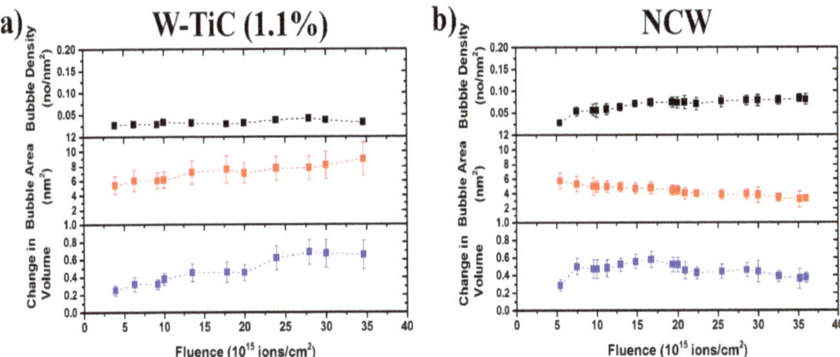

Figure 4. (Color online) Helium bubble density, average area, and the total change in volume in the grain matrices of (**a**) W-TiC and (**b**) NCW as a function of He$^+$ implantation fluence. (For interpretation of the references to color in this figure legend, the reader is referred to the web version of this article.)

3.3. Damage Evolution Comparison

While the final total change in volume in the NCW is about half the change in volume in the W-TiC, indicating higher radiation tolerance of NCW to He$^+$ implantation, the damage evolution (and thus, the bubble density profile) is quite different. In the W-TiC, the change in volume followed the average bubble area profile (whereas the bubble number density was saturated). Therefore, bubble damage in this case is due to the growth of existing He bubbles through the absorption of vacancy, He, and/or He–vacancy complexes. The small bubble density in W-TiC compared to the NCW bubble density, in this case, indicates greater defect accumulation (vacancies and possible He–vacancy complexes with low migration energies) of the former. In addition to the role of high densities of grain boundaries in limiting irradiation or implantation damage, TiC dispersoids have been shown to assist in the defect recombination and absorption of excess vacancies acting as vacancy-biased sinks and to contribute to a higher radiation tolerance of the grain matrices to irradiation damage [12,14]. The same W-TiC material has previously been demonstrated, under heavy ion irradiation, to have higher irradiation resistance to cavity and loop formation (in the grain matrices) compared to commercial coarse-grained tungsten [14]. In previous work [15], nanocrystalline and ultrafine tungsten (where both elongated nanocrystalline and ultrafine grains coexist—SPD material) was implanted under the same conditions used in the current work. Bubble density in the grain matrices of the W-TiC was five times higher than in the SPD material but the average bubble area was ~2.5 times lower in the former. However, the total change in volume was the same. Relatively lower defect mobility, due to the impurities in the W-TiC, could potentially lead to more bubble nucleation sites compared to pure tungsten.

The NCW demonstrated a high density of very small bubbles in the grain matrices which was unexpected. It also showed a decrease in the average bubble size as a function of He$^+$ fluence. The decrease in average size could be due to the nucleation of new small bubbles, which can decrease the average bubble size. However, from the histograms of bubble sizes at four different He$^+$ fluences shown in Figure 5, the average bubble size decrease was shown to be due to a decrease in the sizes of the individual bubbles. Moreover, the average bubble size continued to decrease even after the bubble density plateaued. Such a decrease in bubble size has to be associated with the balance of interstitial and vacancy fluxes against the bubbles in addition to any possible thermal emissions of vacancies at 1223 K (the latter can be difficult due to the high binding energies of vacancies and He) [24] and internal pressure changes due to the He concentration change in the bubble. Both a higher level of interstitial absorption and thermal vacancy emission (the latter of which being suppressed as mentioned above) can lead to the shrinkage of the bubbles. Moreover, nanoporosity from magnetron deposited films can be responsible for higher densities of bubble nucleation sites but this porosity can reduce with irradiation [29] due to adatom accumulation occurring before reaching a stable bubble size. Nanoporous materials were shown to enhance irradiation tolerance in different materials [30–32]. Nanoporosity and possible impurities can also limit the mean free path of defect mobility and thus explain the high density of He bubbles in the NCW. Dislocation loops formed in the NCW were shown to be highly mobile [13]. The interaction of these loops with He bubbles may cause them to shrink. There can also be an effect of residual stress on defect absorption, which is yet to be understood in nanocrystalline materials. Under heavy ion irradiation with no gas introduction, the NCW grades demonstrated higher irradiation resistance to void formation (but with higher void density) compared to the SPD material and commercial coarse-grained tungsten [13]. It was shown in the same study that interstitial and vacancy defects have a higher recombination probability at smaller grain sizes, meaning that NCW would have less cavity damage. In the current work, the bubble sizes at the grain boundaries in the NCW (~8–10 nm as shown in inset in Figure 2h) were smaller than those in the W-TiC grade (~20–25 nm as shown in the inset in Figure 3h). They were also smaller than the bubbles (~10 nm) at the grain boundaries in the SPD material at the same He$^+$ fluence [15]. This represents lower overall bubble damage, which is indicative of a lower He–vacancy complex (with low migration energies) and vacancy transport to the grain boundaries (presumably due to higher levels of recombination). Moreover, the high density of grain boundaries can lead to lower defect transport (to grain boundaries) per boundary. It should be mentioned that the very large bubble sizes at the grain boundaries in the W-TiC material could be due to the TiC impurities at the grain boundaries, which can change the grain boundary sink efficiency and have been shown to be vacancy-biased, as previously discussed. The SPD material in reference [33] showed one order of magnitude higher He$^+$ fluence threshold for fuzz formation compared to commercial tungsten. The NCW in the current work (with its lower total bubble damage in the grain matrices and grain boundaries) is similarly expected to demonstrate an improved fuzz resistance. Such a material needs to be examined under plasma conditions to assess its mechanical property behavior to evaluate its suitability to fusion conditions.

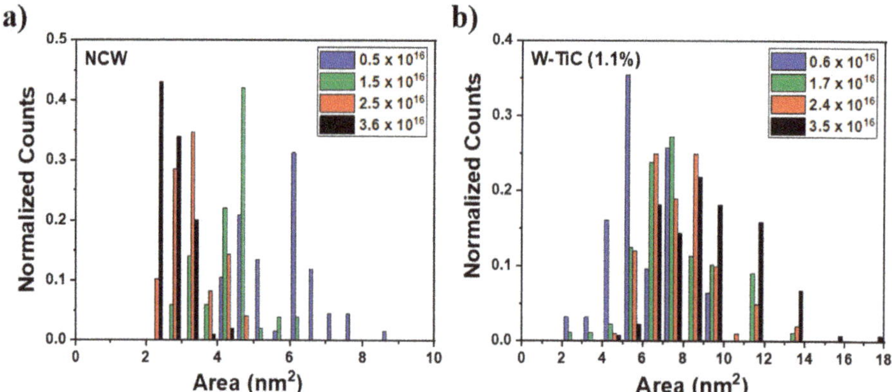

Figure 5. (Color online) normalized bar graphs of bubble size distributions in the grain matrices in (**a**) NCW and (**b**) W-TiC as a function of implantation He$^+$ fluence. (For interpretation of the references to color in this figure legend, the reader is referred to the web version of this article.)

4. Conclusions

In summary, He bubble damage evolution was studied using in-situ TEM as a function of He$^+$ implantation fluence for two strong fusion material candidates (NCW and ultrafine W-TiC alloy). The results have been compared to previously studied SPD nanocrystalline and ultrafine tungsten [15]. Grain boundaries were shown to be decorated with large bubbles, indicating efficient He trapping. He bubble evolution, profiles, and ov]erall bubble damage (total change in volume in grain matrices and grain boundaries) were found to be different in both grades. The NCW showed a higher density of small bubbles but a lower overall change in volume than the W-TiC grade and the previously published SPD tungsten. The high density of small bubbles in the NCW was unexpected but several phenomena can govern this observation. Since SPD tungsten possessed a He^+ fluence threshold for fuzz formation that was one of order of magnitude higher than coarse-grained tungsten, the NCW is then expected to have an even higher He^+ fluence threshold (for fuzz formation). This is true if the fuzz formation mechanism depends on high bubble densities near the surface, as illustrated in previous literature works. These results indicate that while these materials may have improved fuzz resistance under plasma exposures and an enhanced tolerance to He damage due to neutron transmutation reactions, further studies of these materials under fusion conditions need to be performed to develop materials that can withstand these extreme conditions.

Supplementary Materials: The following are available online at http://www.mdpi.com/1996-1944/13/3/794/s1, Figure S1: (Color online) Helium bubble density, average size, and total change in volume in the grain matrices of NCW and W-TiC as functions of He^+ fluence. (a) shows the original and the adjustment data.

Author Contributions: Conceptualization, O.E.A. and S.M.; Methodology, J.H., G.G. and O.E.A.; Formal analysis, K.U., W.S.C. and S.F.; Writing—original draft preparation, O.E.A.; Writing—review and editing, O.E.A., K.U., W.S.C., S.F, J.H., G.G. and S.M. All authors have read and agreed to the published version of the manuscript.

Funding: Research presented in this article was supported by the Laboratory Directed Research and Development program of Los Alamos National Laboratory under project number 20160674PRD3. We gratefully acknowledge the support of the U.S. Department of Energy through the LANL/LDRD Program and the G. T. Seaborg Institute for this work. The authors would also like to thank EPSRC for the funding of the MIAMI facility through grant EP/M028283/1.

Conflicts of Interest: The authors declare no conflict of interest. The funders had no role in the design of the study; in the collection, analyses, or interpretation of data; in the writing of the manuscript, or in the decision to publish the results.

References

1. Zinkle, S.J.; Snead, L.L. Designing radiation resistance in materials for fusion energy. *Annu. Rev. Mater. Res.* **2014**, *44*, 241–267. [CrossRef]
2. Katoh, Y.; Snead, L.; Garrison, L.; Hu, X.; Koyanagi, T.; Parish, C.; Edmondson, P.; Fukuda, M.; Hwang, T.; Tanaka, T. Response of unalloyed tungsten to mixed spectrum neutrons. *J. Nucl. Mater.* **2019**, *520*, 193–207. [CrossRef]
3. Kajita, S.; Sakaguchi, W.; Ohno, N.; Yoshida, N.; Saeki, T. Formation process of tungsten nanostructure by the exposure to helium plasma under fusion relevant plasma conditions. *Nucl. Fusion* **2009**, *49*, 095005. [CrossRef]
4. Hasegawa, A.; Fukuda, M.; Tanno, T.; Nogami, S. Neutron irradiation behavior of tungsten. *Mater. Trans.* **2013**, *54*, 466–471. [CrossRef]
5. Zinkle, S.J.; Was, G. Materials challenges in nuclear energy. *Acta Mater.* **2013**, *61*, 735–758. [CrossRef]
6. Kajita, S.; Yoshida, N.; Yoshihara, R.; Ohno, N.; Yamagiwa, M. TEM observation of the growth process of helium nanobubbles on tungsten: Nanostructure formation mechanism. *J. Nucl. Mater.* **2011**, *418*, 152–158. [CrossRef]
7. Zhang, T.; Wang, Y.; Xie, Z.; Liu, C.; Fang, Q. Synergistic effects of trace ZrC/Zr on the mechanical properties and microstructure of tungsten as plasma facing materials. *Nucl. Mater. Energy* **2019**, *19*, 225–229. [CrossRef]
8. Shen, T.D. Radiation tolerance in a nanostructure: Is smaller better? *Nucl. Instrum. Methods Phys. Res. Sect. B* **2008**, *266*, 921–925. [CrossRef]
9. Shen, T.D.; Feng, S.; Tang, M.; Valdez, J.A.; Wang, Y.; Sickafus, K.E. Enhanced radiation tolerance in nanocrystalline $MgGa_2 O_4$. *Appl. Phys. Lett.* **2007**, *90*, 263113–263115. [CrossRef]
10. Chookajorn, T.; Murdoch, H.A.; Schuh, C.A. Design of stable nanocrystalline alloys. *Science* **2012**, *337*, 951–954. [CrossRef]
11. Kitsunai, Y.; Kurishita, H.; Kayano, H.; Hiraoka, Y.; Igarashi, T.; Takida, T. Microstructure and impact properties of ultra-fine grained tungsten alloys dispersed with TiC. *J. Nucl. Mater.* **1999**, *271*, 423–428. [CrossRef]

12. Kesternich, W.; Rothaut, J. Reduction of helium embrittlement in stainless steel by finely dispersed TiC precipitates. *J. Nucl. Mater.* **1981**, *104*, 845–852. [CrossRef]

13. El-Atwani, O.; Esquivel, E.; Aydogan, E.; Martinez, E.; Baldwin, J.; Li, M.; Uberuaga, B.; Maloy, S. Unprecedented irradiation resistance of nanocrystalline tungsten with equiaxed nanocrystalline grains to dislocation loop accumulation. *Acta Mater.* **2019**, *165*, 118–128. [CrossRef]

14. El-Atwani, O.; Cunningham, W.; Esquivel, E.; Li, M.; Trelewicz, J.; Uberuaga, B.; Maloy, S. In-situ irradiation tolerance investigation of high strength ultrafine tungsten-titanium carbide alloy. *Acta Mater.* **2019**, *164*, 547–559. [CrossRef]

15. El-Atwani, O.; Hinks, J.; Greaves, G.; Allain, J.; Maloy, S. Grain size threshold for enhanced irradiation resistance in nanocrystalline and ultrafine tungsten. *Mater. Res. Lett.* **2017**, *5*, 343–349. [CrossRef]

16. Greaves, G.; Mir, A.; Harrison, R.; Tunes, M.; Donnelly, S.; Hinks, J. New Microscope and Ion Accelerators for Materials Investigations (MIAMI-2) system at the University of Huddersfield. *Nucl. Instrum. Methods Phys. Res. Sect. A Accel. Spectrometers Detect. Assoc. Equip.* **2019**, *931*, 37–43. [CrossRef]

17. Ziegler, J.F.; Ziegler, M.D.; Biersack, J.P. SRIM–The stopping and range of ions in matter (2010). *Nucl. Instrum. Methods Phys. Res. Sect. B Beam Interact. Mater. At.* **2010**, *268*, 1818–1823. [CrossRef]

18. ASTM E521-96 e1. *Standard Practice for Neutron Radiation Damage Simulation by Charged Particle Radiation*; American Society of Testing and Material: Wes Conshohocken, PA, USA, 2009.

19. Jenkins, M. Characterisation of radiation-damage microstructures by TEM. *J. Nucl. Mater.* **1994**, *216*, 124–156. [CrossRef]

20. El-Atwani, O.; Nathaniel, J.; Leff, A.; Muntifering, B.; Baldwin, J.; Hattar, K.; Taheri, M. The role of grain size in He bubble formation: Implications for swelling resistance. *J. Nucl. Mater.* **2017**, *484*, 236–244. [CrossRef]

21. Li, M.; Kirk, M.; Baldo, P.; Xu, D.; Wirth, B. Study of defect evolution by TEM with in situ ion irradiation and coordinated modeling. *Philos. Mag.* **2012**, *92*, 2048–2078. [CrossRef]

22. Iwakiri, H.; Yasunaga, K.; Morishita, K.; Yoshida, N. Microstructure evolution in tungsten during low-energy helium ion irradiation. *J. Nucl. Mater.* **2000**, *283*, 1134–1138. [CrossRef]

23. Wilson, W.; Bisson, C.; Baskes, M. Self-trapping of helium in metals. *Phys. Rev. B* **1981**, *24*, 5616. [CrossRef]

24. Reed, D. A review of recent theoretical developments in the understanding of the migration of helium in metals and its interaction with lattice defects. *Radiat. Eff.* **1977**, *31*, 129–147. [CrossRef]

25. Balluffi, R.W. Vacancy defect mobilities and binding energies obtained from annealing studies. *J. Nucl. Mater.* **1978**, *69*, 240–263. [CrossRef]

26. González, C.; Iglesias, R. Migration mechanisms of helium in copper and tungsten. *J. of Mater. Sci.* **2014**, *49*, 8127–8139. [CrossRef]

27. Caspers, L.; Fastenau, R.; Van Veen, A.; Van Heugten, W. Mutation of vacancies to divacancies by helium trapping in molybdenum effect on the onset of percolation. *Phys. Status Solidi A* **1978**, *46*, 541–546. [CrossRef]

28. Sefta, F.; Hammond, K.D.; Juslin, N.; Wirth, B.D. Tungsten surface evolution by helium bubble nucleation, growth and rupture. *Nucl. Fusion* **2013**, *53*, 073015. [CrossRef]

29. Li, J.; Fan, C.; Li, Q.; Wang, H.; Zhang, X. In situ studies on irradiation resistance of nanoporous Au through temperature-jump tests. *Acta Mater.* **2018**, *143*, 30–42. [CrossRef]

30. Chen, Y.; Yu, K.Y.; Liu, Y.; Shao, S.; Wang, H.; Kirk, M.; Wang, J.; Zhang, X. Damage-tolerant nanotwinned metals with nanovoids under radiation environments. *Nat. Commun.* **2015**, *6*, 1–8. [CrossRef]

31. Gao, E.; Doerner, R.; Williams, B.; Ghoniem, N.M. Low-energy helium plasma effects on textured micro-porous tungsten. *J. Nucl. Mater.* **2019**, *517*, 86–96. [CrossRef]

32. Bringa, E.M.; Monk, J.; Caro, A.; Misra, A.; Zepeda-Ruiz, L.; Duchaineau, M.; Abraham, F.; Nastasi, M.; Picraux, S.; Wang, Y. Are nanoporous materials radiation resistant? *Nano Lett.* **2011**, *12*, 3351–3355. [CrossRef] [PubMed]

33. El-Atwani, O.; Gonderman, S.; Efe, M.; De Temmerman, G.; Morgan, T.; Bystrov, K.; Klenosky, D.; Qiu, T.; Allain, J. Ultrafine tungsten as a plasma-facing component in fusion devices: Effect of high flux, high fluence low energy helium irradiation. *Nucl. Fusion* **2014**, *54*, 083013. [CrossRef]

Article

Effect of Helium on Dispersoid Evolution under Self-Ion Irradiation in A Dual-Phase 12Cr Oxide-Dispersion-Strengthened Alloy

Hyosim Kim [1,*], Tianyao Wang [2], Jonathan G. Gigax [1], Shigeharu Ukai [3], Frank A. Garner [2] and Lin Shao [2,*]

[1] Los Alamos National Laboratory, P.O. Box 1663, Los Alamos, NM 87544, USA; jgigax@lanl.gov
[2] Department of Nuclear Engineering, Texas A&M University, College Station, TX 77843-3128, USA; wty19920822@tamu.edu (T.W.); frank.garner@dslextreme.com (F.A.G.)
[3] Materials Science and Engineering, Faculty of Engineering, Hokkaido University, N13, W-8, Kita-ku, Sapporo, Hokkaido 060-8628, Japan; s-ukai@eng.hokudai.ac.jp
* Correspondence: hkim@lanl.gov (H.K.); lshao@tamu.edu (L.S.)

Received: 14 September 2019; Accepted: 11 October 2019; Published: 14 October 2019

Abstract: As one candidate alloy for future Generation IV and fusion reactors, a dual-phase 12Cr oxide-dispersion-strengthened (ODS) alloy was developed for high temperature strength and creep resistance and has shown good void swelling resistance under high damage self-ion irradiation at high temperature. However, the effect of helium and its combination with radiation damage on oxide dispersoid stability needs to be investigated. In this study, 120 keV energy helium was preloaded into specimens at doses of 1×10^{15} and 1×10^{16} ions/cm^2 at room temperature, and 3.5 MeV Fe self-ions were sequentially implanted to reach 100 peak displacement-per-atom at 475 °C. He implantation alone in the control sample did not affect the dispersoid morphology. After Fe ion irradiation, a dramatic increase in density of coherent oxide dispersoids was observed at low He dose, but no such increase was observed at high He dose. The study suggests that helium bubbles act as sinks for nucleation of coherent oxide dispersoids, but dispersoid growth may become difficult if too many sinks are introduced, suggesting that a critical mass of trapping is required for stable dispersoid growth.

Keywords: oxide-dispersion-strengthened (ODS); ion irradiation; He implantation; dual-phase; ferritic-martensitic; self-ion

---

1. Introduction

Oxide-dispersion-strengthened (ODS) alloys are one class of promising candidate alloys for Gen IV and fusion reactors due to their superior high temperature strength and creep resistance [1–7]. Among various ODS alloys developed worldwide, a dual-phase ferritic/martensitic (F/M) 12Cr ODS alloy has shown good high-temperature oxidation resistance and corrosion resistance [3]. This alloy has specifically controlled excess oxygen and titanium contents to control residual alpha ferrite volume for superior creep rupture strength [1]. A high level of tempered martensite (TM) volume in the matrix contributes to good void swelling resistance [4]. Recent studies have shown good grain stability at 800 displacement-per-atom (dpa) after 600 °C self-ion irradiation, and good swelling resistance (less than 2% and 0.06% at 475 °C for the ferrite and TM phases, respectively) [8,9]. These studies also show that oxide dispersoid size changes saturate with increasing dose. It has been suggested that the increased dispersoid density upon irradiation may improve the swelling resistance by providing more recombination sites for point defects [10,11]. A previous study on the same alloy using the same irradiation condition employed in this study showed that the dispersoid density increased in both ferrite and TM phases due to ballistic dissolution by irradiation and thermodynamic homogenous nucleation at high temperature [10,12]. Due to these properties, the 12Cr ODS alloy is promising for applications in future reactor designs where the operation conditions are more severe with higher damage levels and higher temperatures [13].

However, to deploy this alloy into a fast reactor environment, the effect of helium addition produced by (n, α) transmutation and subsequent He embrittlement needs to be further studied [14]. There have been a few He implantation studies on various ODS alloys [11,15–18]. Yamamoto et al. observed small-size He bubbles trapped on YTiO clusters during simultaneous neutron and He irradiation of MA957 [15]. Edmondson et al. also reported similar results on 14YWT after He implantation alone at high temperature [16]. Yutani et al. investigated the He effect on ODS alloys using single and dual-ion irradiations at high temperature and reported that smaller nano-sized particles can induce a high density of small He bubbles and thereby reduce swelling [11]. Heintze et al. showed that radiation behaviors of ODS alloys are different for single, dual, and sequential He and Fe ion irradiation [17]. Lu et al. reported that dispersoid densities increase after sequential He and Fe ion irradiation of 9Cr ODS alloys [18].

Although many He ion implantation studies have been conducted, there is still a need to further investigate the He effect, particularly in the 12Cr ODS alloy. First, the majority of previous studies focused on swelling, with little attention on dispersoid stability. Dispersoid morphology changes play a significant role in influencing creep resistance, which was one of the main motivations for introducing ODS alloys into nuclear applications. Second, there is no previous report on the He effect on the TM phase of ODS alloys and is unclear if the effect is similar to that reported in ferritic and austenitic ODS alloys. The TM phase has demonstrated much better swelling resistance than the ferrite phase. Thereby TM phase-dominated ODS alloys are very attractive to further improve swelling resistance.

2. Materials and Experimental Procedure

The dual-phase 12Cr ODS alloy was fabricated using a mechanical alloying (MA) process in Hokkaido University (Sapporo, Japan). The MA processed powders were consolidated using a spark plasma sintering method at 1100 °C and were hot-rolled afterward. The final steps involved normalization at 1050 °C for 1 h and tempering at 800 °C for 1 h. More details on the fabrication process can be found in [3]. The chemical composition of the alloy is provided in Table 1. The 12Cr ODS alloy was cut into 3 mm × 6 mm × 1.5 mm pieces. Samples were then mechanically polished by using SiC paper up to p-4000 fine grit, and further polished with 0.25 μm diamond suspension and 0.04 μm silica suspension to remove surface deformation. The final sample thickness was ~0.7 mm.

Table 1. Chemical composition of as-received dual-phase 12Cr oxide-dispersion-strengthened (ODS) alloy (wt %).

Fe	Cr	W	Ni	Ti	C	N	Ar	Y_2O_3	Excess O
Balance	11.52	1.44	0.36	0.28	0.16	0.007	0.006	0.36	0.144

The experiment design is presented in Figure 1. Two pristine specimens were irradiated using a rastered 120 keV He^+ ion beam to a dose of 1×10^{15} and 1×10^{16} ions/cm^2, respectively, both at room temperature. The specimens were then irradiated using a 3.5 MeV Fe^{2+} defocused ion beam to reach 100 peak dpa at 475 °C using a 1.7 MV tandem accelerator. The average dpa rate was 1.74×10^{-3} dpa/s for the Fe ion beam and the ion fluence yielding 100 peak dpa was 9.83×10^{16} ions/cm^2. Note that the Fe irradiation utilized a multiple beam deflection technique to filter out carbon and other contaminants [19–22]. The target chamber vacuum was at 4.0×10^{-8}–6.0×10^{-8} torr during the irradiation. Additionally, liquid nitrogen cold trapping in the target chamber was applied [19–22]. For He implantation, a raster beam was used to guarantee beam uniformity over a large irradiation area. The rastering effect was not a concern for He implantation since the implantation was used only to introduce He bubbles. For the Fe irradiation, however, a static defocused beam was used [23]. Our study further included control samples, some irradiated by He ions only, and others by Fe ions only.

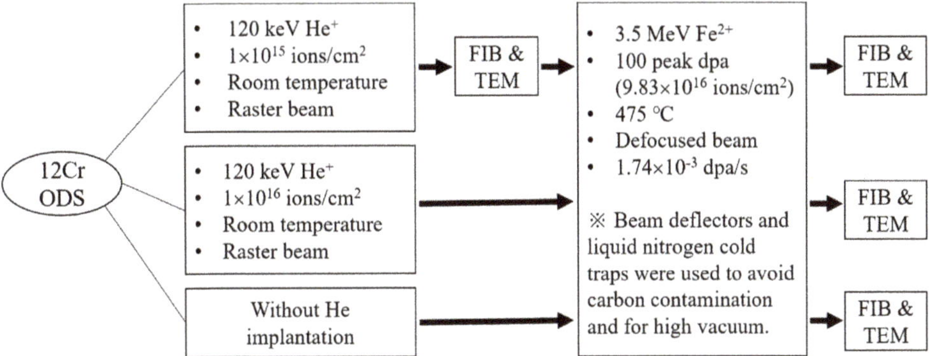

Figure 1. Diagram of experimental steps.

All samples were characterized using a Tescan Lyra-3 transmission electron microscope (TEM). The focused ion beam (FIB) lift-out technique was used to prepare TEM lamella specimens [24], using a FEI Tecnai G2 F20 Super-Twin (FEI, Hillsboro, OR, USA). The TEM specimen thickness was measured using electron energy loss spectroscopy (EELS) on a FEI Tecnai F20. Bright-field (BF) and weak-beam dark-field (WBDF) imaging techniques were used to characterize dispersoid coherency and distribution. High-resolution TEM (HRTEM) and fast Fourier transform (FFT) were used to confirm the crystal structure of the dispersoids. The WBDF imaging is a diffraction-contrast imaging using a weakly excited beam. It is widely used for imaging sub-nano-size features because of better accuracy on position than normal dark-field (DF). While the desired g (for this study, g110 of matrix) is on the optical axis and used for a regular DF, the sample is tilted to make the 3g diffraction pattern brightest to make s_g (excitation error) large. This allows diffracting planes to bend locally back into the Bragg-diffracting orientation to give more intensity in the DF image [25–27].

The average dispersoid diameter was calculated by averaging diameters measured one-by-one from BF and DF micrographs. Dispersoids were counted only when they were present in both BF and DF images. The dispersoid density was calculated by dividing the number of counted coherent dispersoids with (area × specimen thickness) in areas where dispersoids were counted.

The Stopping and Range of Ions in Matter (SRIM) 2013 code was used to calculate damage profiles and implant distributions. The dpa calculation used the Kinchin–Pease mode and the Fe displacement threshold energy was chosen to be 40 eV [28,29]. Figure 2 shows the SRIM calculation of the 120 keV He implant profiles for doses of 1×10^{15} and 1×10^{16} ions/cm² and the dpa profile for 3.5 MeV Fe 100 peak dpa irradiation. The 3.5 MeV Fe beam penetrates to a maximum depth of 1.5 μm below the incident surface, and the dpa peak is ~1 μm deep. In the case of 120 keV He, the maximum penetration depth is approximately 0.5 μm from the surface. The dashed line refers to the 1×10^{15} He/cm² case with a He peak at ~600 appm, while the dotted line refers to the 1×10^{16} He/cm² case with a He peak at ~6000 appm. The He/dpa ratio at the He peak depth is 14.9 appm/dpa for the 1×10^{15} He/cm² and 148.8 He/dpa for the 1×10^{16} He/cm². An energy of 120 keV is selected for He implantation to avoid the free surface effect and to minimize the injected interstitial effect during subsequent Fe ion irradiation [30]. Since void swelling by 3.5 MeV ion irradiation is usually maximized in the front half of the projected range [30], it is ideal to introduce He into the depth region of 300–400 nm. Note that the average local dpa for this region is about 43 dpa.

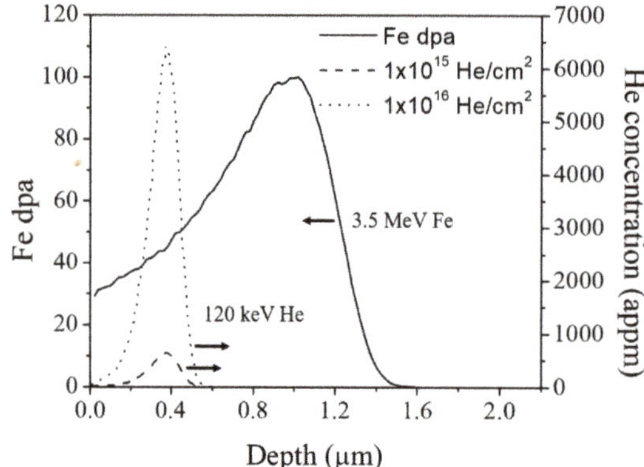

Figure 2. Stopping and Range of Ions in Matter (SRIM) calculations of damage profiles resulting from 3.5 MeV Fe ion irradiation and the He distributions resulting from 120 keV He ion implantation.

3. Results

3.1. He Implantation Only

TEM images were taken from the 350 nm depth region where the He ion peak is located to see the effect of He implantation on oxide dispersoid size and density. Figure 3a,b shows BF and DF images of the He implanted sample using (g, 3 g) condition with the g110 direction excited as indicated by the inset of the diffraction image shown in Figure 3b [24]. The diffraction pattern was obtained from the same grain where the BF and DF images were taken. A total of 33 coherent oxide dispersoids were counted in the 300 to 400 nm depth region. The oxide dispersoids have dark contrast in the BF image regardless of coherency, while only coherent dispersoids appear bright in DF imaging. Other features such as small dislocation loops or dislocation lines can also appear bright if they are aligned with g110. However, dislocation lines and dislocation loops are distinguishable from oxide dispersoids in their shape and contrast under the BF imaging mode. For example, a narrow and long dark contrast line is a dislocation line as shown in the bottom left corner of Figure 3a, which is clearly distinguishable from dispersoids. Small dislocation loops (≤4 nm) can be difficult to separate from small dispersoids. In this study, three factors were considered when distinguishing loops. First, if the shape is not circular, it was not counted as a dispersoid. Second, if it appears as a black dot rather than as a slightly dark contrast, it is considered as a loop or dislocation core. Small dispersoids mostly exhibit a coherent relationship with matrix, yielding slightly darker contrast than the matrix. Third, if the feature has a black-white contrast, it is a loop lying close to the foil surface [27]. By comparing BF and DF images, coherent and incoherent dispersoids can be counted separately. Note that only the TM phase was analyzed in this study for two reasons. First, the TM phase accounts for 80% of the volume of the alloy. Second, the ferrite phase shows a dramatic dispersoid density change even without He implantation (as shown in the previous study reported in [10]). In the present study only coherent dispersoids were counted because newly nucleated oxide dispersoids are prone to take a coherent relationship with the matrix to achieve lower surface energy [31,32].

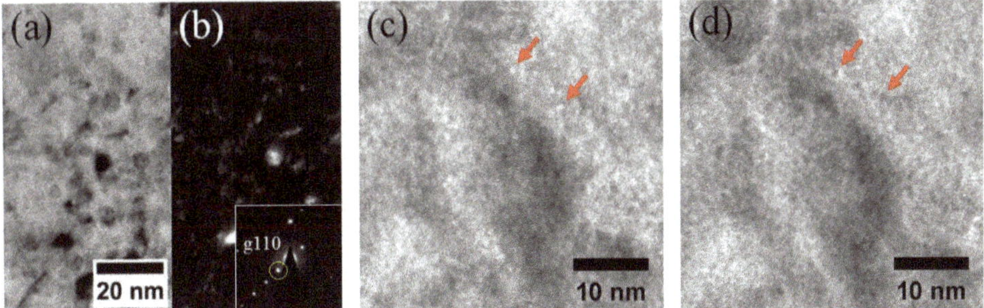

Figure 3. (**a**) Bright-field, (**b**) dark-field, (**c**) under-focused, and (**d**) over-focused TEM images taken from 300–400 nm depth region, after 1×10^{15} He ions/cm^2 implantation. The diffraction pattern used to obtain BF and DF images is superimposed on (**b**) with g110 indexed. He bubbles appear bright and dark in under-focused and over-focused images, respectively.

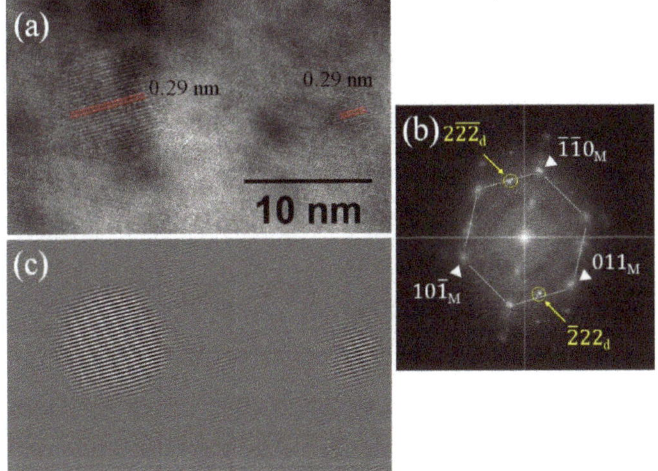

Figure 4. (**a**) High-resolution TEM (HRTEM) of two oxide dispersoids, (**b**) fast Fourier transform (FFT) patterns from the HRTEM, and (**c**) inverse FFT image of dispersoids using FFT patterns in yellow circles. The *d*-spacings measured from HRTEM and FFT patterns are 0.29 nm, which agree with (222) of pyrochlore $Y_2Ti_2O_7$. The scale bar in (a) applies to (c).

Figure 3c,d shows TEM BF under-focused and over-focused images, respectively, taken from the 300 to 400 nm depth region in the 1×10^{15} He ions/cm^2 implanted specimen. In the under-focused image, He bubbles appear white, while they appear dark in the over-focused image, as indicated by red arrows. The He bubble size was measured to be less than 1 nm diameter at this depth. Due to the small size, it is challenging to obtain size and density from the TEM image. The size and density of oxide dispersoids were counted from this region, and the results are discussed in Section 3.4.

Figure 4a shows a HRTEM image of two different size oxide dispersoids obtained from the He implanted region. The HRTEM image was taken at the $[\bar{1}1\bar{1}]$ zone axis of the matrix. Figure 4b shows the FFT patterns from the matrix and dispersoids that are indexed separately using white triangles and yellow arrows, respectively. The *d*-spacings of dispersoids measured from HRTEM and FFT patterns are 0.29 nm, which agree with (222) plane spacing of pyrochlore $Y_2Ti_2O_7$. Previous studies on this alloy also showed that the oxide dispersoids are either orthorhombic Y_2TiO_5 or pyrochlore $Y_2Ti_2O_7$ [8,9].

FFT patterns from the dispersoids (yellow circles in Figure 4b) were used to generate an inverse FFT image as shown in Figure 4c to highlight the dispersoids in Figure 4a.

3.2. Fe Irradiation Only

Figure 5 shows TEM BF and DF images taken from the Fe-irradiated sample at selected depths. Images were taken from six different depths, including the out-of-ion range region (2000 nm). The 2000 nm region was used as a reference point, since it is free from direct ion bombardment (although it does have a possible thermal annealing effect during the irradiation). The same WBDF condition utilized for Figure 3b imaging was used for consistency. Figure 5g shows SRIM damage profile and red arrows are used to show the depths of the TEM characterization. More than 70 coherent oxide dispersoids were counted from each set of micrographs. As shown in Figure 5, large incoherent dispersoids are observed at 2000 nm depth, while only small-size, mostly coherent dispersoids are observed within the ion range ($\leq$1000 nm).

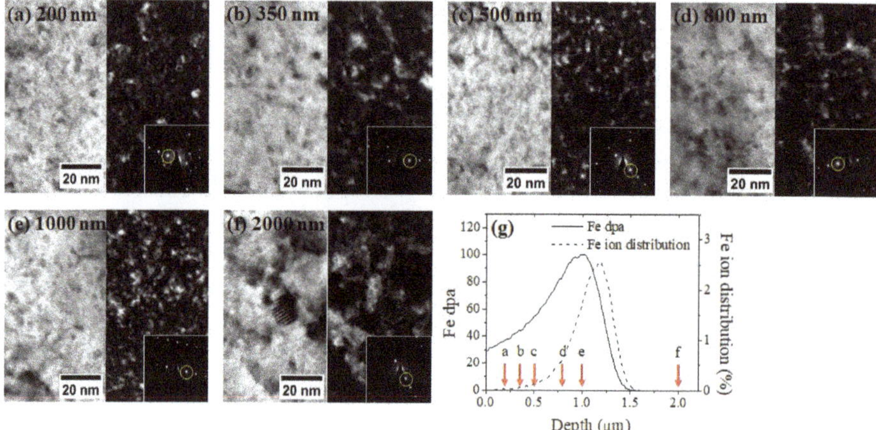

Figure 5. TEM bright-field and dark-field images of a Fe irradiated sample, taken from depths of (**a**) 200, (**b**) 350, (**c**) 500, (**d**) 800, (**e**) 1000, and (**f**) 2000 nm. Diffraction patterns obtained from the same grain are superimposed on each DF image with g110 marked by a yellow circle. (**g**) SRIM Fe damage profile is shown with red arrows indicating the depths of characterization. The dispersoid size decreased and the density increased within the ion range.

3.3. He Implantation Followed by Fe Irradiation

Figure 6 shows TEM BF and DF micrograph sets of the 1×10^{15} He preimplanted and Fe irradiated sample taken from 200, 350, 550, 800, and 1000 nm depths and SRIM calculations of He implant distribution and Fe damage. The same WBDF condition employed in Figure 5 was used. At a depth of 350 nm, corresponding to the He peak location, the TEM DF image in Figure 6b shows the highest density of coherent dispersoids. This is evidence that nucleation of coherent oxide dispersoids is promoted there.

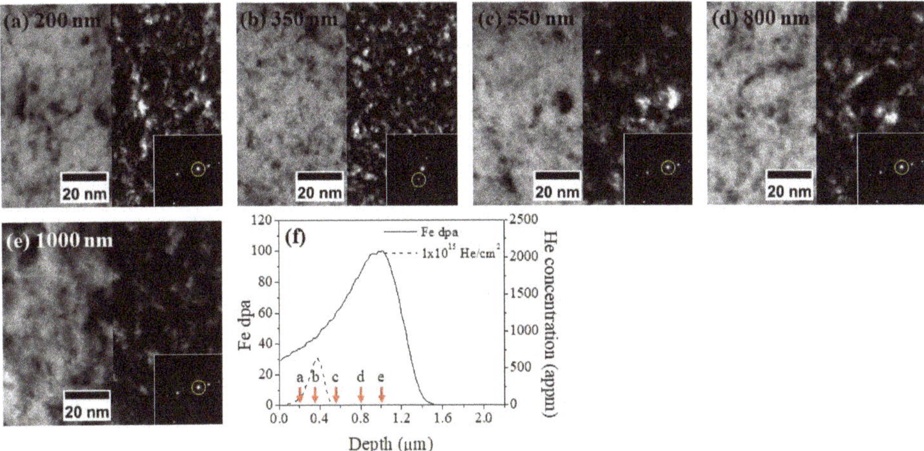

Figure 6. TEM bright-field and dark-field images of the 1×10^{15} He preimplanted and Fe irradiated sample at depths of (**a**) 200, (**b**) 350, (**c**) 550, (**d**) 800, and (**e**) 1000 nm. Diffraction patterns obtained from the same grain are superimposed on each DF image with g110 marked by a yellow circle. (**f**) SRIM calculations of Fe, dpa, and He implant distributions are included, with red arrows indicating the depths of characterization. The dispersoid density was greatly increased in (**b**).

Figure 7 shows TEM BF images of 1×10^{16} He + Fe irradiated sample taken from 300–500 nm depth region. The average diameter of He bubbles was measured to be 3.4 ± 0.9 nm. A high number of He bubbles were observed on the grain boundaries, as shown in Figure 7a, suggesting a sink effect of the grain boundaries. The He bubble density at this depth was measured to be $1.6 \times 10^{23} \pm 1.3 \times 10^{22}$ bubbles/m^3. A He bubble denuded zone was observed near grain boundaries. Figure 7b,c shows under-focused and over-focused images of He bubbles at higher magnification, respectively, taken from the same region. He bubbles appear bright with enhanced edge contrast in under-focus images while they appear dark with a vague boundary in over-focus images. Examples are indicated by the red arrows in Figure 7b,c.

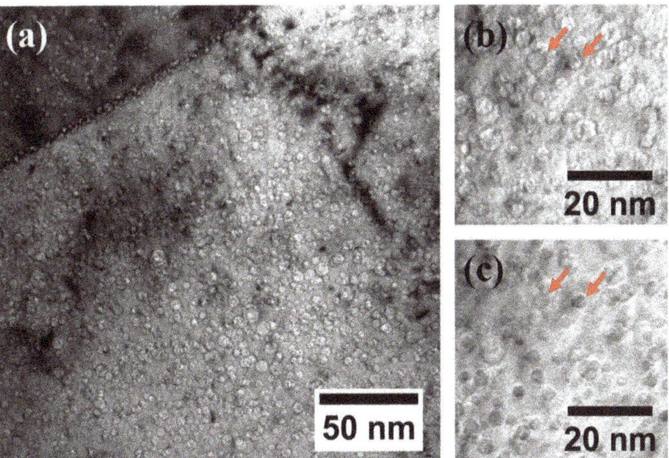

Figure 7. (**a**) Bright-field TEM micrograph, (**b**) under-focused image, and (**c**) over-focused image from the 300–500 nm depth region of the 1×10^{16} He preimplanted and Fe irradiated specimen.

Figure 8 shows TEM BF and DF micrographs of the 1×10^{16} He preimplanted and Fe irradiated sample taken from 200, 350, 550, 800, and 1000 nm depths, along with the SRIM calculations. Unlike the DF image at 350 nm of the 1×10^{15} He + Fe specimen, the DF image of the 1×10^{16} He + Fe sample does not show many bright features, suggesting that the higher dose of He preimplantation did not lead to more nucleation of oxide dispersoids during Fe irradiation. Further analysis on the size and density of dispersoids is given in Section 3.4.

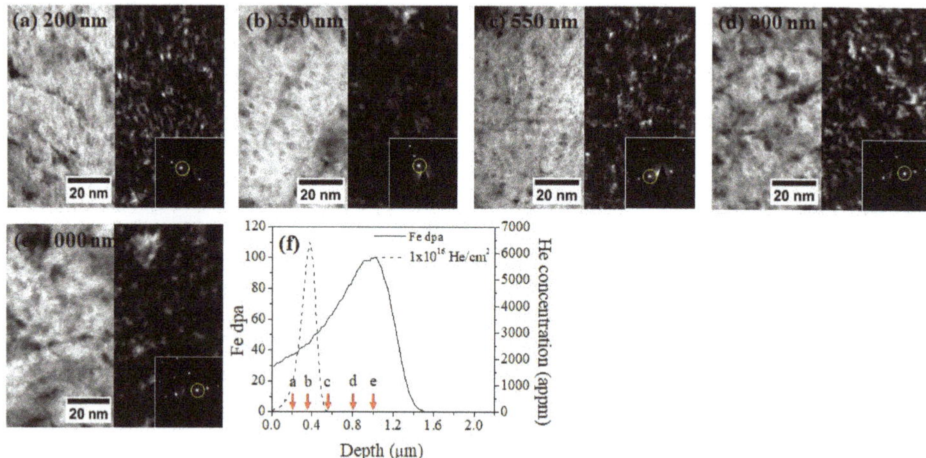

Figure 8. TEM bright-field and dark-field micrographs of the 1×10^{16} He preimplanted and Fe irradiated specimen taken from depths of (**a**) 200, (**b**) 350, (**c**) 550, (**d**) 800, and (**e**) 1000 nm. Diffraction patterns obtained from the same grain are superimposed on each DF image with g110 marked by a yellow circle. (**f**) SRIM calculations are provided to show the depths of TEM characterization. A dispersoid density increase was not observed in (**b**).

3.4. Oxide Dispersoid Size and Density Comparison

Figure 9 shows the average oxide dispersoid diameters of different irradiation conditions as a function of depth. The superimposed dashed and dotted lines refer to the 1×10^{15} and 1×10^{16} 120 keV He ion distributions, respectively, and the black solid line shows the Fe damage profile. The Fe irradiation sample (hollow circles) shows that the dispersoid sizes within the ion range are roughly the same considering the error bars. This uniformity can be explained by the fact that oxide dispersoid size is not dependent on local dpa rates as shown in a previous study [24]. Compared with the out-of-ion range region (2000 nm), oxide dispersoid sizes within the ion range are reduced after Fe irradiation due to ballistic dissolution [8,9,24].

The size of dispersoids after room temperature 1×10^{15} He implantation, indicated by the star in Figure 9, is almost the same as the size of oxide dispersoids at 2000 nm depth, which suggests that He implantation itself does not affect dispersoid morphology. The 1×10^{15} He + Fe irradiated sample shows slightly larger dispersoid sizes within the heavy ion range, but this is actually not related to the He preimplantation effect, since the size is still widely distributed at regions >600 nm. Therefore, considering the error bars, we believe it is a statistical fluctuation varying from sample to sample or grain to grain. The 1×10^{16} He + Fe irradiated specimen indicated by solid diamonds shows that dispersoid sizes are similar to that of the Fe irradiated specimen. Overall, there is no clear trend that He preimplantation affects dispersoid size after subsequent Fe ion irradiation.

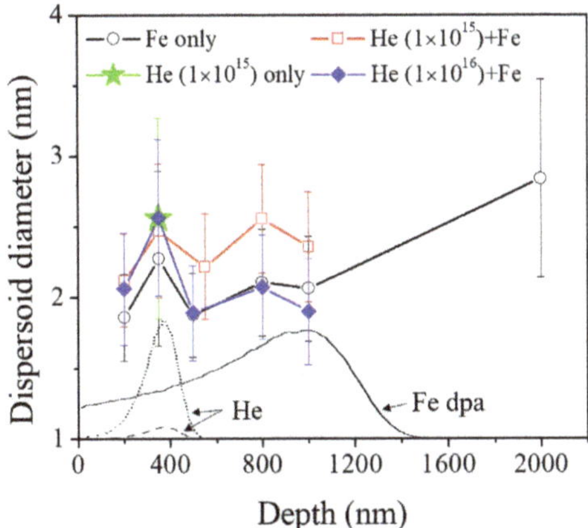

Figure 9. Average dispersoid diameters of Fe irradiated only, He implanted only, and He + Fe irradiated samples as a function of depth. The SRIM calculations are superimposed. The average size decreased within the ion range, but no significant difference was observed between the specimens.

Figure 10 shows the dispersoid densities as a function of depth, superimposed with the SRIM calculations. The He implanted-only sample (indicated by stars) shows a density similar to that of the out-of-ion-range of the Fe irradiated sample (2000 nm), which implies that He implantation alone does not affect the dispersoid density, at least when implanted at room temperature at 120 keV energy.

The coherent dispersoid densities, however, are dramatically increased after Fe irradiation as indicated by circles, and there is a factor of two density increase at 350 nm depth, in comparison with the point at 2000 nm. On the other hand, the 1×10^{15} He preimplanted + Fe irradiated sample (indicated by squares) shows the highest coherent dispersoid density at the He ion peak location (350 nm). The enhancement is a factor of 2.7 in comparison with the 2000 nm reference point. The coherent oxide dispersoid size and density for each case at 350 nm depth and 2000 nm depth are summarized in Table 2.

For the 1×10^{16} He preimplanted + Fe irradiated specimen (indicated by solid diamonds), the densities of dispersoid are systematically lower than those of the Fe irradiated sample within the ion range ($\leq$1000 nm). This suggests that the large-size He bubbles do not affect dispersoid nucleation under Fe ion irradiation, and there is a possibility that the large bubbles even suppress the nucleation process. It appears that small-size He bubbles in the matrix assist coherent oxide dispersoid nucleation during Fe irradiation, resulting in a higher density of coherent dispersoids at the He peak region. However, this effect occurs only when the He bubble size is small.

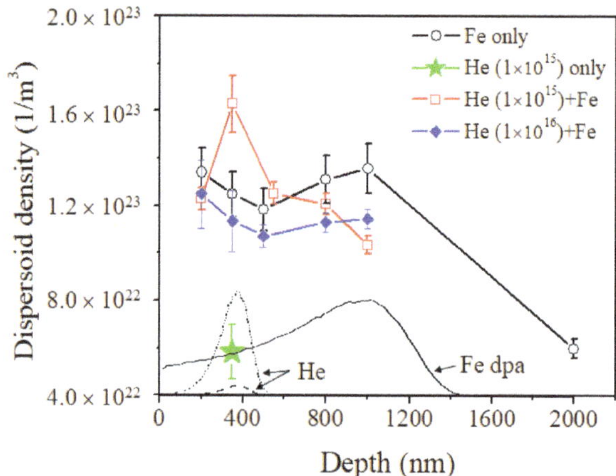

Figure 10. Dispersoid densities of Fe irradiated, He implanted and He + Fe irradiated samples as a function of depth. The He ion profiles and Fe damage curve are superimposed. The densities increased within the ion range for all cases, and the highest density was observed after the 1×10^{15} He + Fe irradiation.

Table 2. Summary of coherent oxide dispersoid size and density for each irradiation case.

Specimen	Diameter (nm)	Density (Particles/m³)
Fe (350 nm)	2.3 ± 0.6	$1.2 \times 10^{23} \pm 9.6 \times 10^{21}$
He (350 nm)	2.6 ± 0.7	$5.9 \times 10^{22} \pm 1.1 \times 10^{22}$
1×10^{15} He + Fe (350 nm)	2.5 ± 0.5	$1.6 \times 10^{23} \pm 1.2 \times 10^{22}$
1×10^{16} He + Fe (350 nm)	2.6 ± 0.6	$1.1 \times 10^{23} \pm 1.3 \times 10^{22}$
Fe (2000 nm)	2.8 ± 0.6	$6.0 \times 10^{22} \pm 4.0 \times 10^{21}$

Figure 11 shows dispersoid size distributions of Fe irradiated, He implanted, 1×10^{15} He + Fe irradiated, and 1×10^{16} He + Fe irradiated specimens taken from the 350 nm depth region. Black solid lines superimposed on each figure are the dispersoid size distribution of the 2000 nm depth region from the Fe irradiated sample for comparison. With an exception of the He implanted-only sample, the size distributions follow a gaussian distribution. Figure 11a shows that the dispersoid size of the 350 nm depth of the Fe irradiated specimen is skewed toward a lower depth, due to ballistic dissolution. Figure 11b suggests a bi-modal distribution after 1×10^{15} He implantation, with a high density of dispersoids of both ~3 nm and <2 nm observed.

The 1×10^{15} He + Fe sample shows a size distribution similar to the out-of-range reference point, but with fewer large dispersoids and a higher frequency of small dispersoids. This means that large dispersoids dissolved under Fe irradiation, and small coherent dispersoids nucleated. Figure 11d shows the size distribution of 1×10^{16} He + Fe specimen, and appears very similar to that of Figure 11a. This means that a high dose of He preimplantation at room temperature does not affect the dispersoid size or size distribution. As only coherent dispersoids were counted in TM phase, the dispersoid size distribution remains in the same range (1.5–5.5 nm diameter) for all cases.

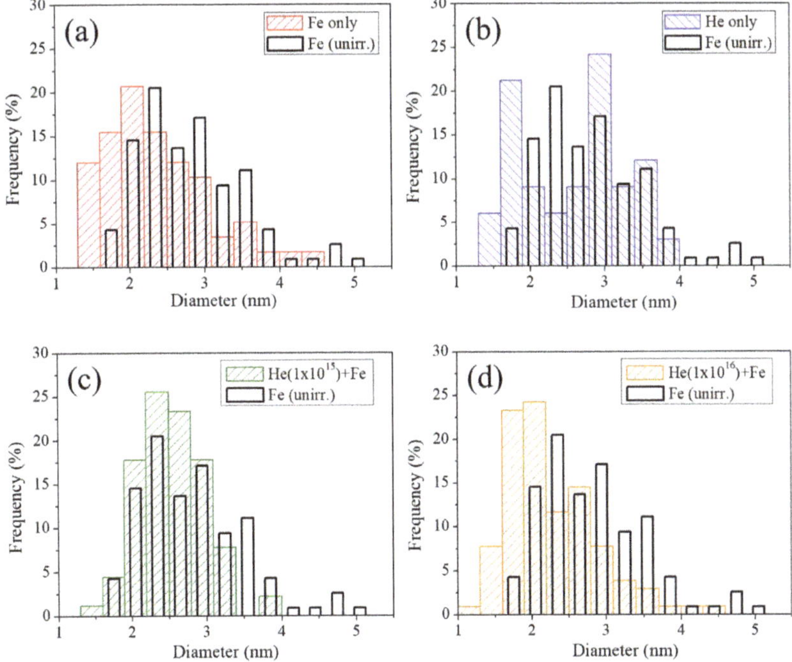

Figure 11. Dispersoid size distributions of (**a**) Fe irradiated, (**b**) He implanted-only, (**c**) He (1×10^{15}) + Fe, and (**d**) He (1×10^{16}) + Fe irradiated samples from the 350 nm depth. Black lines superimposed on each figure are the dispersoid size distribution at 2000 nm depth of Fe irradiated specimen.

4. Discussion

As its migration energy is low at 0.078 eV, helium can migrate easily and become trapped at defects such as vacancies, dislocations, grain boundaries, and precipitate surfaces [33]. When He is trapped by a vacancy, He-vacancy clusters are formed in the matrix [34–38], and these He-vacancy clusters can attract oxygen due to their high vacancy-oxygen affinity. According to a previous density functional theory (DFT) study on the vacancy mechanism of high oxygen solubility and nucleation of stable oxygen enriched clusters in Fe, oxygen in an interstitial position shows a high affinity for vacancies due to a weak bonding with the Fe matrix [39]. This O-vacancy mechanism further enables the nucleation of O-enriched nanoclusters, and it further attracts solutes like Ti and Y with high oxygen affinities. When trapped elements reach certain concentrations, dispersoids are nucleated. In a similar mechanism, when oxide dispersoids and He bubbles are present in the matrix, dispersoids will dissolve under irradiation through ballistic dissolution. Then, oxygen from dispersoids and matrix will form O-enriched nanoclusters at He-vacancy clusters which further attracts Ti and Y to form oxide dispersoids. We believe that an increase of coherent dispersoid density in the 1×10^{15} He + Fe irradiation case is due to this mechanism.

It is unclear at this stage why a higher He ion fluence at 1×10^{16} ions/cm^2 does not lead to further increases in dispersoid density. We speculate that at much higher He fluence, the number density of small cavities or He-vacancy clusters is larger. Thus the amount of O/Ti/Y trapped per cavity may be reduced significantly. If there is a critical trapping mass required to reach a stable dispersoid, in a way similar to formation of stable void nuclei, dispersoid growth might be difficult if the trap density is too large.

The observed opposite trend of He effect on dispersoids appears to be quite similar to the often-observed He effect on swelling. In previous helium co-implantation studies on various non-ODS alloys, it has been reported that helium does not always increase swelling [40–43]. Void swelling increased for low He/dpa levels and decreased for high He/dpa levels. Helium promotes cavity nucleation. At low He/dpa levels, cavities are defect sinks for vacancies, and interstitial loops/dislocations are biased sinks for interstitials. However, if the cavity density is too high at high He/dpa levels, cavities become the dominant defect sinks. Since cavities are thought to be neutral sinks for both interstitials and vacancies, defect recombination is promoted instead of biased vacancy trapping. This leads to limited and saturated void growth [43].

5. Conclusions

A dual-phase F/M 12Cr ODS alloy was preloaded with He atoms by using 120 keV He^+ ions at room temperature at two different fluences of 1×10^{15} and 1×10^{16} ions/cm^2. He was first implanted and followed by irradiation with 3.5 MeV Fe^{2+} ion at 475 °C to 100 peak dpa. The He preimplantation effect on the oxide dispersoids in the TM phase was investigated for the first time in this study. The coherent oxide dispersoids' size and density were characterized as a function of depth. The 1×10^{15} ions/cm^2 He implantation itself did not change dispersoid sizes. However, the dispersoid density was increased after sequential He + Fe ion irradiation for the He dose of 1×10^{15} ions/cm^2. On the other hand, such enhancement was not observed for the higher He dose of 1×10^{16} ions/cm^2, suggesting that additional bubble and vacancy cluster trapping may dilute local solute concentrations, which makes the initial nucleation stage difficult.

Author Contributions: Conceptualization, L.S. and H.K.; analysis, H.K.; investigation, H.K., T.W., J.G.G.; resources, S.U.; writing—original draft preparation, H.K.; writing—review and editing, L.S., F.A.G.; visualization, H.K.; supervision, L.S.; project administration, L.S.; funding acquisition, L.S.

Funding: This research was funded by the US National Science Foundation, grant number 1708788.

Conflicts of Interest: The authors declare no conflict of interest.

References

1. Ukai, S.; Ohtsuka, S. Nano-mesoscopic structure control in 9Cr ODS ferritic steels. *Energy Mater.* **2007**, *2*, 26–35. [CrossRef]
2. Ukai, S. Microstructure and high-temperature strength of 9Cr ODS ferritic steel. In *Metal, Ceramic and Polymeric Composites for Various Uses*; Cuppoletti, J., Ed.; In Tech: Rijeka, Croatia, 2011; pp. 283–302.
3. Ukai, S.; Kudo, Y.; Wu, X.; Oono, N.; Hayashi, S.; Ohtsuka, S.; Kaito, T. Residual ferrite formation in 12Cr ODS steels. *J. Nucl. Mater.* **2014**, *455*, 700–703. [CrossRef]
4. Ohtsuka, S.; Kaito, T.; Tanno, T.; Yano, Y.; Koyama, S.; Tanaka, K. Microstructure and high-temperature strength of high Cr ODS tempered martensitic steels. *J. Nucl. Mater.* **2013**, *442*, S89–S94. [CrossRef]
5. Klueh, R.L.; Maziasz, P.J.; Kim, I.S.; Heatherly, L.; Hoelzer, D.T.; Hashimoto, N.; Kenik, E.A.; Miyahara, K. Tensile and creep properties of an oxide dispersion-strengthened ferritic steel. *J. Nucl. Mater.* **2002**, *307–311*, 773–777. [CrossRef]
6. Klueh, R.L.; Shingledecker, J.P.; Swindeman, R.W.; Hoelzer, D.T. Oxide dispersion-strengthened steels: A comparison of some commercial and experimental alloys. *J. Nucl. Mater.* **2005**, *341*, 103–114. [CrossRef]
7. Ageev, V.S.; Vil'danova, N.F.; Kozlov, K.A.; Kochetkova, T.N.; Nikitina, A.A.; Sagaradze, V.V.; Safronov, B.V.; Tsvelev, V.V.; Chukanov, A.P. Structure and thermal creep of the oxide-dispersion-strengthened EP-450 reactor steel. *Phys. Met. Metallogr.* **2008**, *106*, 318–325. [CrossRef]
8. Chen, T.; Aydogan, E.; Gigax, J.G.; Chen, D.; Wang, J.; Wang, X.; Ukai, S.; Garner, F.A.; Shao, L. Microstructural changes and void swelling of a 12Cr ODS ferritic-martensitic alloy after high-dpa self-ion irradiation. *J. Nucl. Mater.* **2015**, *467*, 42–49. [CrossRef]
9. Chen, T.; Gigax, J.G.; Price, L.; Chen, D.; Ukai, S.; Aydogan, E.; Maloy, S.A.; Garner, F.A.; Shao, L. Temperature dependent dispersoid stability in ion-irradiated ferritic-martensitic dual-phase oxide-dispersion-strengthened alloy: Coherent interfaces vs. incoherent interfaces. *Acta Mater.* **2016**, *116*, 29–42. [CrossRef]
10. Kim, H.; Gigax, J.G.; Chen, T.; Ukai, S.; Garner, F.A.; Shao, L. Oxide dispersoid coherency study on dual-phase 12Cr oxide-dispersion-strengthened alloy under self-ion irradiation. *J. Nucl. Mater.* **2019**. submitted.

11. Yutani, K.; Kishimoto, H.; Kasada, R.; Kimura, A. Evaluation of helium effects on swelling behavior of oxide dispersion strengthened ferritic steels under ion irradiation. *J. Nucl. Mater.* **2007**, *367-370*, 423–427. [CrossRef]

12. Wharry, J.P.; Swenson, M.J.; Yano, K.H. A review of the irradiation evolution of dispersed oxide nanoparticles in the b.c.c. Fe-Cr system: Current understanding and future directions. *J. Nucl. Mater.* **2017**, *486*, 11–20. [CrossRef]

13. Zinkle, S.J.; Was, G.S. Materials challenges in nuclear energy. *Acta Mater.* **2013**, *61*, 735–758. [CrossRef]

14. Odette, G.R.; Alinger, M.J.; Wirth, B.D. Recent developments in irradiation-resistant steels. *Annu. Rev. Meter. Res.* **2008**, *38*, 471–503. [CrossRef]

15. Yamamoto, T.; Odette, G.R.; Miao, P.; Hoelzer, D.T.; Bentley, J.; Hashimoto, N.; Tanigawa, H.; Kurtz, R.J. The transport and fate of helium in nanostructured ferritic alloys at fusion relevant He/dpa ratios and dpa rates. *J. Nucl. Mater.* **2007**, *367–370*, 399–410. [CrossRef]

16. Edmondson, P.D.; Parish, C.M.; Zhang, Y.; Hallen, A.; Miller, M.K. Helium bubble distributions in a nanostructured ferritic alloy. *J. Nucl. Mater.* **2013**, *434*, 210–216. [CrossRef]

17. Heintze, C.; Bergner, F.; Hernandez-Mayoral, M.; Kogler, R.; Muller, G.; Ulbricht, A. Irradiation hardening of Fe-9Cr-based alloys and ODS Eurofer: Effect of helium implantation and iron-ion irradiation at 300 °C including sequence effects. *J. Nucl. Mater.* **2016**, *470*, 258–267. [CrossRef]

18. Lu, C.; Lu, Z.; Wang, X.; Xie, R.; Li, Z.; Higgins, M.; Liu, C.; Gao, F.; Wang, L. Enhanced radiation-tolerant oxide dispersion strengthened steel and its microstructure evolution under helium-implantation and heavy-ion irradiation. *Sci. Rep.* **2017**, *7*, 40343. [CrossRef] [PubMed]

19. Gigax, J.G.; Kim, H.; Aydogan, E.; Garner, F.A.; Maloy, S.; Shao, L. Beam-contamination-induced compositional alteration and its neutron-atypical consequences in ion simulation of neutron-induced void swelling. *Mater. Res. Lett.* **2018**, *5*, 478–495. [CrossRef]

20. Shao, L.; Gigax, J.; Chen, D.; Kim, H.; Garner, F.A.; Wang, J.; Toloczko, M.B. Standardization of accelerator irradiation procedures for simulation of neutron induced damage in reactor structural materials. *Nucl. Inst. Meth. Phys. Res. B* **2017**, *409*, 251–254. [CrossRef]

21. Shao, L.; Gigax, J.; Kim, H.; Garner, F.A.; Wang, J.; Toloczko, M.B. Carbon contamination, its consequences and its mitigation in ion-simulation of neutron-induced swelling of structural metals. In Proceedings of the 18th International Conference on Environmental Degradation of Materials in Nuclear Power Systems—Water Reactors, Portland, OR, USA, 13–17 August 2017; Springer: Cham, Switzerland, 2018; pp. 681–693.

22. Gigax, J.G.; Kim, H.; Aydogan, E.; Price, L.M.; Wang, X.; Maloy, S.A.; Garner, F.A.; Shao, L. Impact of composition modification induced by ion beam Coulomb-drag effects on the nanoindentation hardness of HT9. *Nucl. Inst. Meth. Phys. Res. B* **2019**, *444*, 68–73. [CrossRef]

23. Gigax, J.G.; Aydogan, E.; Chen, T.; Chen, D.; Shao, L.; Wu, Y.; Lo, W.Y.; Yang, Y.; Garner, F.A. The influence of ion beam rastering on the swelling of self-ion irradiated pure iron at 450 °C. *J. Nucl. Mater.* **2015**, *465*, 343–348. [CrossRef]

24. Kim, H.; Gigax, J.G.; Chen, T.; Ukai, S.; Garner, F.A.; Shao, L. Dispersoid stability in ion irradiated oxide-dispersion-strengthened alloy. *J. Nucl. Mater.* **2018**, *509*, 504–512. [CrossRef]

25. Cockayne, D.J.H. A theoretical analysis on the weak-beam method of electron microscopy. *J. Phys. Sci.* **1972**, *27*, 452–460. [CrossRef]

26. Cockayne, D.J.H. Weak-beam electron microscopy. *Ann. Rev. Mater. Sci.* **1981**, *11*, 75–95. [CrossRef]

27. Jenkins, M.L. *Characterization of Radiation Damage by Transmission Electron Microscopy*, 1st ed.; Taylor & Francis Group: Boca Raton, FL, USA, 2000; pp. 7–26.

28. Ziegler, J.F.; Ziegler, M.D.; Biersack, J.P. SRIM-the stopping and range of ions in matter. *Nucl. Instrum. Methods Phys. Res. B* **2010**, *268*, 1818–1823. [CrossRef]

29. Stroller, R.E.; Toloczko, M.B.; Was, G.S.; Certain, A.G.; Dwaraknath, S.; Garner, F.A. On the use of SRIM for computing radiation damage exposure. *Nucl. Instrum. Methods Phys. Res. B* **2013**, *310*, 75–80. [CrossRef]

30. Shao, L.; Wei, C.-C.; Gigax, J.; Aitkaliyeva, A.; Chen, D.; Sencer, B.H.; Garner, F.A. Effect of defect imbalance on void swelling distributions produced in pure iron irradiated with 3.5 MeV self-ions. *J. Nucl. Mater.* **2014**, *453*, 176–181. [CrossRef]

31. Tang, Q.; Ukai, S.; Oono, N.; Hayashi, S.; Leng, B.; Sugino, Y.; Han, W.; Okuda, T. Oxide particle refinement in 4.5 mass% Al Ni-based ODS superalloys. *Mater. Trans.* **2012**, *53*, 645–651. [CrossRef]

32. Zhang, L.; Ukai, S.; Hoshino, T.; Hayashi, S.; Qu, X. Y_2O_3 evolution and dispersion refinement in Co-base ODS alloys. *Acta Mater.* **2009**, *57*, 3671–3682. [CrossRef]
33. Kemp, R.; Cottrell, G.; Bhadeshia, H.K.D. Classical thermodynamic approach to void nucleation in irradiated materials. *Energy Mater.* **2006**, *1*, 103–105. [CrossRef]
34. Morishita, K.; Sugano, K.; Wirth, B.D.; Diaz de la Rubia, T. Thermal stability of helium-vacancy clusters in iron. *Nucl. Instrum. Methods Phys. Res. B* **2003**, *202*, 76–81. [CrossRef]
35. Fu, C.L.; Willaime, F. *Ab initio* study of helium in α-Fe: Dissolution, migration, and clustering with vacancies. *Phys. Rev. B* **2005**, *72*, 064117. [CrossRef]
36. Ventelon, L.; Wirth, B.D.; Domain, C. Helium-self-interstitial atom interaction in α-iron. *J. Nucl. Mater.* **2006**, *351*, 119–132. [CrossRef]
37. Borodin, V.A.; Vladimirov, P.V.; Möslang, A. Lattice kinetic Monte-Carlo modelling of helium-vacancy cluster formation in bcc iron. *J. Nucl. Mater.* **2007**, *367–370*, 286–291. [CrossRef]
38. Gao, N.; Victoria, M.; Chen, J.; Van Swygenhoven, H. Formation of dislocation loops during He clustering in bcc Fe. *J. Phys. Condens. Matter* **2011**, *23*, 245403. [CrossRef]
39. Fu, C.L.; Krčmar, M.; Painter, G.S.; Chen, X. Vacancy mechanism of high oxygen solubility and nucleation of stable oxygen-enriched clusters in Fe. *Phys. Rev. Lett.* **2007**, *99*, 225502. [CrossRef] [PubMed]
40. Getto, E.; Jiao, Z.; Monterrosa, A.M.; Sun, K.; Was, G.S. Effect of preimplanted helium on void swelling evolution in self-ion irradiated HT9. *J. Nucl. Mater.* **2015**, *462*, 458–469. [CrossRef]
41. Katoh, Y.; Kohno, Y.; Kohyama, A. Dual-ion irradiation effects on microstructure of austenitic alloys. *J. Nucl. Mater.* **1993**, *205*, 354–360. [CrossRef]
42. Stoller, R.E. The influence of helium on microstructural evolution: Implications for DT fusion reactors. *J. Nucl. Mater.* **1990**, *174*, 289–310. [CrossRef]
43. Bhattacharya, A.; Meslin, E.; Henry, J.; Décamps, B.; Barbu, A. Dramatic reduction of void swelling by helium in ion-irradiated high purity α-iron. *Mater. Res. Lett.* **2010**, *6*, 372–377. [CrossRef]

Article

Swelling and Helium Bubble Morphology in a Cryogenically Treated FeCrNi Alloy with Martensitic Transformation and Reversion after Helium Implantation

Feifei Zhang [1], Lynn Boatner [2], Yanwen Zhang [2,3,*], Di Chen [4], Yongqiang Wang [4] and Lumin Wang [1,*]

1 Department of Nuclear Engineering and Radiological Sciences, University of Michigan, Ann Arbor, MI 48108, USA
2 Materials Science and Technology Division, Oak Ridge National Laboratory, Oak Ridge, TN 37831, USA
3 Department of Materials Science and Engineering, University of Tennessee-Knoxville, Knoxville, TN 37996, USA
4 Materials Science and Technology Division, Los Alamos National Laboratory, Los Alamos, NM 87545, USA
* Correspondence: zhangy1@ornl.gov (Y.Z.); lmwang@umich.edu (L.W.)

Received: 30 June 2019; Accepted: 20 August 2019; Published: 2 September 2019

Abstract: A cryo-quenched 70 wt % Fe-15 wt% Cr-15 wt% Ni single-crystal alloy with fcc (face centered cubic), bcc (body centered cubic), and hcp (hexagonal close packed) phases was implanted with 200 keV He^+ ions up to 2×10^{17} ions·cm^{-2} at 773 K. Surface-relief features were observed subsequent to the He^+ ion implantation, and transmission electron microscopy was used to characterize both the surface relief properties and the details of associated "swelling effects" arising cumulatively from the austenitic-to-martensitic phase transformation and helium ion-induced bubble evolution in the single-crystal ternary alloy. The bubble size in the bcc phase was found to be larger than that in the fcc phase, while the bubble density in the bcc phase was correspondingly lower. The phase boundaries with misfit dislocations formed during the martensitic transformation and reversion processes served as helium traps that dispersed the helium bubble distribution. Swelling caused by the phase transformation in the alloy was dominant compared to that caused by helium bubble formation due to the limited depth of the helium ion implantation. The detailed morphology of helium bubbles formed in the bcc, hcp, and fcc phases were compared and correlated with the characters of each phase. The helium diffusion coefficient under irradiation at 773 K in the bcc phase was much higher (i.e., by several orders of magnitude) than that in the fcc phase and led to faster bubble growth. Moreover, the misfit phase boundaries were shown to be effective sites for the diffusion of helium atoms. This feature may be considered to be a desirable property for improving the radiation tolerance of the subject, ternary alloy.

Keywords: FeCrNi alloy; helium bubble; bubble swelling; ion irradiation; phase transformation

---

1. Introduction

FeCrNi alloys are often considered as model austenitic stainless steels, i.e., systems that are important structural materials for use in present-day and future nuclear reactors. In particular, such materials that are able to withstand extreme irradiation doses are of general importance for applications in advanced nuclear-energy systems [1,2].

Macroscopic radiation damage effects in structural components of nuclear devices, such as fission or fusion reactors, are the consequence of two fundamentally different types of interactions between the energetic particles in the irradiation process and the atoms in the material lattice, i.e., atomic displacements result in vacancy and self-interstitial type lattice defects and nuclear reactions result in the introduction of foreign elements. The defect agglomerates that are produced contribute to swelling, hardening, amorphization, and embrittlement, i.e., properties that may accelerate the materials' degradation and potentially lead to failure. In particular, the (n, α) transmutation reaction caused by neutron irradiation produces large amounts of helium atoms that may lead to the precipitation and formation of helium bubbles [3,4]. This creation of helium atoms in metals is considered to be of particular concern, since their precipitation into large bubbles can substantially deteriorate the mechanical properties of the material [5].

Different strategies continue to be explored to enhance the irradiation tolerance of metals and alloys, including the tolerance of the materials to helium build-up during neutron irradiation. In particular, the introduction of a high density of grain boundaries and interphase boundaries is one approach for increasing the radiation resistance that is currently being explored [6–10]. Most grain boundaries and interfaces are effective sinks for point defects that can potentially lead to enhanced irradiation tolerance. Atomistic simulations have, in fact, confirmed the ability of grain boundaries to remove collision cascade-induced point defects in both fcc and bcc metals [11,12]. Additionally, studies of multilayered composites have shown that heterophase interfaces are also excellent point defect sinks [13].

Recently, Boatner et al. have reported that large-size single crystal Fe-15Cr-15Ni alloys can be used as a decorative steel after the austenitic-to-martensitic phase transformation is induced by a cryo-quenching method [14]. Extensive research has also been carried out previously on the formation of bcc α'-martensite and hcp ε-martensite in Fe-Cr-Ni system alloys [15–18]. For high-temperature applications, the reverse transformation of these phases to γ-austenite should be considered. This transformation is of interest both in structural mechanics and radiation resistance considering the high density of dislocations and phase boundaries formed during the transformation. According to previous researchers, phase reversion is also significant in relation to the behavior of certain stainless steels used in reactor applications [19]. The formation and co-existence of α'-martensite, hcp ε-martensite, and γ-austenite phases in FeCrNi alloys (along with their different radiation behavior) suggest that it may be possible to markedly improve the irradiation resistance of many materials by designing or controlling the composition of the constituent phases.

In the present work, we report the properties of the helium bubble evolution in cryo-quenched single crystal Fe-15Cr-15Ni alloys after He$^+$ irradiation at 773 K. Here, we provide evidence that radiation tolerance can be improved by misfit phase boundaries between the different phases in the Fe-15Cr-15Ni alloy. Specifically, the typical features of helium bubbles formed in the fcc and bcc phases were examined in detail and correlated with the characteristics of each phase.

2. Experimental

The Fe-Cr-Ni alloy specimen used here was produced by first growing a single crystal of a 70 wt % Fe, 15 wt% Cr, 15 wt% Ni composition, referred to as Fe-15Cr-15Ni in this work. High-purity elemental components (e.g., 99.99% to 99.999% pure Fe, Ni, and Cr) were used to form the alloy. Laue back-reflection X-ray methods were used to orient Fe-15Cr-15Ni austenitic alloy single crystals that were then cut using an electric-arc discharge machine to expose the desired crystal surface. These surface cuts were chosen perpendicular to the (100), (110), and (111) planes of the different single crystals, respectively. After shaping of the item that was to be produced by the metal-removal method, the work piece with a crystallographically oriented surface was lapped and then given a final high polish. The material was then cryogenically quenched at 77 K by immersion in liquid nitrogen followed by a return to room temperature. Cryogenic quenching results in the formation of a surface relief on the ternary single-crystal alloy. Additional details on processing this alloy are available elsewhere [14].

The crystal samples with different surface orientations were affixed on a copper block using a high-temperature silver paste. A thermocouple was attached to the block to monitor the sample temperature during the He$^+$ irradiation. The crystals were then irradiated with 200 keV He$^+$ ions at 773 K to a fluence of 2×10^{17} ions cm^{-2} at a flux of approximately 2×10^{13} ions cm$^{-2} \cdot$s^{-1} using the Danfysik Research Ion Implanter (Danfysik A/S, Taastrup, Denmark) at the Ion Beam Materials Laboratory in Los Alamos National Laboratory. During the irradiation process, the target chamber was maintained at a typical vacuum level of 10^{-7} Torr. The irradiation-induced damage profile and helium concentration were predicted using the Stopping and Range of Ions in Matter (SRIM) code by choosing the calculation type "Ion distribution and quick calculation of damage" [20,21]. The composition and density of the target in the calculation was the Fe-15Cr-15Ni alloy with a density of 7.987 g/cm^3. Likewise, a value of 40 eV was set for the displacement energy threshold energy for Fe, Cr, and Ni [22].

The irradiated samples were analyzed using transmission electron microscopy (TEM) and scanning transmission electron microscopy (STEM) in cross-section. All of the TEM samples were prepared by focused ion beam (FIB) lift-out techniques. The TEM samples were examined using JEOL 3011 (JEOL, Tokyo, Japan) and JEOL 3100RS (JEOL, Tokyo, Japan) transmission electron microscopes operated using 300 keV electrons. The helium bubbles formed during the irradiation process were imaged by conventional TEM with a Fresnel contrast mechanism, and the crystallographic structure was analyzed by high-resolution TEM (HRTEM) and selected area diffraction (SAD) techniques.

3. Results

3.1. Surface Morphology after Helium Irradiation at 773 K

After He$^+$ irradiation at 773 K, a relief pattern formed on the surface of the Fe-15Cr-15Ni single crystal. SEM results in Figure 1 show the surface relief patterns formed on {001}, {011}, and {111} surfaces. Since the samples were polished before the He$^+$ ion irradiation, the surface-relief features after He$^+$ ion irradiation may be induced by either a phase transformation at the elevated temperature during the implantation or helium bubble swelling or a combination of both. Similar structures were also reported by Breedis [23], in grains of polycrystal of a Fe-16Cr-12Ni alloy (previously quenched to 173 K) when annealing the samples at 673 to 1173 K.

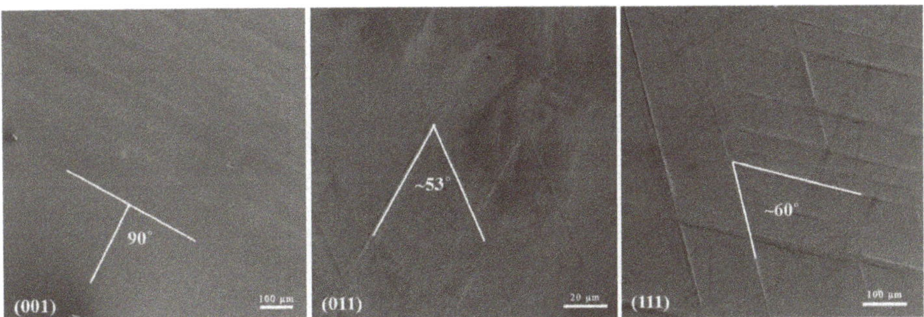

Figure 1. Optical micrographs of the surface morphology of the cryo-quenched Fe-15Cr-15Ni single-crystal alloy after He$^+$ irradiation to 2×10^{17} ions cm^{-2} at 500 °C (irradiated surfaces are well polished prior to He$^+$ irradiation).

3.2. Microstructural Observations of γ-Austenite, α′-Martensite, and ε-Martensite

The decorative surface structures produced from cooling an Fe-15Cr-15Ni alloy single crystal have been reported previously by Boatner et al. [14] as consisting primarily of two differing crystalline phases (i.e., austenite and retained martensite) after cycling through the phase transition. In Fe-15Cr-15Ni austenitic stainless steels, two martensitic phases form after cooling below Ms. temperature (the martensite start temperature). The ε martensitic phase has an hcp structure and forms as plates, while α′ martensite with a bcc structure forms as laths having a <110>$_\gamma$ long direction, lying in bands bounded by {111}$_\gamma$ planes [23,24].

Helium irradiation was performed at 773 K, which means the samples were effectively undergoing heat treatment that might induce some phase alteration effects (e.g., a martensite reverse transformation). Various microstructural features of reversed γ have been reported [24–26], including stacking faults, fine twins, and areas with a high dislocation density and subgrains.

Figures 2 and 3 show results for TEM samples that were lifted out from the surface relief and depression regions, respectively. Figure 2 shows an area containing hcp regions. Figure 2a shows the hcp phase at one of its zone axes, and Figure 2b shows the corresponding electron diffraction pattern. Figure 2c,d show the most common morphology of the ε-martensite phase, that is a nano-layer hcp phase in this case. Figure 3a shows an area containing bcc regions, with a lath-like band crossing over the entire cross-section of the sample. The crystallographic relationships between the austenite (γ) and bcc phases were examined in the electron microscope by placing an appropriate diffraction aperture across a common boundary to allow the superposition of both grain orientations in the same selected area diffraction (SAD) pattern (Figure 3a). The results are exemplified in the composite shown in Figure 3 that illustrates the following orientation relationships: $\{111\}_\gamma//\{011\}_{\alpha'}$, $[101]_\gamma//[111]_\alpha$. The fcc/bcc orientation relationships were consistent along with the Kurdjumov–Sachs (K–S) relationship. Additional evidence via high-resolution TEM (HRTEM) is shown in Figure 3c,d, which confirms the fcc and bcc structure.

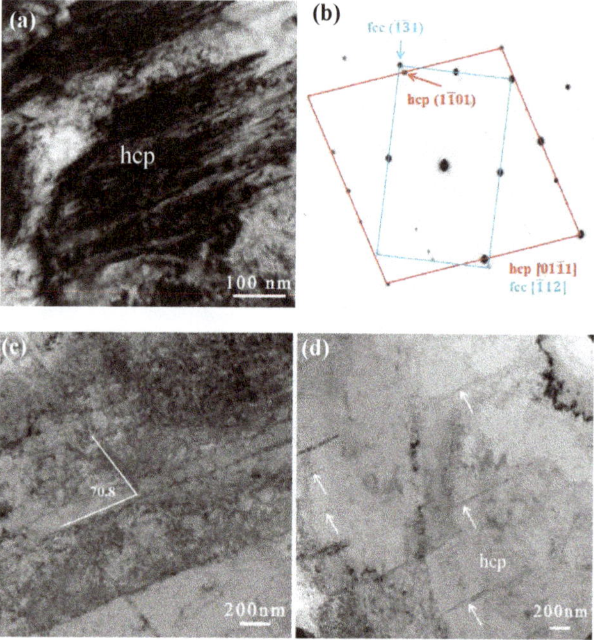

Figure 2. TEM images of the microstructure of the Fe-15Cr-15Ni alloy cryogenically treated at 77 K and heated to 773 K for a 2×10^{17} ions cm^{-2} He$^+$ irradiation—showing ε martensite in the matrix. (**a**) ε-martensite (hcp phase) and (**b**) corresponding electron diffraction pattern, (**c**) and (**d**) nano-layers of ε-martensite.

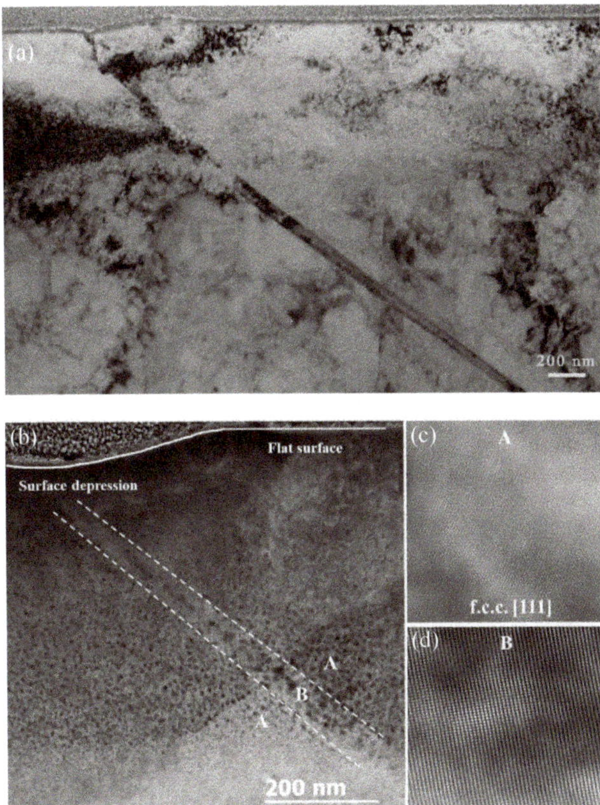

Figure 3. TEM images of the microstructure of the Fe-15Ni-15Cr alloy cryogenically treated at 77 K and heated to 773 K for a 2×10^{17} ions cm^{-2} He$^+$ irradiation—showing α' martensite lath that formed with the <110> direction. The surface depression is related to the martensitic reverse transformation. (**a**) TEM micrograph showing a martensite lath in matrix, (**b**) STEM micrograph show helium bubbles (black dots) in martensite lath and matrix, (**c**) and (**d**) is inversed FFT showing fcc and bcc structure of martensite lath and matrix.

3.3. Irradiation Damage of Cryo-Quenched Fe-15Cr-15Ni Alloy

The dose and helium concentration as a function of sample depth, as predicted by SRIM for the sample irradiated to a fluence of 2×10^{17} ions cm^{-2}, are shown in Figure 4a. The dose shows the predicted damage profile extending from the surface to a depth of 700 nm with a peak damage of ~5 dpa at 500 nm. The predicted helium concentration profile reveals a relatively sharp peak, extending from 400 to 700 nm with a peak of 1×10^5 appm of helium at a depth of 550 nm.

Cross-section TEM micrographs recorded in "under-focused" conditions for the sample irradiated to 2×10^{17} ions cm^{-2} are shown in Figure 4b. The resulting Fresnel contrast observed here is indicative of cavities (either bubbles or voids) in the matrix. In this particular material, these cavities are pressurized helium bubbles. The discernible bubbles extend from the surface to a depth of ~800 nm. As expected, the bubbles in the regions of the higher helium concentrations are, on average, larger than those in the lower-concentration regions—although significant variations in the sizes were observed with increasing distance from the implantation surface. Larger bubbles are observed in the center of the band of bubbles, and smaller bubbles are found on either side of these regions.

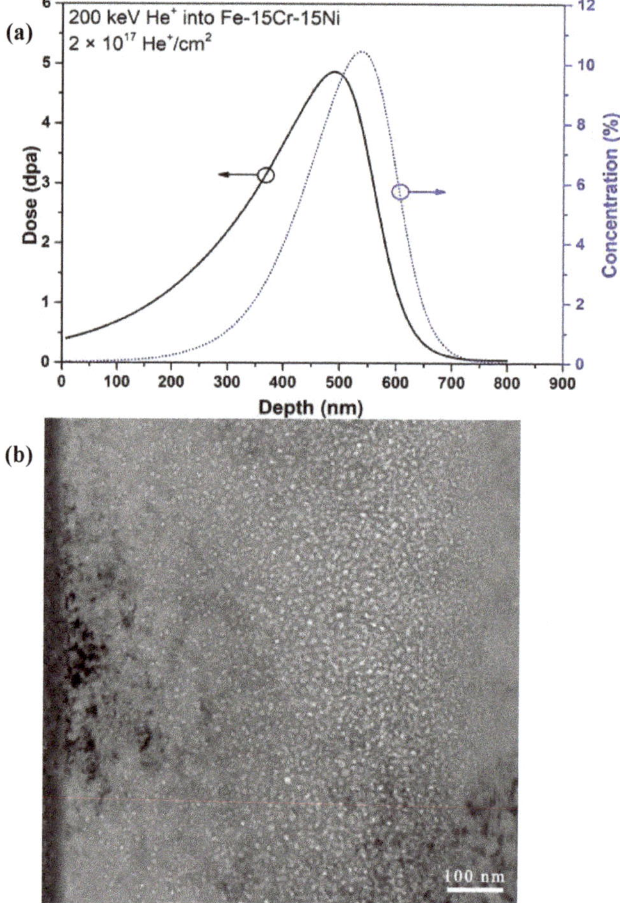

Figure 4. (**a**) The damage distribution in dose (dpa) and helium concentration for 200 keV He$^+$ irradiation to 2×10^{17} ions cm^{-2}, as predicted by SRIM. (**b**) A bright-field Fresnel-type image recorded in an under-focused condition where the bubbles are shown as bright regions.

3.3.1. Helium Bubble Distribution in the Cryo-Quenched Fe-15Cr-15Ni Alloy

Cryo-quenching to 77 K induced a phase transformation in the Fe-15Cr-15Ni alloy single crystal [14]. Diffractometer evidence from single crystals and polycrystalline aggregates shows that three structures (fcc, hcp, bcc) occur in varying proportions. To investigate the helium bubble distribution in cryo-quenched Fe-15Cr-15Ni alloy, samples with $[110]_\gamma$, $[100]_\gamma$, $[111]_\gamma$, and $[112]_\gamma$ zone axes parallel to the incident electron beam were selected.

In order to study the radiation-induced defects as well as the helium bubble formation and growth processes, cross-section TEM characterization of the irradiated bulk materials was carried out. TEM was employed to investigate both the helium bubble morphology and phase transformation. Figure 5 shows the helium bubble distribution and irradiation-induced features at low magnification for samples with the (001)-, (011)-, and (111)-orientations. No channeling effect was observed, and the helium distribution matches the SRIM predictions.

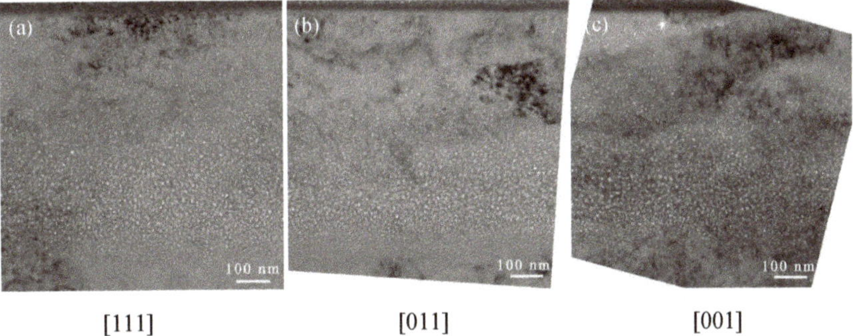

Figure 5. Low magnification of 200 keV He$^+$ irradiated Fe-15Cr-15Ni samples with different orientations (**a**) [111], (**b**) [011], (**c**) [001]. No channeling effect was observed.

Due to the high-temperature irradiation, faceted helium bubbles were observed in all of the irradiated samples. Figure 6 shows the details of the faceted helium bubbles at higher magnification. As indicated by the lines and inserted diffraction patterns in Figure 6a–c, although the bubble size distribution is broad, the shape of the bubbles is faceted. There are preferential orientations for the facets (viewed along the zone axis of [011]): The edges of all of the faceted bubbles are parallel to [001], [11–1], or [–11–1].

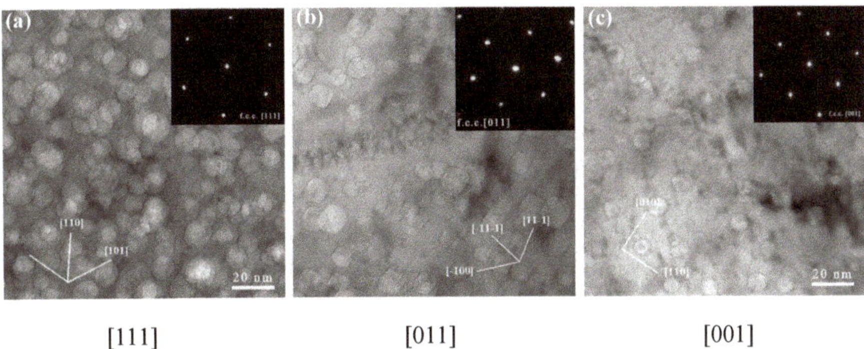

Figure 6. Comparison of the experimental observations of helium bubbles with the projected shape of an octahedron for the different zone axes: (**a**) [111], (**b**) [011], and (**c**) [001] in the fcc matrix.

Based on these observations, a polyhedron composed of {111} planes is suggested to surround the bubble. Geometric construction of a polyhedron consisting of {111} planes shows an octahedral shape. In most cases, a truncated octahedron or octahedron with rounded edges and corners is more energetically favorable [27]. Since it is embedded in the substrate and consists of {111} substrate planes, the octahedron has the same symmetry as the substrate. In order to minimize the overall free energy, it is expected that the facets with the lowest surface energy will occupy most of the surface. In the fcc structure, the lowest surface energy is found on the most closely packed {111} planes. In our case, {100} planes occupy the corners of the octahedral composed of {111} planes, which means that there is a {111} octahedron with six {100} truncations. The total surface area of a faceted bubble with the {111} surface decreases when the bubble is truncated by the {100} planes. A truncated octahedron or rounded corners are more energetically favorable.

Figure 3 represents the process of phase identification from a [111]$_\gamma$ zone axis. As shown in Figure 3a, a long, narrow trace passes through the entire cross-section of the image, illustrating the existence of the α-structure. It shows the α-phase with a thickness of ~20 nm. At higher magnification, as shown in Figure 7a, faceted bubbles in the bcc phase are observed. Figure 7b shows the indexed SAD pattern, indicating that the fcc and bcc phases are associated with the zone axis of $[111]_\gamma//[011]_\alpha$. The observed orientation relationship is consistent with the well-known K–S (Kurdjumov–Sachs) relationship, i.e., the $[111]_\gamma//[011]_\alpha$ of fcc and bcc lattices.

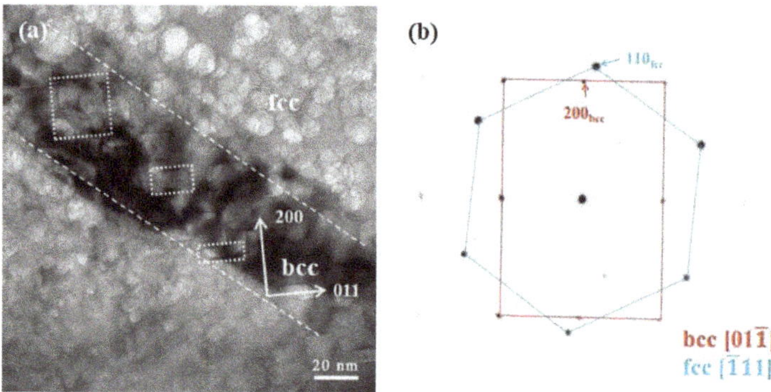

Figure 7. (a) Faceted helium bubbles in an α'-martensite lath (bcc phase). The orientation relationship between γ-austenite and α'-martensite was determined by (b) the SAD pattern, $\{111\}_\gamma//\{011\}_{\alpha'}$, $[101]_\gamma//[111]_\alpha$

Also shown in Figure 7 are the different helium bubble morphologies in the fcc and bcc phases. In the fcc phase, the helium bubbles are somewhat rounded. In the bcc phase, however, the helium bubbles are less rounded and more "plate-like" in (and near) the helium peak area. The facets of the helium bubbles in the bcc phase are composed of $\{011\}_\alpha$ and $\{001\}_\alpha$ planes.

The TEM micrographs show narrow (nm wide) traces that are parallel to $\{111\}_\gamma$ in different irradiated samples with different orientations, i.e., [112] and [011]. Figure 8 is a TEM image that shows the microstructures observed at a depth of 300 nm from the sample surface. Clearly, the BF TEM image and SAD patterns reveal the formation of overlapping fcc and hcp phases. The SAD pattern was recorded by tilting the specimen along the beam direction of [211] γ- austenite. Figure 8b shows the indexed SAD pattern, comprised of fcc phase and hcp phases. The low-energy surfaces in the fcc phase are (111), (100), and (011), and the low energy surfaces in the hcp phase are (01–11), (0001), (01–10), and (11–20). Faceted helium bubbles are observed with two different morphologies, following the minimized energy rule in the fcc and hcp phases.

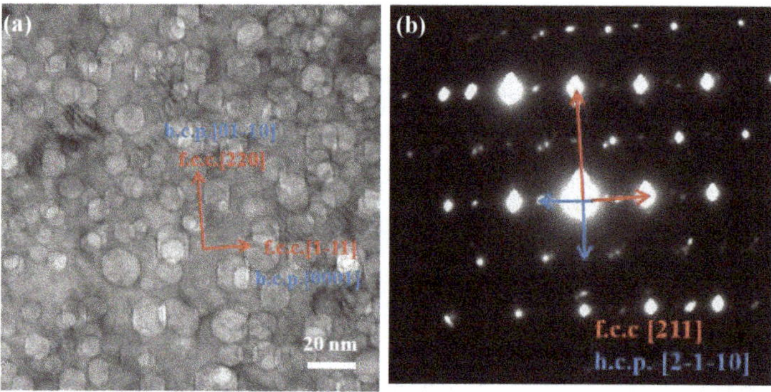

Figure 8. (a) TEM micrograph and (b) SAED pattern show the evidence of the co-existence of fcc and hcp phases along with helium bubbles in the irradiated area.

3.3.2. Helium Bubble-Induced Swelling and Phase Transformation-Induced Surface Relief

Cross-section samples were prepared by the FIB lift-out technique from the surface relief areas. Figure 9 shows the cross-section TEM sample from a surface relief area where a second phase is observed that is consistent with the surface "upheaval" or dilation feature. At higher magnification, Figure 10a,b show helium bubbles at the helium peak concentration region in the fcc and bcc phases. The helium bubbles are much larger but less dense in the bcc phase than in the fcc phase. In order to determine the qualitative effect of the two different phases, the number of the helium bubbles in the peak area was characterized. In the peak helium concentration region, the helium bubbles exhibit the greatest variation in bubble size in the different phases. The plotted results are shown in Figure 10c. The average sizes of helium bubbles in the bcc phase and fcc phases are about ~18 and ~9 nm in diameter, respectively. The density of helium bubbles in the bcc phase is two times lower than that in the fcc phase. The bubble swelling is ~8% in the bcc phase and ~4% in the fcc phase. Because these two areas are adjacent with the same implanted helium concentration, the difference in helium bubble formation can be directly observed and compared. According to the difference in the bubble swelling between the fcc and bcc phases, the surface dilation would only be several nanometers in height. The lattice parameter ratio, $a_\gamma / a_{\alpha'}$, is 1.244, as calculated from the diffraction pattern in Figure 8b. The "swelling" induced by the bcc phase is about 3.7%, according to the natural lattice differentials. This result cannot yield values of the surface relief that span a range of tens to hundreds of micrometers for our bulk samples. To evaluate the phase transformation contribution to the surface relief, the TEM sample containing both surface "upheaval" and depression was characterized. From the TEM results (Figure 9), it is directly observed that the α'-martensite phase is strongly related to surface relief.

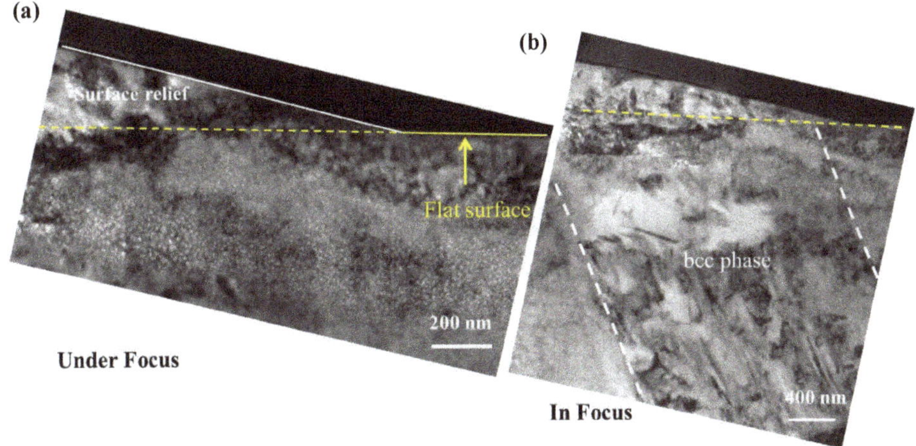

Figure 9. TEM images of the microstructure of the Fe-15Cr-15Ni alloy cryogenically treated at 77 K and heated to 773 K for a 2×10^{17} ions cm^{-2} He$^+$ irradiation—showing the α'-martensite phase (transformation) that is related to the surface relief at (**a**) high magnification and (**b**) low magnification.

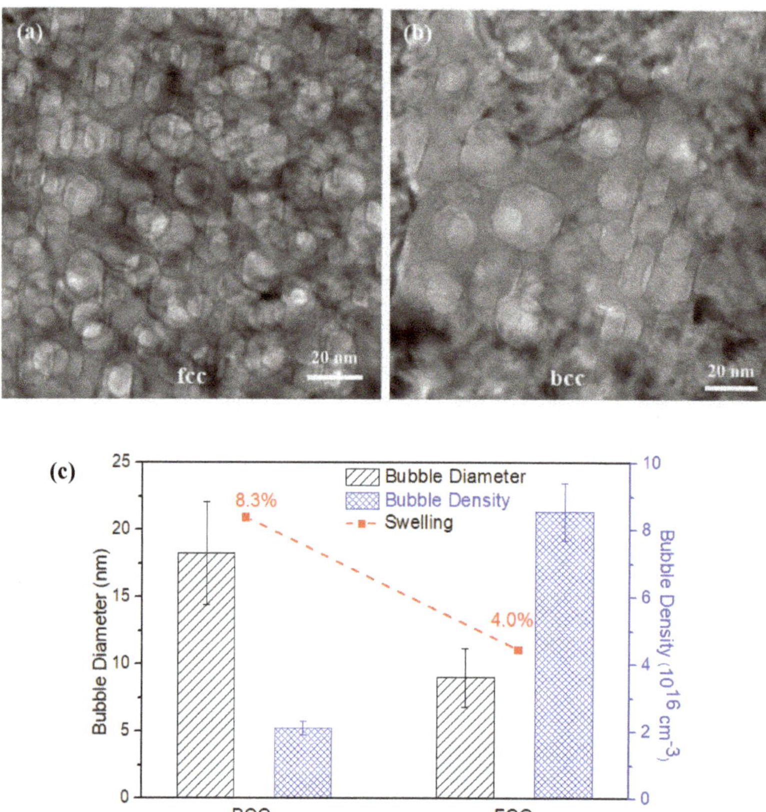

Figure 10. Surface dilation and helium bubbles corresponding to the fcc and bcc phases, (**a**,**b**), respectively. (**c**) Statistical data for the helium bubbles formed in the bcc and fcc phases—including the bubble size, bubble density, and bubble swelling.

3.3.3. Helium Bubbles along Phase Boundaries

As illustrated in Figure 11, a TEM micrograph shows a helium bubble distribution along a coherent twin boundary that has a {111} boundary plane. Figure 11a shows that a large number of helium bubbles with a small size were formed in the vicinity of the coherent fcc–hcp phase boundary after irradiation at 773 K. The distribution of these helium bubbles is roughly uniform, and there are no obvious bubble-denuded zones along the coherent twin boundary. This observation is consistent with previous investigations on irradiated Cu, i.e., the defect agglomerates formed around coherent twin boundaries are similar as those in the interior of the grains. Statistical results are plotted in Figure 11b and the corresponding microstructure is shown in Figure 11a. Compared with the matrix, helium bubbles at the coherent twin boundary are distributed with average diameters that are smaller and the bubble density is higher.

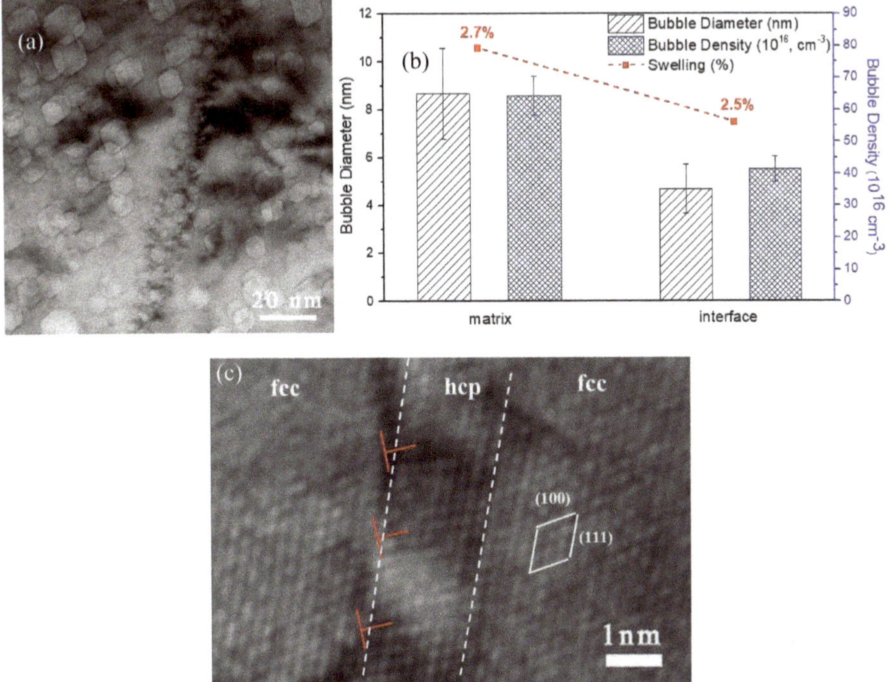

Figure 11. TEM micrograph (**a**) and statistical data (**b**) for the helium bubbles along the fcc–lamellae hcp boundary, including the helium bubble size, helium bubble density, and bubble swelling. (**c**) High-resolution TEM confirms the presence of coherent phase boundaries between the hcp and fcc phases.

4. Discussion

4.1. Phase Transformation in the Fe-15Cr-15Ni Alloy

4.1.1. Reversion of bcc Martensite in Fe-15Cr-15Ni Alloy

As shown in Figure 1, the surface relief spans a range of 100 μm after He^+ irradiation at 773 K. The same phenomenon was observed by Breedis and Roberson [23]. They reported that the surface relief was produced after reheating the cryo-quenched Fe-16Cr-12Ni sample. Subsequent transformation produces surface upheavals (dilation) or depressions at the identical sites of the previous transformation. West [25] indicated that the martensite reverse transformation that occurs between A_s (austenite initiation temperature) and A_f (austenite formation final temperature) of Fe-Cr-Ni austenitic alloys is very different depending on the alloy composition. The A_s and A_f temperatures for the α' to γ transformation is 813 and 923 K for the Fe-18Cr-8Ni, and 743 K and 883 K for the Fe-18Cr-12Ni composition. The Fe-18Cr-12Ni alloy has the higher Ni content that lowers the M_s as well as the A_s temperatures [24]. From the previous results, one may speculate that the A_s temperature of Fe-15Cr-15Ni should be around 773 K, i.e., the temperature of the He implantation. This means the α' to γ transformation potentially occurred during the He^+ irradiation.

4.1.2. Phases in Cryo-Quenched-Reheated Single Crystal Fe-15Cr-15Ni Alloy

We observed bcc, hcp, and fcc phases in our Fe-15Cr-15Ni alloy. Extensive research has been carried out on the formation of hcp and bcc martensites in austenitic stainless steels, based on the model Fe-Cr-Ni system. In the work of Coleman, West, and Breedis, all three phases were found in a cryo-quenched-reheated FeCrNi alloy [28]. Observations have been reported of reversed austenite nucleating either within the bcc phase or being formed athermally from whole bcc laths. They indicated that two alternative explanations for the appearance of an hcp structure in the Fe-Cr-Ni alloy. One explanation is that the increase in volume due to the formation of the bcc structure produces extensive faulting of the austenite phase. The second possible explanation is attributed to the formation of the hcp structure as an intermediate structure in the transformation sequence.

4.2. Helium Bubbles Evolution in fcc, bcc, and hcp Phases

4.2.1. Faceted Bubble Formation

Based on the present observations, a polyhedron composed of {111} planes is expected to surround the bubbles. As noted above, geometric construction of a polyhedron consisting of {111} planes yields an octahedral shape. Nelson determined that the surface energy sequence for fcc iron is $E(111) < E(100) < E(110)$ [29]. In our case, the {100} planes occupy the corners of the octahedron composed of {111} planes, yielding a {111} octahedron with a six {100}-plane truncation. The total surface area of a faceted bubble with the {111} surface decreases when the bubble is truncated by the {100} planes so the truncated octahedron or rounded corners are therefore more energetically favorable.

The bubble formation characteristics in a bcc structure have received relatively less attention. In Figure 8, the faceted helium bubbles are composed of $\{100\}_{bcc}$ and $\{110\}_{bcc}$ planes. It is interesting to note that helium bubbles in the lower helium concentration range are elongated along the $\{100\}_{bcc}$ surface, while they become more "square" in shape in the He peak range. The distinct $\{100\}_{bcc}$ facets are consistent with the results in the literature [30–33]. In order to minimize the overall free energy, the facets with the lowest surface energy will occupy most of the bubble surface, and in the bcc structure, the low surface energy is {100}, {110}, and {111}.

Helium bubbles in the hcp phase have a polyhedral geometry bounded by {0001}, {0111}, and {0110} facets, as determined by viewing the voids along the <2110> zone axes. Given the fact that the surface energies of the {0001} and {0111} surfaces are similar, and slightly lower than that of {0110} [34], the equilibrium bubble shape should be approximately equiaxed [35].

4.2.2. Helium Bubble Size in the bcc and fcc Phases

In the present study, the helium concentration and thermal treatment were the same in the bcc and fcc phases. The most important factor that affects the helium bubble nucleation and growth should be the helium atom diffusion process. Because helium has an extremely low solubility in metals, its diffusion is a basic requirement for bubble nucleation and growth [36]. It is the result of random jumps of helium atoms from one stable lattice site to another or diffusion of helium as interstitials. The dominant positions for helium atoms in a lattice are interstitial and substitutional sites, and the preferential position and dominant migration mode depend on the temperature as well as on the irradiation conditions. For helium diffusion under ion irradiation, atomic displacements and resulting vacancies, self-interstitial atoms (SIAs), and clusters of these defects play an important role. Singh also indicated that the dominant mechanism of helium atom diffusion is the "replacement mechanism" between 0.2 and 0.5 T_m [3]. Bubble growth by the dissolution and re-trapping mechanism requires the transport of helium atoms and vacancies. Variation of the size, $\bar{r}$, and density, ρ, of bubbles as a function of the annealing time is described by the relations [37]:

$$\bar{r} \sim (D_{He})^{1/n}, \tag{1}$$

$$\rho \sim 1/D_{He}, \tag{2}$$

where D_{He} is the diffusion mobility of the helium in the matrix, and the exponent, n, depends on the bubble-growth mechanism ($n = 2\sim6$ [38]). In the present study, the temperature of helium implantation was at 773 K (0.2 T_m < 773 K < 0.5 T_m), which corresponds to the application range of this mechanism (i.e., dissolution and re-trapping).

An analysis of Equations (1) and (2) above shows that as bubbles grow by the dissolution and re-trapping method, the smaller the value of D_{He}, the smaller the bubbles and the higher their density should be. Other researchers have indicated that the diffusion coefficient of helium atoms in bcc steel is higher than that in fcc steel by several orders of magnitude when it is in the intermediate temperature range (600 K < T < 1200 K) [39]. This is comparable to the results in our study where the helium bubbles in the fcc phase are larger and their density is smaller than in the bcc phase. Bubble nucleation, under identical irradiation and annealing conditions, occurs in fcc metals much earlier than in bcc metals [38,40], and during He^+ implantation, a helium atom injected into an Fe lattice would be expected to be located at an interstitial (i.e., tetrahedral) site.

4.2.3. Phase Transformation Comparison with Helium Bubble Swelling

Reversion of the austenite phase in a Fe-15Cr-15Ni alloy, as pointed out above, induces surface relief features up to several micrometers in size. According to the measurement and calculation of helium bubble swelling effects, e.g., ~8% in the bcc phase and ~4% in the fcc phase in Figure 9c, the bubble swelling-induced changes on the surface relief height would only be about 16 and 8 nm in the bcc phase and the fcc phases, respectively, due to the limited depth range affected by the helium implantation (e.g., up to ~800 nm as shown in Figure 4). Thus, bubble swelling in both bcc and fcc phases is not the primary reason responsible for the measured surface relief in the sample studied here. However, "swelling" induced by the bcc phase transformation is only about 3.7%, based on the lattice differentials. Therefore, the bubble-produced swelling would be much more significant than the surface relief caused by the phase transformation if the helium concentration were constant throughout the sample thickness.

4.3. *Helium Bubbles Evolution along the Phase Boundary with Misfit Dislocations*

Grain/phase boundaries are considered as effective sites for trapping helium. Generally, possible effects of helium bubble formation on mechanical properties are embrittlement where intergranular fracture is induced by helium bubbles' growth and coalescence. In the past decades, people have made the effort to find interfaces that are excellent in trapping helium and susceptible to helium embrittlement simultaneously.

In this work, as well as some previously reported investigations [41,42], it is clear that the misfit dislocation structure plays a crucial role in helium nucleation. Helium bubbles on the phase boundary with misfit dislocations, described above, are smaller and denser. These observations indicate that boundaries with misfit dislocations are the preferred trapping sites for helium and will also induce helium dispersion. In the investigation of helium at grain boundaries in austenitic steels, Lane and Goodhew [41] also found that helium bubble densities were greatest at grain boundaries containing misfit dislocations and that these dislocations were, in fact, preferred bubble nucleation sites.

Based on the previous investigations [43] and this study, we may begin to envision new ways for how interfaces may be engineered to control the properties of helium implanted into structural materials. For example, by controlling the distribution of phase boundaries with misfit dislocation, helium bubbles or bubble nuclei may be templated with desired nucleation sites, bubble density, and size.

Helium is highly mobile in most metals [44,45], and is an important factor on controlling the voids' nucleation and growth. When atomic displacement occurs, both helium atoms and the radiation-induced defects are essential for the formation and growth of helium cavities [46]. Wang reported the influence of the helium concentration on the void formation of ferritic martensitic steels under very high radiation damage (up to 500 dpa) and their work also indicated that at certain helium concentrations, helium can suppress void swelling [47]. This is caused by trapped helium or the helium–vacancy complex serves as the nucleation site for voids. Wei reported that in a fcc/bcc multilayer system, the change in hardness is negligible when the layer thickness is less than 10 nm following helium implantation [9]. If the helium or helium–vacancy cluster can be dispersed by tailoring material with misfit phase/grain boundaries, it may effectively suppress the void growth. This may be a new method to design nuclear materials with high helium tolerance and less helium embrittlement bias.

5. Conclusions

(1) Single crystals of a 70Fe-15Cr-15Ni alloy transform from a fcc- to a bcc-structured martensitic phase on cooling to the liquid nitrogen temperature and a reverse transformation when the temperature was returned to room temperature and continued to evolve during He$^+$ ion irradiation at 773 K along with an accompanying intermedium phase of the hcp structure. The reverse transformation also induced significant surface relief.

(2) The size of the ion implantation-induced helium bubbles and the corresponding swelling of the bcc phase were significantly larger than that in the fcc phase at the peak helium concentration depth. The helium diffusion coefficient in the bcc phase was higher than that in the fcc phase by several orders of magnitude at the irradiation temperature (773 K). This lead to faster bubble growth in the bcc phase.

(3) Compared to the volume change induced during the phase transformation (fcc to bcc), the bubble-induced swelling was actually much more significant even though the surface relief caused by the phase transformation in the samples investigated in this study was dominant because of the limited depth of the helium implantation.

(4) The faceted bubbles preferred to grow in a shape consisting of the most closely packed planes in both the fcc and bcc phases where the surface energy is a minimum, consistent with previous experimental observations.

(5) Helium bubbles formed along the "plate-like" intermediate hcp phase and its boundary with the fcc matrix were smaller and had a higher density compared to that in the matrix. The misfit phase boundaries can be an effective site for the dispersion of helium atoms, and this property may help to increase the swelling resistance of the material.

Author Contributions: Conceptualization, L.B. and Y.Z.; Investigation, F.Z., L.W., L.B. and Y.Z.; Resources, Y.W. and D.C.; Writing-Original Draft Preparation, F.Z.; Writing-Review & Editing, F.Z., L.W., Y.Z., Y.W. and L.B.; Project Administration, Y.Z.; Funding Acquisition, Y.Z.

Funding: This work was supported as part of the Energy Dissipation to Defect Evolution (EDDE) center, an Energy Frontier Research Center funded by the US Department of Energy (DOE), Office of Science, Basic Energy Sciences.

Acknowledgments: Authors are grateful to EDDE center for supporting this work. Helium ion implantations were supported by the Center for Integrated Nanotechnologies (CINT), a DOE Office of Science User Facility jointly operated by Los Alamos and Sandia National Laboratories. Electron microscopy analysis was conducted at the Michigan Center for Material Characterization (MC2).

Conflicts of Interest: The authors declare no conflict of interest.

References

1. Nastasi, M.; Mayer, J.; Hirvonen, J.K. *Ion-Solid Interactions: Fundamentals and Applications*; Cambridge University Press: Cambridge, UK, 1996.
2. Zinkle, S.J.; Busby, J.T. Structural materials for fission & fusion energy. *Mater. Today* **2009**, *12*, 12–19.
3. Trinkaus, H.; Singh, B. Helium accumulation in metals during irradiation—Where do we stand? *J. Nucl. Mater.* **2003**, *323*, 229–242. [CrossRef]
4. Zhang, F.; Wang, X.; Wierschke, J.B.; Wang, L. Helium bubble evolution in ion irradiated Al/B4C metal metrix composite. *Scr. Mater.* **2015**, *109*, 28–33. [CrossRef]
5. Schroeder, H.; Kesternich, W.; Ullmaier, H. Helium effects on the creep and fatigue resistance of austenitic stainless steels at high temperatures. *Nucl. Eng. Des. Fusion* **1985**, *2*, 65–95. [CrossRef]
6. Zhuo, M.; Fu, E.; Yan, L.; Wang, Y.; Zhang, Y.; Dickerson, R.; Uberuaga, B.; Misra, A.; Nastasi, M.; Jia, Q. Interface-enhanced defect absorption between epitaxial anatase TiO$_2$ film and single crystal SrTiO$_3$. *Scr. Mater.* **2011**, *65*, 807–810. [CrossRef]

7. Rose, M.; Balogh, A.; Hahn, H. Instability of irradiation induced defects in nanostructured materials. *Nucl. Instrum. Methods Phys. Res. Sect. B Beam Interact. Mater. Atoms* **1997**, *127*, 119–122. [CrossRef]

8. Li, N.; Fu, E.; Wang, H.; Carter, J.; Shao, L.; Maloy, S.; Misra, A.; Zhang, X.; Maloy, S. He ion irradiation damage in Fe/W nanolayer films. *J. Nucl. Mater.* **2009**, *389*, 233–238. [CrossRef]

9. Wei, Q.; Li, N.; Mara, N.; Nastasi, M.; Misra, A. Suppression of irradiation hardening in nanoscale V/Ag multilayers. *Acta Mater.* **2011**, *59*, 6331–6340. [CrossRef]

10. Singh, B.N. Effect of grain size on void formation during high-energy electron irradiation of austenitic stainless steel. *Philos. Mag.* **1974**, *29*, 25–42. [CrossRef]

11. Samaras, M.; Derlet, P.M.; Van Swygenhoven, H.; Victoria, M. Computer Simulation of Displacement Cascades in Nanocrystalline Ni. *Phys. Rev. Lett.* **2002**, *88*, 125505. [CrossRef]

12. Bai, X.M.; Voter, A.F.; Hoagland, R.G.; Nastasi, M.; Uberuaga, B.P.; Bai, X. Efficient Annealing of Radiation Damage Near Grain Boundaries via Interstitial Emission. *Science* **2010**, *327*, 1631–1634. [CrossRef] [PubMed]

13. Misra, A.; Demkowicz, M.J.; Zhang, X.; Hoagland, R.G. The radiation damage tolerance of ultra-high strength nanolayered composites. *JOM* **2007**, *59*, 62–65. [CrossRef]

14. Boatner, L.; Kolopus, J.; Lavrik, N.V.; Phani, P.S. Cryo-quenched Fe-Ni-Cr alloy single crystals: A new decorative steel. *J. Alloys Compd.* **2017**, *691*, 666–671. [CrossRef]

15. Venables, J.A. The martensite transformation in stainless steel. *Philos. Mag.* **1962**, *7*, 35–44. [CrossRef]

16. Lagneborgj, R. The martensite transformation in 18% Cr-8% Ni steels. *Acta Met.* **1964**, *12*, 823–843. [CrossRef]

17. Fujita, H.; Ueda, S. Stacking faults and fcc (γ) $\rightarrow$ hcp (ε) transformation in 188-type stainless steel. *Acta Metall.* **1972**, *20*, 759–767. [CrossRef]

18. Brooks, J.; Loretto, M.; Smallman, R. In situ observations of the formation of martensite in stainless steel. *Acta Met.* **1979**, *27*, 1829–1838. [CrossRef]

19. Poirier, J.; Dupouy, J. *Proceedings of the International Conference on Irradiation Behavior of Metallic Materials for Fast Reactor Core Components*; CEA: Ajaccio, France; 1979; p. 425.

20. Ziegler, J.F.; Ziegler, M.; Biersack, J. SRIM—The stopping and range of ions in matter. *Nucl. Instrum. Methods Phys. Res. Sect. B Beam Interact. Mater. Atoms* **2010**, *268*, 1818–1823. [CrossRef]

21. Stoller, R.; Toloczko, M.; Was, G.; Certain, A.; Dwaraknath, S.; Garner, F. On the use of SRIM for computing radiation damage exposure. *Nucl. Instrum. Methods Phys. Res. Sect. B Beam Interact. Mater. Atoms* **2013**, *310*, 75–80. [CrossRef]

22. *ASTM E521-16: Standard Practice for Investigating the Effects of Neutron Radiation Damage Using Charged-Particle Irradiation*; ASTM International: West Conshohocken, PA, USA, 2016.

23. Breedis, J.; Robertson, W. The martensitic transformation in single crystals of iron-chromium-nickel alloys. *Acta Metall.* **1962**, *10*, 1077–1088. [CrossRef]

24. Guy, K.B.; Butler, E.P.; West, D.R.F. Reversion of bcc α' martensite in Fe–Cr–Ni austenitic stainless steels. *Met. Sci.* **1983**, *17*, 167–176. [CrossRef]

25. Smith, H.; West, D.R.F. The reversion of martensite to austenite in certain stainless steels. *J. Mater. Sci.* **1973**, *8*, 1413–1420. [CrossRef]

26. Coleman, T.H.; West, D.R.F. The Reversion of Martensite to Austenite in an Fe–16Cr–12Ni Alloy. *Met. Sci.* **1975**, *9*, 342–345. [CrossRef]

27. Wei, Q.; Li, N.; Sun, K.; Wang, L. The shape of bubbles in He-implanted Cu and Au. *Scr. Mater.* **2010**, *63*, 430–433. [CrossRef]

28. Breedis, J.F.; Kaufman, L. The formation of Hcp and Bcc phases in austenitic iron alloys. *Met. Mater. Trans. A* **1971**, *2*, 2359–2371. [CrossRef]

29. Nelson, R.S.; Mazey, D.J.; Barnes, R.S. The thermal equilibrium shape and size of holes in solids. *Philos. Mag.* **1965**, *11*, 91–111. [CrossRef]

30. Luklinska, Z.; Goodhew, P.; Von Bradsky, G. Helium bubble growth in ferritic stainless steel. *J. Nucl. Mater.* **1985**, *135*, 206–214. [CrossRef]

31. Möslang, A.; Preininger, D. Effect of helium implantation on the mechanical properties and the microstructure of the martensitic 12% Cr-steel 1.4914. *J. Nucl. Mater.* **1988**, *155*, 1064–1068. [CrossRef]

32. Fréchard, S.; Walls, M.; Kociak, M.; Chevalier, J.; Henry, J.; Gorse, D. Study by EELS of helium bubbles in a martensitic steel. *J. Nucl. Mater.* **2009**, *393*, 102–107. [CrossRef]

33. Goodhew, P. Shapes of pores in metals. *Met. Sci.* **1981**, *15*, 377–385. [CrossRef]

34. Johansen, C.G.; Huang, H.; Lu, T.-M. Diffusion and formation energies of adatoms and vacancies on magnesium surfaces. *Comput. Mater. Sci.* **2009**, *47*, 121–127. [CrossRef]

35. Jostsons, A.; Farrell, K. Structural damage and its annealing response in neutron irradiated magnesium. *Radiat. Effic.* **1972**, *15*, 217–225. [CrossRef]

36. Kombaiah, B.; Edmondson, P.; Wang, Y.; Boatner, L.; Zhang, Y. Mechanisms of radiation-induced segregation around He bubbles in a Fe-Cr-Ni crystal. *J. Nucl. Mater.* **2019**, *514*, 139–147. [CrossRef]

37. Singh, B.; Trinkaus, H. An analysis of the bubble formation behaviour under different experimental conditions. *J. Nucl. Mater.* **1992**, *186*, 153–165. [CrossRef]

38. Kalin, B.; Chernov, I.; Kalashnikov, A.; Esaulov, M. Characteristic features of the interaction of implanted helium with interstitial and substitution elements in nickel and iron. *Vopr. At. Nauk. Tekh. Ser. Fiz. Radiats Povrezhden. Radiats Mater.* **1997**, *2*, 53–79.

39. Zhang, Y. Helium Migration in Iron. Master's Thesis, University of Cambridge, Cambridge, UK, 2004.

40. Kalin, B.A.; Chernov, I.I.; Kalashnikov, A.N.; Binyukova, S.Y.; Timofeev, A.A.; Dedyurin, A.I. Effect of Doping on Helium Behavior and Bubble Structure Development in Nickel and Vanadium Alloys. *AT Energy* **2002**, *92*, 50–56. [CrossRef]

41. Lane, P.L.; Goodhew, P.J. Helium bubble nucleation at grain boundaries. *Philos. Mag. A* **1983**, *48*, 965–986. [CrossRef]

42. Singh, B.; Leffers, T.; Green, W.; Victoria, M. Nucleation of helium bubbles on dislocations, dislocation networks and dislocations in grain boundaries during 600 MeV proton irradiation of aluminium. *J. Nucl. Mater.* **1984**, *125*, 287–297. [CrossRef]

43. Di, Z.; Bai, X.-M.; Wei, Q.; Won, J.; Hoagland, R.G.; Wang, Y.; Misra, A.; Uberuaga, B.P.; Nastasi, M. Tunable helium bubble superlattice ordered by screw dislocation network. *Phys. Rev. B* **2011**, *84*, 052101. [CrossRef]

44. Stewart, D.; Osetskiy, Y.; Stoller, R. Atomistic studies of formation and diffusion of helium clusters and bubbles in BCC iron. *J. Nucl. Mater.* **2011**, *417*, 1110–1114. [CrossRef]

45. Shu, X.; Tao, P.; Li, X.; Yu, Y. Helium diffusion in tungsten: A molecular dynamics study. *Nucl. Instrum. Methods Phys. Res. Sect. B Beam Interact. Mater. Atoms* **2013**, *303*, 84–86. [CrossRef]

46. Raineri, V.; Coffa, S.; Saggio, M.; Frisina, F.; Rimini, E. Radiation damage–He interaction in He implanted Si during bubble formation and their evolution in voids. *Nucl. Instrum. Methods Phys. Res. Sect. B Beam Interact. Mater. Atoms* **1999**, *147*, 292–297. [CrossRef]

47. Wang, X.; Yan, Q.; Was, G.S.; Wang, L. Void swelling in ferritic-martensitic steels under high dose ion irradiation: Exploring possible contributions to swelling resistance. *Scr. Mater.* **2016**, *112*, 9–14. [CrossRef]

Article

Dual Beam In Situ Radiation Studies of Nanocrystalline Cu

Cuncai Fan [1], Zhongxia Shang [1], Tongjun Niu [1], Jin Li [1], Haiyan Wang [1,2] and Xinghang Zhang [1,*]

[1] School of Materials Engineering, Purdue University, West Lafayette, IN 47907, USA
[2] School of Electrical and Computer Engineering, Purdue University, West Lafayette, IN 47907, USA
* Correspondence: xzhang98@purdue.edu

Received: 25 July 2019; Accepted: 21 August 2019; Published: 25 August 2019

Abstract: Nanocrystalline metals have shown enhanced radiation tolerance as grain boundaries serve as effective defect sinks for removing radiation-induced defects. However, the thermal and radiation stability of nanograins are of concerns since radiation may induce grain boundary migration and grain coarsening in nanocrystalline metals when the grain size falls in the range of several to tens of nanometers. In addition, prior in situ radiation studies on nanocrystalline metals have focused primarily on single heavy ion beam radiations, with little consideration of the helium effect on damage evolution. In this work, we utilized in situ single-beam (1 MeV Kr^{++}) and dual-beam (1 MeV Kr^{++} and 12 keV He^{+}) irradiations to investigate the influence of helium on the radiation response and grain coarsening in nanocrystalline Cu at 300 °C. The grain size, orientation, and individual grain boundary character were quantitatively examined before and after irradiations. Statistic results suggest that helium bubbles at grain boundaries and grain interiors may retard the grain coarsening. These findings provide new perspective on the radiation response of nanocrystalline metals.

Keywords: in situ TEM; dual-beam irradiation; nanocrystalline; grain coarsening; helium bubbles

1. Introduction

Irradiation of metals and alloys produces supersaturated point defects (Frenkel pairs) and defect clusters [1,2], and thus leads to the degradation in their physical and mechanical properties [3,4]. One effective strategy to alleviate radiation damage is to use various types of interfaces [5], such as grain boundaries (GBs) [6,7], twin boundaries (TBs) [8,9], phase boundaries [10,11], and free surfaces [12,13]. These interfaces act as defect sinks and are expected to attract, absorb, and annihilate radiation-induced defects [14]. There are increasing evidences showing that nanostructured materials with high volume fraction of interfaces are more radiation-tolerant than conventional materials [15–19], in terms of lower defect density [20,21], less radiation-induced hardening [22,23], and stronger resistance against amorphization [24]. In spite of their enhanced radiation tolerance, nanostructured materials tend to become thermally unstable because of the extra energy stored at interfaces [25]. For instance, radiation-assisted GB and TB migration, accompanied by grain coarsening and detwinning, were reported in nanocrystalline (NC) [26,27] and nanotwinned (NT) metals [28–31], especially when the grain size or twin spacing reduces to several to tens of nanometers [29,32].

At the core of nuclear reactors, structural materials are exposed to intense fluxes of neutrons at elevated temperatures [33]. In order to investigate such radiation damage in a safe, economic, and efficient way, heavy ion irradiation technique was developed and has been widely adopted as a surrogate for emulating neutron irradiation damage in the past decades [34]. However, there are various challenges for the use of single heavy ion irradiation technique to emulate neutron-radiation-induced damage [35]. For instance, a single type of heavy ion irradiation study often lacks helium (He), inevitably arising from some nuclear reactions, like the D-T nuclear fusion reaction [36]. As an inert gas, He is hardly soluble in solids and plays an important role in microstructure evolution [37–40]. Under irradiation, He can easily combine with excess vacancies and precipitate as bubbles in matrix, dislocations, GBs, or heterointerfaces [41,42]. With increasing neutron fluxes and addition of He atoms, the bubbles may keep growing, leading to void swelling, hardening, and embrittlement [4,43–45]. Therefore, to better simulate the neutron radiation damage with heavy ion irradiation technique, it has been suggested that pre-injection or simultaneous implantation of He may be necessary while conducting regular heavy ion irradiation studies [46].

In this work, we utilize in situ transmission electron microscope (TEM) technique to directly compare the distinctions between single-beam heavy ion irradiation (1 MeV Kr^{++}) and dual-beam irradiation (by 1 MeV Kr^{++} and 12 keV He^+), and combine the recent advance in automated crystal orientation mapping capability in transmission electron microscope, to explore the He effect on ion irradiation-induced grain coarsening in NC Cu at 300 °C. The findings provide new insights for understanding the irradiation response of NC metals and their potential applications in advanced nuclear energy system.

2. Materials and Methods

NC Cu (99.995 at.%) films (~2 μm) were deposited on the HF-etched Si (111) substrates at room temperature (RT) by direct current magnetron sputtering technique. Plan-view TEM specimens were prepared and subsequently irradiated using the Intermediate Voltage Electron Microscope (IVEM) Tandem Facility at the Argonne National Laboratory (Chicago, IL, USA), where an ion accelerator is attached to a Hitachi 900 NAR microscope (Hitachi, Tokyo, Japan) operated at 200 kV. The ion source included 1 MeV Kr^{++} and 12 keV He^+ with the ion beam incidented at 30° from the electron beam and 15° from the foil normal, as schematically illustrated in Figure 1a. To explore the effect of He on radiation damage, two independent irradiation experiments were conducted using the single heavy ion beam of Kr^{++}, and the dual beams of Kr^{++} plus He^+. For single-beam irradiation, the specimen was irradiated by 1 MeV Kr^{++} at 300 °C at a dose rate of 6.25×10^{11} ions cm^{-2} s^{-1} up to a fluence of 1×10^{15} ions cm^{-2}. For dual-beam irradiation, the specimen was first implanted at RT by 12 keV He^+, with a dose rate of 1.25×10^{12} ions cm^{-2} s^{-1} and to a fluence of 6.7×10^{14} ions cm^{-2}. The He-injected specimen was then irradiated simultaneously by 1 MeV Kr^{++} and 12 keV He^+, and their dose rates were 6.25×10^{11} ions cm^{-2} s^{-1} and 2.08×10^{10} ions cm^{-2} s^{-1}, respectively, up to the same fluence of 1×10^{15} ions cm^{-2}.

Irradiation damage was calculated by the Stopping and Range of Ions in Matter (SRIM) with full damage cascades and the displacement energy of 30 eV for Cu [47]. The calculated depth profiles of ion concentration and radiation damage, in unit of displacements-per-atom (dpa), are given in Figure 1b,c. The TEM foil thickness is estimated to be ~100 nm, and the SRIM calculations reveal that most (~95%) of the Kr^{++} transmitted through the TEM foil and caused a high radiation dose of ~5 dpa, while most (~92%) of the He^+ ions were injected into the foil with negligible damage, ~0.01 dpa. The average He concentration is ~1.5 at.%.

All the TEM specimens, as-deposited or irradiated, were characterized by a Thermo Fischer Scientific/FEI Talos 200X microscope equipped with a NanoMEGAS ASTAR precession electron diffraction system that allows for high-resolution crystal orientation mapping [48]. Multiple locations with the same area (1.92×1.92 μm^2) were selected and scanned for each specimen by using ASTAR system. The spot size of electron beam for scanning is ~2 nm, and the scanning step size is ~5 nm.

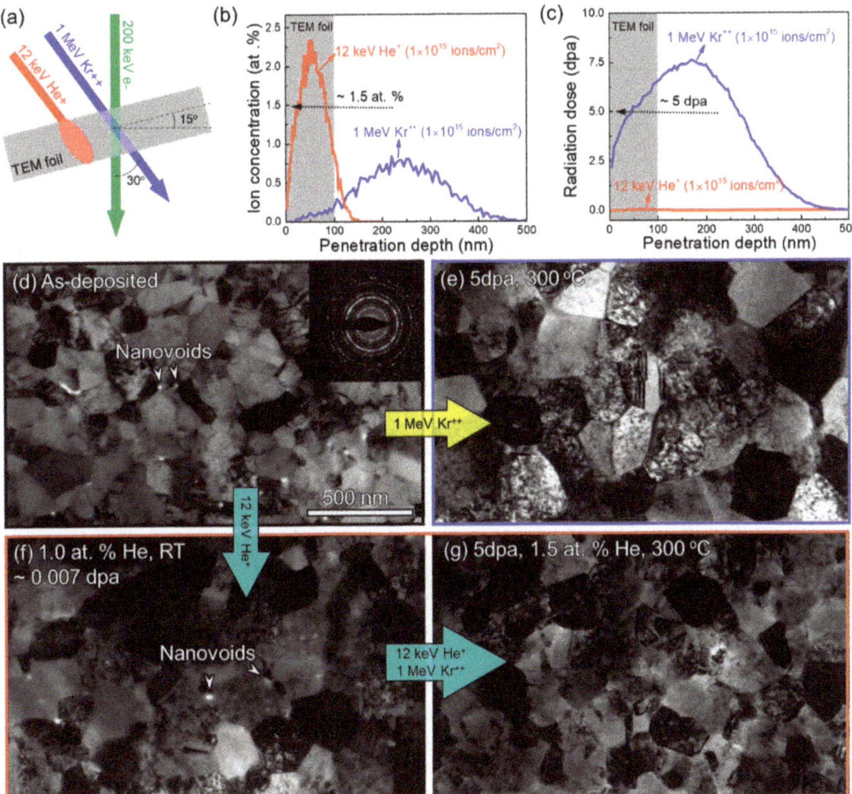

Figure 1. Singe- and dual-beam irradiations on nanocrystalline (NC) Cu at 300 °C. (**a**) Experimental set up of in situ heavy ion TEM irradiations; (**b,c**) SRIM calculations of depth profiles of ion concentration and corresponding radiation dose; (**d**) As-deposited NC Cu with some preexisting nanovoids at grain boundaries (GBs); (**e**) Single-beam (1 MeV Kr^{++}) irradiation at 300 °C for 5 dpa; (**f**) He pre-injection at room temperature (RT); (**g**) Dual-beam (1 MeV Kr^{++} and 12 keV He^{+}) irradiation at 300 °C to 5 dpa.

3. Results

3.1. In Situ Study of Irradiation-Induced Microstructure Evolution

Figure 1d is a bright-field (BF) TEM micrograph of the as-deposited sample with a broad distribution of grain sizes, ranging from tens of nm to a few hundred nm. The inset selected area diffraction (SAD) pattern indicates the formation of polycrystalline metals, and the arrows denote some nanovoids formed along GBs. Figure 1e shows the microstructure after single-beam irradiation by 1 MeV Kr^{++} to 5 dpa at 300 °C. Compared with Figure 1d, grain sizes in Figure 1e have apparently increased and the preexisting nanovoids have disappeared. Figure 1f shows the microstructure after He-injection to a concentration of 1 at.% at RT and a low dose of only 0.007 dpa. Most of the preexisting nanovoids retained. After irradiations with dual beams of 1 MeV Kr^{++} and 12 keV He^{+} to 5 dpa at 300 °C, corresponding to a He concentration of 1.5 at.%, Figure 1g displays that most of the nanovoids have disappeared. In addition, the average grain size after dual-beam irradiation in Figure 1g seems to be between that of as-deposited specimen in Figure 1d and that of single-beam irradiated specimen in Figure 1e.

Figure 2 compares the TEM snapshots of NC Cu subjected to single-beam and dual-beam irradiations to 5 dpa. Frequent GB migrations of small grains were captured in single-beam irradiation in Figure 2a–d, and the shrinkage rate of small grains tended to decrease with increasing grain size. For instance, 7 representative tiny grains that are <100 nm are denoted by 1–7 in Figure 2a, and they all shrank rapidly and disappeared when irradiated to 1.25 dpa, as shown in Figure 2b. In contrast, another large grain marked as 8 in Figure 2a–d remained its triangular shape and shrank gradually. Moreover, several other large grains barely shrank but evolved into polygons. Their initially curved GBs became straight, as marked by the arrows in Figure 2a–c. Meanwhile, the angles between adjacent grains evolved to an equilibrium angle of ~120°, as shown in Figure 2d. In comparison, no obvious GB migrations were observed in dual-beam irradiated Cu shown in Figure 2e–h. Some of the grains slightly rearranged their geometry as shown by a typical outlined grain in Figure 2e–h. He bubbles emerged at 1.25 dpa with a He concentration of ~1.125 at.%, as shown by the inset in Figure 2f.

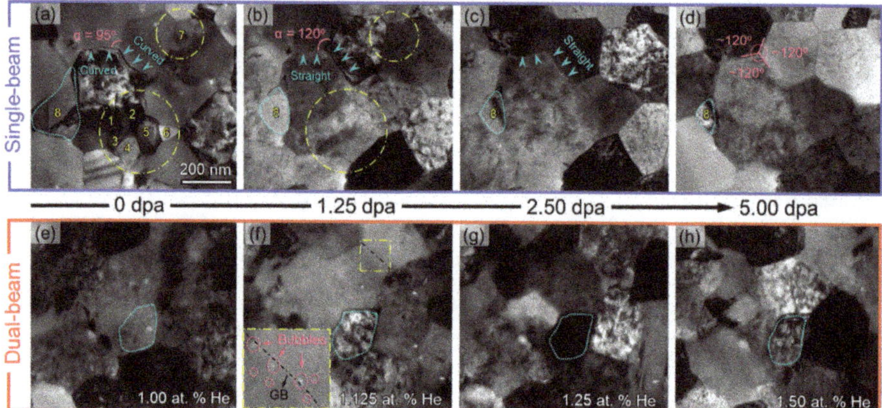

Figure 2. In situ TEM snapshot displaying microstructural evolution of NC Cu under single-beam (**a–d**) and dual-beam irradiation (**e–h**). GB migrations were frequently captured in single-beam irradiation and grain coarsening occurred at the expense of small grains, as evidenced by the shrinkage of several tiny grains marked by number 1–8 in (**a–d**). The arrows in (**a**) mark the curved GBs for a large grain that became straight with increasing dose in (**b, c**). In contrast, the grains under dual-beam irradiation only experienced slight rearrangement of their geometries, as shown by the dotted lines in (**e–h**).

3.2. Post-irradiation Analyses

To better characterize the evolution of GBs, an ASTAR automated crystal orientation mapping system was used to analyze the as-deposited, single-beam irradiated, and dual-beam irradiated Cu samples. Comparison of the orientation maps in Figure 3a–c clearly demonstrates prominent grain growth in irradiated specimens. In addition, the grain boundary maps in Figure 3d,f reveal a large fraction of Σ3 coherent TBs (yellow lines) in all samples. The enlarged view in Figure 3g reveals that the as-deposited NC Cu is characterized by irregular and curved GBs. In comparison, the single-beam irradiated NC Cu contains a significant number of straight boundaries often forming angles of 120° at triple junctions, as shown in Figure 3h. The dual-beam irradiated NC Cu, on the other hand, maintains curved GBs that are decorated with abundant He bubbles as shown in Figure 3i.

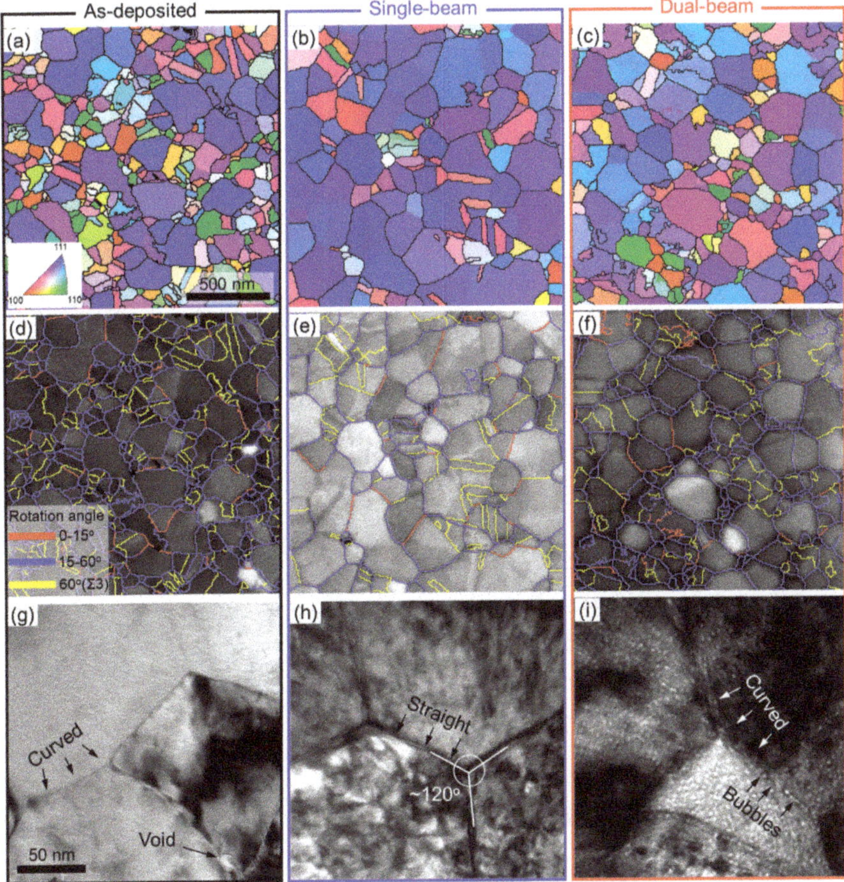

Figure 3. Microstructural characterization of as-deposited (**a,d,g**), single-beam irradiated (**b,e,h**), and dual-beam irradiated samples (**c,f,i**). (**a–c**) Crystal orientation maps obtained from ASTAR system. (**d–f**) Grain boundary maps superimposed upon image quality maps. (**g–i**) TEM micrographs showing enlarged views of representative GBs.

The statistics of grain size evolutions were derived from a study of 1646 grains for the as-deposited sample, 552 and 696 grains for the single-beam and dual-beam irradiated samples, respectively. The grain size is quantified by equivalent diameter D that equals to $2\sqrt{(A/\pi)}$, where A refers to the individual grain area. The grain size histograms for the three samples are shown in Figure 4a, and their corresponding cumulative probabilities P are plotted in Figure 4b as a function of D. Note that the probability curves shift rightward after irradiations due to grain growth, and the red curve for dual-beam irradiated sample in Figure 4b is between that of as-deposited (black) and single-beam irradiated sample (blue). The fraction of smaller grains that are less than 100 nm drops from 83% to 60 and 47% after dual-beam and single-beam irradiation, respectively. The corresponding median grain size $D_{0.5}$ ($P = 0.5$) increases from 56 ± 4 nm to 83 ± 2 nm after dual-beam irradiation, and to 103 ± 5 nm after single-beam irradiation. In particular, the lower right inset in Figure 4b reveals that the fraction of grains smaller than 40 nm is around 30% in the as-deposited sample, and it decreases drastically after either dual-beam or single-beam irradiation.

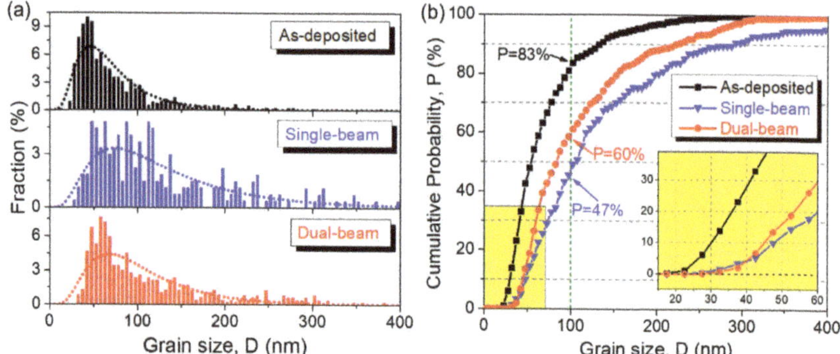

Figure 4. (**a**) Grain size (equivalent diameter) histograms for as-deposited (black), single-beam irradiated (blue), and dual-beam irradiated (red) sample. (**b**) Cumulative probability versus grain size.

The misorientation angle (θ) distributions for the three specimens are compared in Figure 5a. The GBs have been divided into three major categories according to their misorientation angle: low-angle GBs ($\theta < 15°$), high-angle GBs ($15° < \theta < 60°$), and special $\Sigma 3$ coherent TBs ($\theta = 60°$). TBs account for one-quarter of all boundaries for the three specimens. The statistic results in Figure 5b,c show that all GBs decreased in length after irradiations, and it was found that the high-angle GBs ($15° < \theta$) in single-beam irradiated sample reduced the most. The detailed statistics on evolutions of grain size and GB misorientation angles are summarized in Table 1.

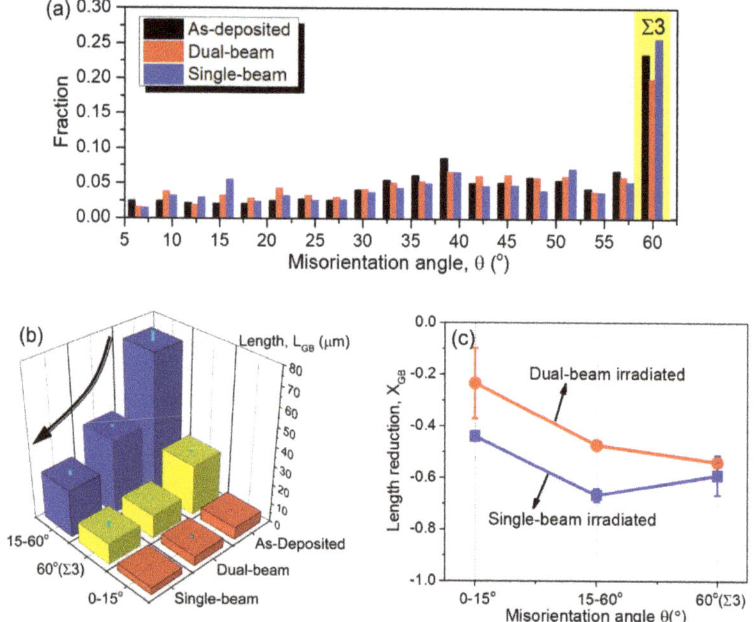

Figure 5. (**a**) Misorientation angle (θ) histograms for as-deposited (black), single-beam irradiated (blue), and dual-beam irradiated (red) samples. (**b**) Boundary length for low-angle (red bars, $\theta = 0–15°$) and high-angle (blue bars, $\theta = 15–60°$) GBs, as well as $\Sigma 3$ twin boundaries (TBs) (yellow bars, $\theta = 60°$). (**c**) GB length reduction for dual-beam (red) and single-beam (blue) irradiated samples, relative to the as-deposited sample.

Table 1. Summary of grain size and GB characters for as-deposited, dual-beam irradiated, and single-beam irradiated NC Cu, collected from a large area (11.06 μm^2) at three different locations. $D_{0.5}$: the median grain size when $P = 0.5$.

Sample	Number of Grains	Grain Size $D_{0.5}$ (nm)	Grain Boundary Length, L_{GB} (μm)		
			0–15°	15–60°	60° (Σ3)
As-deposited	1646	56 ± 4	6.2 ± 0.1	73.8 ± 4.4	26.2 ± 0.7
Dual-beam	696	83 ± 2	4.7 ± 0.2	38.9 ± 0.7	12.1 ± 0.1
Single-beam	552	103 ± 5	3.5 ± 0.1	24.6 ± 1.9	10.8 ± 2.0

4. Discussion

Grain coarsening arises from GB migration. The migration velocity v of an isolated boundary in one dimension can be described by [49]:

$$v = -M\frac{\partial \mu}{\partial x} \tag{1}$$

where M is the GB mobility and increases with increasing temperature, and $\partial \mu / \partial x$ is the driving force and increases with decreasing grain size due to the boundary curvature effect. There are increasing experimental evidences that show GB migration velocity is accelerated considerably under irradiation [27,32,50,51], and the irradiation-enhanced grain coarsening occurs even at room temperature when thermal activation makes little contribution [26]. To describe the radiation effects on grain coarsening, a thermal spike (damage cascade) model was proposed [26,49], according to which the radiation-assisted GB migration occurs within thermal spikes through atomic jumps that are biased by local GB curvature. Moreover, the radiation effects on GB structure and its migration were also well studied by atomistic simulations [6,50,52–55]. It was found that the free volume in GBs can accommodate extra interstitials [6,52], which makes interstitial-loaded GBs so unstable that they frequently migrate to annihilate vacancy clusters nearby [53,54]. In addition, for small grains with dimensions comparable to the thermal spike volume, their boundary area may overlap with thermal spikes, so small grains may undergo drastic grain growth through disorder-driven mechanism [32,55]. These prior studies suggest that irradiation-induced grain growth is often determined by two major factors: the GB curvature and the interaction between damage cascades and GBs. In the current study, we found that the radiation-assisted grain coarsening can also be influenced by the He bubbles, and the underlying mechanism will be discussed later in detail.

The interaction between damage cascades and GBs can be simply divided into two scenarios that are sink-dominated or recombination-dominated. For the former scenario, the majority of radiation-induced defects, including equal numbers of vacancies and interstitials, are trapped and annihilated by defect sinks; whereas for the latter case, defects are mostly eliminated through vacancy-interstitial recombination. It was proposed that a dimensionless parameter E that considers the defect recombination and fluxes into defect sinks can be used to evaluate which mechanism is dominating [26]. In the single-beam irradiation, the parameter E_1 can be written as:

$$E_1 = \frac{\left(k_{GB}^2 D_i\right)\left(k_{GB}^2 D_v\right)}{4 K_0 K_{iv}} \tag{2}$$

where D_i and D_v are the diffusion coefficients for interstitials and vacancies, respectively. k_{GB}^2 is the GB sink strength, estimated as [56]:

$$k_{GB}^2 = \frac{60}{D^2} \tag{3}$$

where D is the grain size. K_0 is the displacement rate (~0.003 dpa/s in current study), and K_{iv} is the constant for interstitial-vacancy recombination rate and is given by:

$$K_{iv} = \frac{4\pi(D_i + D_v)}{\Omega} = K_{iv_0}(D_i + D_v) \tag{4}$$

where Ω is the atomic volume, and K_{iv_0}, in cm^{-2}, is a material constant, ~6.98 × 10^{16} cm^{-2} for Cu [26]. Substituting Equations (3) and (4) into Equation (2) yields:

$$E_1 = \frac{D^4 K_0 K_{iv_0}}{900}\left(\frac{1}{D_i} + \frac{1}{D_v}\right) \tag{5}$$

As $D_i \gg D_v$, for irradiation of Cu at 300 °C, Equation (5) is simplified to:

$$E_1 = \frac{D^4 K_0 K_{iv_0}}{900 D_v} \tag{6}$$

Under dual-beam irradiation, He bubbles act as extra defect sinks that compete with GBs in absorbing point defects. The effective parameter E_2 in the presence of He bubbles is thus modified as:

$$E_2 = \frac{\left(k_{GB}^2 D_i\right)\left(k_{GB}^2 D_v\right)}{4K_0 K_{iv} + \left(k_B^2 D_i\right)\left(k_B^2 D_v\right)} \tag{7}$$

where k_B^2 is the sink strength for bubbles and is given by [57]:

$$k_B^2 = \frac{4\pi\rho R^2}{a} \tag{8}$$

where ρ is the bubble density, R is the bubble radius, and a is the lattice parameter (~0.3615 nm for Cu). Combining Equations (3)–(5) and (8), Equation (7) can be simplified into:

$$E_2 = \frac{900 D_v a^2}{a^2 K_0 K_{iv_0} + 4\pi R^4 \rho^2 D_v D^4} \tag{9}$$

Post-irradiation TEM study in Figure 3i shows that the bubble radius R is ~1 nm, and the bubble density is around 0.0005 nm^{-3}. The vacancy diffusivity D_v is estimated by [57]:

$$D_v = a^2 \nu \exp\left(-\frac{E_m^v}{kT}\right) \tag{10}$$

where ν is the Debye frequency (~10^{13} s^{-1}), E_m^v is the vacancy activation migration energy (0.8 eV for Cu) [57], k is the Boltzmann constant, and T is the temperature (300 °C in current study). Substituting all the parameters into in Equations (6) and (9) and plotting the values of E_1 and E_2 as a function of grain size D result in Figure 6.

The physical meaning of E in our calculations refers to the magnitude of radiation-induced defects that can be trapped by preexisting GBs relative to the defects removed through recombination or He bubbles. According to previous studies, it is plausible to assume that only when sufficient defects diffuse into GBs, GBs can experience structure change and instability, followed by GB migration and grain coarsening [53]. In principle, a value of $E \gg 1$ indicates the radiation–GB interaction is sink-dominated, whereas $E \ll 1$ indicates the interaction is recombination-dominated. The curves obtained in Figure 6 suggest the interaction transits from sink-dominated regime to recombination-dominated regime with increasing grain size D. There is a critical grain size, below which the interaction is sink-dominated and the radiation-assisted grain coarsening is most likely to occur. Note that the red curve of dual-beam irradiation is below the blue curve, and the critical grain size shifts from 25 nm to 50 nm, when He bubbles are present. The plot also suggests that grain size D should be kept slightly larger than the transition value, so that the nanograins can effectively enhance radiation tolerance while retaining their structural stability.

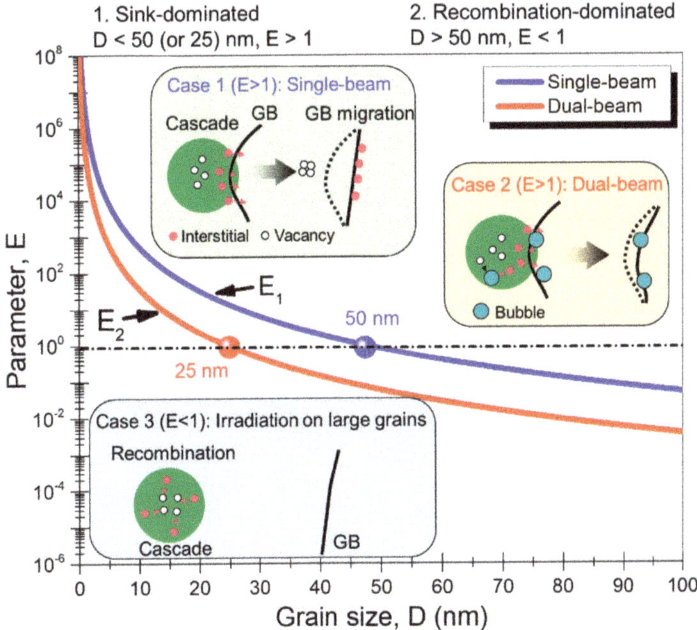

Figure 6. *E* parameter (logarithmic scale) versus grain size for single-beam (blue curve) and dual-beam (red curve) irradiated Cu at 300 °C. Case 1: Single-beam irradiation on small grains. Case 2: Dual-beam irradiation on small grains. Case 3: Irradiation of large grains. See the text for more details.

It is worth pointing out that the two curves in Figure 6 show little difference when the grain size D is less than 5 nm. Our in situ observations in Figure 2 and post-irradiation statistics in Figure 4 also indicate that nanograins rapidly disappeared under single-beam or dual-beam irradiation. The damage cascade size D^* of 1 MeV Kr^{++} irradiation on Cu is estimated to be 7 nm (see Appendix A). Therefore, the direct overlap of damage cascade with nanograins of similar dimension may lead to drastic grain coarsening at the expense of fine grains through disorder-driven mechanism [32,55].

Next, we compare the radiation-assisted GB migration and grain coarsening under single-beam and dual-beam irradiations. For simplicity, we divide the results into three cases according to the grain size D, as schematically illustrated in Figure 6.

Case 1: Radiation-assisted grain coarsening for small grains under single-beam irradiation. This case applies to sink-dominated interaction ($E > 1$), and damage cascade can entirely or partially overlap with small grains. Sufficient point defects can diffuse into GBs, leading to GB structure change and migration through disorder-driven [32,55] or thermal spike mechanism [26,49].

Case 2: Radiation-assisted grain coarsening for small grains under dual-beam irradiation. In this case, the radiation-GB interaction is still sink-dominated ($E > 1$). However, due to the formation of high-density He bubbles at GBs or grain interiors, the fraction of defects contributing to GB structure change and migration is reduced. Compared with single-beam irradiation, the radiation-assisted GB migration in dual-beam irradiated specimen is inhibited for two reasons: first, the bubbles at grain interiors can capture more point defects; second, the bubbles at GBs can exert pinning effect on GBs.

Case 3: Irradiation on large grains with recombination-dominated interactions ($E < 1$). In this case, most of the radiation-induced defects are removed through vacancy-interstitial recombination, so they make little contribution to GB migrations. For single-beam irradiation, damage cascade may occasionally occur near GBs. However, as the grain sizes are rather large, the local boundary curvature change due to cascade may be too small to drive prominent GB migrations [54]. The GB migrations may stop when GBs become straight, as shown in Figure 2. Meanwhile, the triple junctions can reach a stable state with three equal angles of ~120°, as shown in Figure 3h. For dual-beam irradiation, the local GBs can hardly move because of the pinning effect arising from He bubbles. As a result, GBs remain curved after irradiation, as shown in Figure 3i.

Finally, it should be emphasized that, in addition to grain size, radiation-GB interaction and GB migration may also be influenced by GB inherent structures [58]. For instance, compared with regular high-angle GBs, coherent TBs (CTBs) exhibit remarkable thermal stability after annealing to 800 °C [59], and they can retain their structural integrity during heavy irradiation [29,60]. Our post-irradiation analyses in Figure 3 also reveal a large fraction of surviving Σ3 CTBs in single-beam and dual-beam irradiated samples. A previous study on He ion irradiation response by Demkowicz et al. [61] suggested that TBs are not effective defect sinks, based on the observation that there were no defect denuded zones near TBs in He ion irradiated NT Cu. A thorough analysis may be beneficial to compare the density and dimension of He bubbles in bulk Cu irradiated to the same dose. Interestingly the atomistic simulations by Demkowicz et al. revealed that Σ3 CTBs can promote Frenkel pair recombination and decrease point defect production rate [61]. There are numerous prior studies that show TBs are effective defect sinks [8,31,62,63]. For instance, in Kr ion irradiated NT Ag, the density of stacking fault tetrahedrons is less in NT Ag with smaller twin spacing [62]. In other words, a clear size effect exists in irradiated NT metals. Recently, it was reported that less He bubbles are produced in NT Cu with nanovoids than annealed coarse-grained Cu, when subjected to identical He ion irradiation conditions at RT [63]. Similar phenomenon was also reported in the Fe ion radiation studies on NT austenitic stainless steel [8]. Meanwhile, there are increasing in situ radiation studies that show TBs engage, interact, and eliminate radiation-induced defect clusters [64,65]. For instance, an in situ radiation study on NT Ag showed that there is indeed a defect denuded zone near TBs based on the statistics of time accumulated defect cluster density [65]. These observations and current study may offer a new strategy for improving radiation tolerance while maintaining microstructural stability through the coupling of TB architectures with other defects sinks [60,66].

5. Conclusions

Nanocrystalline Cu films were irradiated with single-ion beam (1 MeV Kr^{++}) and dual-ion beams (1 MeV Kr^{++} and 12 keV He^{+}). Substantial GB migration and grain coarsening were captured in irradiated samples. The irradiation-induced GB migration is attributed to the interaction between damage cascade and preexisting GBs. With increasing grain size, the interaction transits from sink-dominated to recombination-dominated regime, and the radiation-assisted grain coarsening occurs in sink-dominated region at the expense of small grains. In situ radiation experiments also show that grain coarsening in dual-beam irradiated sample was retarded when He bubbles were introduced at GBs and grain interiors.

Author Contributions: C.F. designed and conducted the preliminary experiments with Z.S., T.N., and J.L.; Z.S. aided in ASTAR analysis; C.F. analyzed the data and wrote the manuscript under the supervision of X.Z. and H.W.

Funding: This research was funded by National Science Foundation, Civil, Mechanical and Manufacturing Innovation, under grant number 1728419. This work was also supported by National Science Foundation, Division of Materials Research, Metallic Materials and Nanostructures Program under grant number 1611380.

Acknowledgments: We acknowledge Meimei Li, Pete Baldo, and Wei-Ying Chen at Argonne National Laboratory (Chicago, IL, USA) for their help in our radiation experiments.

Conflicts of Interest: The authors declare no conflict of interest.

Appendix A Estimation of radiation cascade size D^*

The radiation cascade size, D^*, is determined by incident particle energy, and the average cascade volume V is given by [57]:

$$V = \frac{4}{3}\pi\left(\frac{1}{2}D^*\right)^3 = \frac{E_D}{NU_a} \tag{A1}$$

where N is the atom density, ~8.5×10^{22} atoms/cm^3 for Cu, and U_a is the energy per atom that can be estimated from the melting temperature of the target, ~0.3 eV for Cu. E_D in Equation (A1) refers to the damage energy stored in a cascade, given by [67]:

$$E_D = \frac{E_T}{1 + fg(\varepsilon)} \tag{A2}$$

and the inelastic energy loss is calculated using a numerical approximation to the universal function $g(\varepsilon)$:

$$g(\varepsilon) = 3.4008\varepsilon^{\frac{1}{6}} + 0.40244\varepsilon^{\frac{3}{4}} + \varepsilon \tag{A3}$$

$$f = 0.1337\, Z_1^{\frac{1}{6}}\left(\frac{Z_1}{A_1}\right)^{\frac{1}{2}} \tag{A4}$$

The ε in Equation (A3) refers to the reduced energy described as:

$$\varepsilon = \frac{A_2 E_T}{A_1 + A_2} \frac{A}{Z_1 Z_2 e^2} \tag{A5}$$

$$A = A_0\left(\frac{9\pi^2}{128}\right)^{\frac{1}{3}}\left(Z_1^{\frac{2}{3}} + Z_2^{\frac{2}{3}}\right) \tag{A6}$$

where A_0 is the Bohr radius (~0.053 nm), e is the electronic charge, Z_1 and Z_2 are the atomic numbers of the projectile and target, and A_1 and A_2 are the mass numbers of the atoms.

The term E_T in Equation (A2) is the transferred energy to primary knock-on atom (PKA). For the incident particles of 1 MeV Kr^{++}, classified as heavy slow ions, the inverse square potential is proper for calculating the average transferred energy, E_T, to primary knock-on atom (PKA) in the Cu lattice [57]. The description of E_T is given by:

$$E_T = \sqrt{\gamma E_i T_{min}} \tag{A7}$$

where E_i is the incident energy (1 MeV), T_{min} is the minimum transferred energy that is equal to Cu displacement energy E_d, ~30 eV, and γ is a mass ratio defined by atomic masses of incident particle m and lattice atom M in the form of:

$$\gamma = \frac{4mM}{(m + M)^2} \tag{A8}$$

where m is 83.80 u for Kr, and M is 63.55 u for Cu.

Combining the Equations (A2)–(A8) yields $E_D = 41.13$ keV, substituting which into Equation (A1) gives the average thermal spike volume $V = 162$ nm^3 and thermal spike size $D^* = 7$ nm.

References

1. Zinkle, S. Radiation-Induced Effects on Microstructure. In *Comprehensive Nuclear Materials*, 1st ed.; Konings, R., Allen, T.R., Stoller, R.E., Yamanaka, S., Eds.; Elsevier: Oxford, UK, 2012; Chapter 1.03; pp. 65–98.
2. Kiritani, M. Microstructure evolution during irradiation. *J. Nucl. Mater.* **1994**, *216*, 220–264. [CrossRef]
3. Zinkle, S.J.; Matsukawa, Y. Observation and analysis of defect cluster production and interactions with dislocations. *J. Nucl. Mater.* **2004**, *329*, 88–96. [CrossRef]

4. Barnes, R. Embrittlement of stainless steels and nickel-based alloys at high temperature induced by neutron radiation. *Nature* **1965**, *206*, 1307. [CrossRef]

5. Han, W.; Demkowicz, M.J.; Mara, N.A.; Fu, E.; Sinha, S.; Rollett, A.D.; Wang, Y.; Carpenter, J.S.; Beyerlein, I.J.; Misra, A. Design of radiation tolerant materials via interface engineering. *Adv. Mater.* **2013**, *25*, 6975–6979. [CrossRef] [PubMed]

6. Bai, X.-M.; Voter, A.F.; Hoagland, R.G.; Nastasi, M.; Uberuaga, B.P. Efficient annealing of radiation damage near grain boundaries via interstitial emission. *Science* **2010**, *327*, 1631–1634. [CrossRef] [PubMed]

7. Sun, C.; Zheng, S.; Wei, C.; Wu, Y.; Shao, L.; Yang, Y.; Hartwig, K.; Maloy, S.; Zinkle, S.; Allen, T. Superior radiation-resistant nanoengineered austenitic 304L stainless steel for applications in extreme radiation environments. *Sci. Rep.* **2015**, *5*, 7801. [CrossRef] [PubMed]

8. de Bellefon, G.M.; Robertson, I.; Allen, T.; van Duysen, J.-C.; Sridharan, K. Radiation-resistant nanotwinned austenitic stainless steel. *Scr. Mater.* **2019**, *159*, 123–127. [CrossRef]

9. Li, J.; Xie, D.; Xue, S.; Fan, C.; Chen, Y.; Wang, H.; Wang, J.; Zhang, X. Superior twin stability and radiation resistance of nanotwinned Ag solid solution alloy. *Acta Mater.* **2018**, *151*, 395–405. [CrossRef]

10. Demkowicz, M.; Hoagland, R.; Hirth, J. Interface structure and radiation damage resistance in Cu-Nb multilayer nanocomposites. *Phys. Rev. Lett.* **2008**, *100*, 136102. [CrossRef]

11. Fan, Z.; Fan, C.; Li, J.; Shang, Z.; Xue, S.; Kirk, M.A.; Li, M.; Wang, H.; Zhang, X. An in situ study on Kr ion–irradiated crystalline Cu/amorphous-CuNb nanolaminates. *J. Mater. Res.* **2019**, *34*, 1–11. [CrossRef]

12. Bringa, E.M.; Monk, J.; Caro, A.; Misra, A.; Zepeda-Ruiz, L.; Duchaineau, M.; Abraham, F.; Nastasi, M.; Picraux, S.; Wang, Y. Are nanoporous materials radiation resistant? *Nano Lett.* **2011**, *12*, 3351–3355. [CrossRef] [PubMed]

13. Li, J.; Fan, C.; Li, Q.; Wang, H.; Zhang, X. In situ studies on irradiation resistance of nanoporous Au through temperature-jump tests. *Acta Mater.* **2018**, *143*, 30–42. [CrossRef]

14. Beyerlein, I.J.; Demkowicz, M.J.; Misra, A.; Uberuaga, B. Defect-interface interactions. *Prog. Mater. Sci.* **2015**, *74*, 125–210. [CrossRef]

15. Beyerlein, I.; Caro, A.; Demkowicz, M.; Mara, N.; Misra, A.; Uberuaga, B. Radiation damage tolerant nanomaterials. *Mater. Today* **2013**, *16*, 443–449. [CrossRef]

16. Misra, A.; Demkowicz, M.; Zhang, X.; Hoagland, R. The radiation damage tolerance of ultra-high strength nanolayered composites. *JOM* **2007**, *59*, 62–65. [CrossRef]

17. Zhang, X.; Hattar, K.; Chen, Y.; Shao, L.; Li, J.; Sun, C.; Yu, K.; Li, N.; Taheri, M.L.; Wang, H. Radiation damage in nanostructured materials. *Prog. Mater. Sci.* **2018**, *96*, 217–321. [CrossRef]

18. Nita, N.; Schaeublin, R.; Victoria, M. Impact of irradiation on the microstructure of nanocrystalline materials. *J. Nucl. Mater.* **2004**, *329*, 953–957. [CrossRef]

19. Wurster, S.; Pippan, R. Nanostructured metals under irradiation. *Scr. Mater.* **2009**, *60*, 1083–1087. [CrossRef]

20. Edwards, D.J.; Simonen, E.P.; Bruemmer, S.M. Evolution of fine-scale defects in stainless steels neutron-irradiated at 275 C. *J. Nucl. Mater.* **2003**, *317*, 13–31. [CrossRef]

21. Song, M.; Wu, Y.; Chen, D.; Wang, X.; Sun, C.; Yu, K.; Chen, Y.; Shao, L.; Yang, Y.; Hartwig, K. Response of equal channel angular extrusion processed ultrafine-grained T91 steel subjected to high temperature heavy ion irradiation. *Acta Mater.* **2014**, *74*, 285–295. [CrossRef]

22. Yu, K.; Liu, Y.; Sun, C.; Wang, H.; Shao, L.; Fu, E.; Zhang, X. Radiation damage in helium ion irradiated nanocrystalline Fe. *J. Nucl. Mater.* **2012**, *425*, 140–146. [CrossRef]

23. Cheng, G.; Xu, W.; Wang, Y.; Misra, A.; Zhu, Y. Grain size effect on radiation tolerance of nanocrystalline Mo. *Scr. Mater.* **2016**, *123*, 90–94. [CrossRef]

24. Shen, T.D.; Feng, S.; Tang, M.; Valdez, J.A.; Wang, Y.; Sickafus, K.E. Enhanced radiation tolerance in nanocrystalline MgGa$_2$O$_4$. *Appl. Phys. Lett.* **2007**, *90*, 263115. [CrossRef]

25. Lu, K. Stabilizing nanostructures in metals using grain and twin boundary architectures. *Nat. Rev. Mater.* **2016**, *1*, 16019. [CrossRef]

26. Kaoumi, D.; Motta, A.; Birtcher, R. A thermal spike model of grain growth under irradiation. *J. Appl. Phys.* **2008**, *104*, 073525. [CrossRef]

27. Radiguet, B.; Etienne, A.; Pareige, P.; Sauvage, X.; Valiev, R. Irradiation behavior of nanostructured 316 austenitic stainless steel. *J. Mater. Sci.* **2008**, *43*, 7338–7343. [CrossRef]

28. Fan, C.; Li, J.; Fan, Z.; Wang, H.; Zhang, X. In Situ Studies on the Irradiation-Induced Twin Boundary-Defect Interactions in Cu. *Metall. Mater. Trans. A* **2017**, *48*, 1–9. [CrossRef]

29. Fan, C.; Xie, D.; Li, J.; Shang, Z.; Chen, Y.; Xue, S.; Wang, J.; Li, M.; El-Azab, A.; Wang, H. 9R phase enabled superior radiation stability of nanotwinned Cu alloys via in situ radiation at elevated temperature. *Acta Mater.* **2019**, *167*, 248–256. [CrossRef]

30. Yu, K.Y.; Bufford, D.; Khatkhatay, F.; Wang, H.; Kirk, M.A.; Zhang, X. In situ studies of irradiation induced twin boundary migration in nanotwinned Ag. *Scr. Mater.* **2013**, *69*, 385. [CrossRef]

31. Chen, Y.; Li, J.; Yu, K.; Wang, H.; Kirk, M.; Li, M.; Zhang, X. In situ studies on radiation tolerance of nanotwinned Cu. *Acta Mater.* **2016**, *111*, 148–156. [CrossRef]

32. Zhang, Y.; Aidhy, D.S.; Varga, T.; Moll, S.; Edmondson, P.D.; Namavar, F.; Jin, K.; Ostrouchov, C.N.; Weber, W.J. The effect of electronic energy loss on irradiation-induced grain growth in nanocrystalline oxides. *Phys. Chem. Chem. Phys.* **2014**, *16*, 8051–8059. [CrossRef] [PubMed]

33. Zinkle, S.J.; Was, G. Materials challenges in nuclear energy. *Acta Mater.* **2013**, *61*, 735–758. [CrossRef]

34. Was, G.; Jiao, Z.; Getto, E.; Sun, K.; Monterrosa, A.; Maloy, S.; Anderoglu, O.; Sencer, B.; Hackett, M. Emulation of reactor irradiation damage using ion beams. *Scr. Mater.* **2014**, *88*, 33–36. [CrossRef]

35. Was, G.S. Challenges to the use of ion irradiation for emulating reactor irradiation. *J. Mater. Res.* **2015**, *30*, 1158–1182. [CrossRef]

36. Knaster, J.; Moeslang, A.; Muroga, T. Materials research for fusion. *Nat. Phys.* **2016**, *12*, 424. [CrossRef]

37. Mansur, L.; Coghlan, W. Mechanisms of helium interaction with radiation effects in metals and alloys: A review. *J. Nucl. Mater.* **1983**, *119*, 1–25. [CrossRef]

38. Zinkle, S.; Wolfer, W.; Kulcinski, G.; Seitzman, L. II. Effect of oxygen and helium on void formation in metals. *Philos. Mag. A* **1987**, *55*, 127–140. [CrossRef]

39. Dai, Y.; Odette, G.; Yamamoto, T. The effects of helium in irradiated structural alloys. In *Comprehensive Nuclear Materials*, 1st ed.; Konings, R., Allen, T.R., Stoller, R.E., Yamanaka, S., Eds.; Elsevier: Oxford, UK, 2012; Chapter 1.06; pp. 141–193.

40. Brimbal, D.; Décamps, B.; Henry, J.; Meslin, E.; Barbu, A. Single-and dual-beam in situ irradiations of high-purity iron in a transmission electron microscope: Effects of heavy ion irradiation and helium injection. *Acta Mater.* **2014**, *64*, 391–401. [CrossRef]

41. Odette, G.; Alinger, M.; Wirth, B. Recent developments in irradiation-resistant steels. *Annu. Rev. Mater. Res.* **2008**, *38*, 471–503. [CrossRef]

42. Li, S.-H.; Li, J.-T.; Han, W.-Z. Radiation-induced helium bubbles in metals. *Materials* **2019**, *12*, 1036. [CrossRef]

43. Zinkle, S.J.; Farrell, K. Void swelling and defect cluster formation in reactor-irradiated copper. *J. Nucl. Mater.* **1989**, *168*, 262–267. [CrossRef]

44. Knapp, J.; Follstaedt, D.; Myers, S. Hardening by bubbles in He-implanted Ni. *J. Appl. Phys.* **2008**, *103*, 013518. [CrossRef]

45. Hu, X.; Koyanagi, T.; Fukuda, M.; Kumar, N.K.; Snead, L.L.; Wirth, B.D.; Katoh, Y. Irradiation hardening of pure tungsten exposed to neutron irradiation. *J. Nucl. Mater.* **2016**, *480*, 235–243. [CrossRef]

46. Taller, S.; Woodley, D.; Getto, E.; Monterrosa, A.M.; Jiao, Z.; Toader, O.; Naab, F.; Kubley, T.; Dwaraknath, S.; Was, G.S. Multiple ion beam irradiation for the study of radiation damage in materials. *Nucl. Instrum. Methods Phys. Res. B* **2017**, *412*, 1–10. [CrossRef]

47. Ziegler, J.F. SRIM-2003. *Nucl. Instrum. Methods Phys. Res. B* **2004**, *219*, 1027–1036. [CrossRef]

48. Moeck, P.; Rouvimov, S.; Rauch, E.; Véron, M.; Kirmse, H.; Häusler, I.; Neumann, W.; Bultreys, D.; Maniette, Y.; Nicolopoulos, S. High spatial resolution semi-automatic crystallite orientation and phase mapping of nanocrystals in transmission electron microscopes. *Cryst. Res. Technol.* **2011**, *46*, 589–606. [CrossRef]

49. Alexander, D.E.; Was, G.S. Thermal-spike treatment of ion-induced grain growth: Theory and experimental comparison. *Phys. Rev. B* **1993**, *47*, 2983. [CrossRef]

50. Zhang, Y.; Jiang, W.; Wang, C.; Namavar, F.; Edmondson, P.D.; Zhu, Z.; Gao, F.; Lian, J.; Weber, W.J. Grain growth and phase stability of nanocrystalline cubic zirconia under ion irradiation. *Phys. Rev. B* **2010**, *82*, 184105. [CrossRef]

51. Bufford, D.; Abdeljawad, F.; Foiles, S.; Hattar, K. Unraveling irradiation induced grain growth with in situ transmission electron microscopy and coordinated modeling. *Appl. Phys. Lett.* **2015**, *107*, 191901. [CrossRef]

52. Samaras, M.; Derlet, P.; Van Swygenhoven, H.; Victoria, M. Computer simulation of displacement cascades in nanocrystalline Ni. *Phys. Rev. Lett.* **2002**, *88*, 125505. [CrossRef]

53. Jin, M.; Cao, P.; Yip, S.; Short, M.P. Radiation damage reduction by grain-boundary biased defect migration in nanocrystalline Cu. *Acta Mater.* **2018**, *155*, 410–417. [CrossRef]

54. Jin, M.; Cao, P.; Short, M.P. Mechanisms of grain boundary migration and growth in nanocrystalline metals under irradiation. *Scr. Mater.* **2019**, *163*, 66–70. [CrossRef]

55. Voegeli, W.; Albe, K.; Hahn, H. Simulation of grain growth in nanocrystalline nickel induced by ion irradiation. *Nucl. Instrum. Methods Phys. Res. B* **2003**, *202*, 230–235. [CrossRef]

56. Mansur, L. Theory and experimental background on dimensional changes in irradiated alloys. *J. Nucl. Mater.* **1994**, *216*, 97–123. [CrossRef]

57. Was, G.S. *Fundamentals of Radiation Materials Science: Metals and Alloys*; Springer: Berlin, Germany, 2016.

58. Han, W.; Demkowicz, M.; Fu, E.; Wang, Y.; Misra, A. Effect of grain boundary character on sink efficiency. *Acta Mater.* **2012**, *60*, 6341–6351. [CrossRef]

59. Anderoglu, O.; Misra, A.; Wang, H.; Zhang, X. Thermal stability of sputtered Cu films with nanoscale growth twins. *J. Appl. Phys.* **2008**, *103*, 094322. [CrossRef]

60. Jiao, S.; Kulkarni, Y. Radiation tolerance of nanotwinned metals–An atomistic perspective. *Comput. Mater. Sci.* **2018**, *142*, 290–296. [CrossRef]

61. Demkowicz, M.J.; Anderoglu, O.; Zhang, X.; Misra, A. The influence of $\Sigma 3$ twin boundaries on the formation of radiation-induced defect clusters in nanotwinned Cu. *J. Mater. Res.* **2011**, *26*, 1666–1675. [CrossRef]

62. Yu, K.; Bufford, D.; Sun, C.; Liu, Y.; Wang, H.; Kirk, M.; Li, M.; Zhang, X. Removal of stacking-fault tetrahedra by twin boundaries in nanotwinned metals. *Nat. Commun.* **2013**, *4*, 1377. [CrossRef]

63. Fan, C.; Li, Q.; Ding, J.; Liang, Y.; Shang, Z.; Li, J.; Su, R.; Cho, J.; Chen, D.; Wang, Y. Helium irradiation induced ultra-high strength nanotwinned Cu with nanovoids. *Acta Mater.* **2019**, *177*, 107–120. [CrossRef]

64. Li, J.; Chen, Y.; Wang, H.; Zhang, X. In situ studies on twin-thickness-dependent distribution of defect clusters in heavy ion-irradiated nanotwinned Ag. *Metall. Mater. Trans. A* **2017**, *48*, 1466–1473. [CrossRef]

65. Li, J.; Yu, K.; Chen, Y.; Song, M.; Wang, H.; Kirk, M.; Li, M.; Zhang, X. In situ study of defect migration kinetics and self-healing of twin boundaries in heavy ion irradiated nanotwinned metals. *Nano Lett.* **2015**, *15*, 2922–2927. [CrossRef] [PubMed]

66. Chen, Y.; Yu, K.Y.; Liu, Y.; Shao, S.; Wang, H.; Kirk, M.; Wang, J.; Zhang, X. Damage-tolerant nanotwinned metals with nanovoids under radiation environments. *Nat. Commun.* **2015**, *6*, 7036. [CrossRef] [PubMed]

67. Norgett, M.; Robinson, M.; Torrens, I. A proposed method of calculating displacement dose rates. *Nucl. Eng. Des.* **1975**, *33*, 50–54. [CrossRef]

Communication

Investigating Helium Bubble Nucleation and Growth through Simultaneous In-Situ Cryogenic, Ion Implantation, and Environmental Transmission Electron Microscopy

Caitlin A. Taylor [1], Samuel Briggs [1,2], Graeme Greaves [3], Anthony Monterrosa [1], Emily Aradi [3], Joshua D. Sugar [4], David B. Robinson [4], Khalid Hattar [1,* and Jonathan A. Hinks [3]

[1]   Sandia National Laboratories, Albuquerque, NM 87185, USA
[2]   Nuclear Science and Engineering, Oregon State University, Corvallis, OR 97331, USA
[3]   School of Computing and Engineering, University of Huddersfield, Huddersfield HD1 3DH, UK
[4]   Sandia National Laboratories, Livermore, CA 94551, USA
*   Correspondence: khattar@sandia.gov

Received: 2 July 2019; Accepted: 13 August 2019; Published: 16 August 2019

Abstract: Palladium can readily dissociate molecular hydrogen at its surface, and rapidly accept it onto the octahedral sites of its face-centered cubic crystal structure. This can include radioactive tritium. As tritium β-decays with a half-life of 12.3 years, He-3 is generated in the metal lattice, causing significant degradation of the material. Helium bubble evolution at high concentrations can result in blister formation or exfoliation and must therefore be well understood to predict the longevity of materials that absorb tritium. A hydrogen over-pressure must be applied to palladium hydride to prevent hydrogen from desorbing from the metal, making it difficult to study tritium in palladium by methods that involve vacuum, such as electron microscopy. Recent improvements in in-situ ion implantation Transmission Electron Microscopy (TEM) allow for the direct observation of He bubble nucleation and growth in materials. In this work, we present results from preliminary experiments using the new ion implantation Environmental TEM (ETEM) at the University of Huddersfield to observe He bubble nucleation and growth, in-situ, in palladium at cryogenic temperatures in a hydrogen environment. After the initial nucleation phase, bubble diameter remained constant throughout the implantation, but bubble density increased with implantation time. β-phase palladium hydride was not observed to form during the experiments, likely indicating that the cryogenic implantation temperature played a dominating role in the bubble nucleation and growth behavior.

Keywords: in-situ; helium implantation; environmental transmission electron microscopy; palladium tritide

---

1. Introduction

Helium is insoluble in almost all solids and precipitates into nanometer-sized bubbles that can result in mechanical property degradation and eventually fracture. At very high He concentrations (tens of atomic percent), micrometer scale blisters can form, resulting in exfoliation, gas release, or both [1,2]. Bubble nucleation and growth are sensitive to most environmental conditions, including material composition, crystal structure, and temperature.

Palladium-based materials are under consideration for many applications [3], including H_2 purification, storage, and detection, as well as fuel cell catalysis, due to its ability to easily dissociate molecular H_2 on its surfaces and incorporate H atoms into octahedral sites in its face-centered cubic (fcc) crystal structure as a metal hydride [4–6]. One of these applications is solid-state tritium storage, where ^{3}H will decay to ^{3}He with a half-life of 12.3 years, causing rapid accumulation of ^{3}He in the Pd lattice. Studying ^{3}He evolution in PdT_x is difficult due to the safety constraints of radiological work, though some microscopy has been performed investigating early stage ^{3}He bubble formation (less than one year of aging) [7–9]. Over these short aging times, ^{3}He bubbles reached 1–2 nm in diameter. While He ion implantation has been utilized as an accelerated aging method to study blister formation in Pd metal at high doses [1,2], He implantation into Pd hydride is difficult in most facilities because a constant H_2 over-pressure is required to maintain the hydride structure. If the over-pressure is removed, most H will diffuse out of the material [4–6].

New capabilities in in-situ ion irradiation allow for direct observation of He bubble nucleation and growth as a function of He implantation dose and temperature [10]. In the new Microscope and Ion Accelerators for Materials Investigations (MIAMI-2) facility at the University of Huddersfield [11], in-situ He implantation has been combined with Environmental Transmission Electron Microscopy (ETEM), which allows for imaging in the presence of milliTorr H_2 pressures. We have used this combination of capabilities to investigate whether the presence of H_2 affects the nucleation and early growth of bubbles. We performed much of the work at sub-ambient sample temperatures to increase the H solubility in the sample, but did not observe formation of the concentrated, β-phase Pd hydride, which is difficult to characterize with electron diffraction techniques. α-phase Pd hydride is expected to have formed to some degree under the experimental conditions, but cannot be properly identified using electron diffraction due to its characteristic minute change in lattice parameter Thus, cryogenic temperature likely dominated the observed He bubble nucleation and growth kinetics.

2. Materials and Methods

Specimens and Irradiation Treatment

Palladium wire was purchased from Alfa Aesar (Alfa Aesar, Haverhill, MA, USA) and was annealed prior to TEM sample preparation to cause pre-existing voids identified near the surface to coalesce into larger voids that could not be confused with He bubbles. The wire was annealed at 700 °C for 1.5 h in an evacuated quartz ampoule with a base pressure of 1×10^{-7} Torr at the time of sealing. TEM sample preparation was done using the Focused Ion Beam (FIB) method with a FEI Helios Nanolab 660 (ThermoFisher Scientific, Hillsboro, OR, USA). The resulting lamellae were mounted on Mo FIB grids and thinned to electron transparency, with final cleaning steps utilizing a 5 kV accelerating voltage. Samples were then transported and imaged in the Hitachi H-9500 ETEM (Hitachi High-Technologies, Tokyo, Japan) at the MIAMI-2 facility (University of Huddersfield, West Yorkshire, UK) [11]. Unless otherwise stated, all TEM imaging was conducted in a Bright Field (BF) imaging condition with an accelerating voltage of 300 kV. Initial TEM imaging showed a high density of defects, likely resulting from either the FIB procedure or the original wire extrusion process. Since pre-existing defects will affect ^{4}He bubble nucleation and hydride formation, the specimens were annealed at 400 °C for one hour in vacuum using a Gatan Model 652 double-tilt heating holder (Gatan, Pleasanton, CA, USA) in an attempt to reduce defect density.

Thermodynamic calculations [12], shown in Figure 1a, were used to estimate the temperature required to hydride Pd at ETEM relevant pressures (on the order of 10^{-2} Torr). The α-phase has been characterized by a slight unit cell expansion from the fcc Pd lattice of 3.88 Å to 3.89 Å when H/Pd = 0.03. As H_2 content increases, a new set of fcc lattice reflections form, corresponding to the β-phase, which has a cell constant of 4.02 Å when the α→β transformation is complete (H/Pd~0.57) [13]. This corresponds to a 10% volume expansion, or a 3.6% lattice parameter expansion, compared to Pd metal.

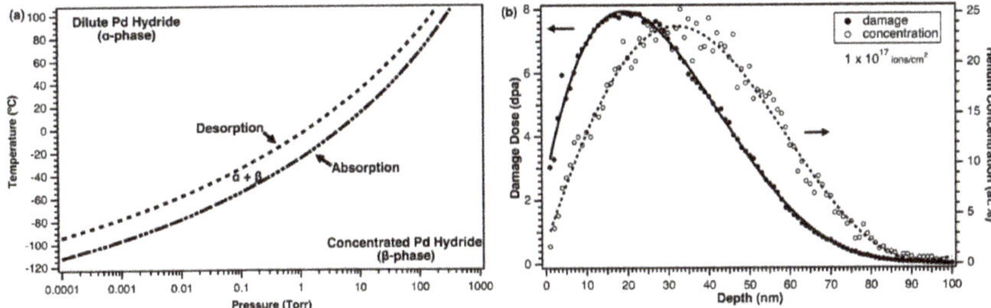

Figure 1. Experimental parameters, including: (**a**) thermodynamic calculations showing when H_2 is expected to absorb and desorb from pure Pd as a function of temperature and pressure, and (**b**) SRIM prediction, shown for a fluence of 10^{17} ions/cm^2, of implantation depth, damage dose, and ^{4}He concentration for 10 keV ^{4}He into Pd at 18.7°. Lines are meant to guide the eye in (**b**).

We expect concentrated hydride, or β-phase, to form below the "absorption" line in Figure 1a, and the dilute, or α-phase, to form above the "desorption" line at a given pressure as the temperature is increased. The crystal structure is expected to remain fcc in all cases. The region between the "desorption" and "absorption" lines in Figure 1a consists of α + β-phases [5,6]. These calculations do not include kinetic aspects of hydride formation, which are not well documented for Pd at cryogenic temperatures and may influence the achievement of the hydride phase and final stoichiometry. Furthermore, the thermodynamic data are extrapolated, and are potentially sample-dependent, so we consider Figure 1a to be only an approximate guide.

Samples were cooled to −100 °C in the ETEM using a Gatan Model 636 (Gatan, Pleasanton, CA, USA) liquid nitrogen-cooled cryogenic holder. Thermodynamic calculations (Figure 1a) show that the pressure must be above 6.4×10^{-4} Torr to form concentrated β-phase at −100 °C. However, experimental isotherms show that the concentrated β-phase forms above ~1×10^{-4} Torr at −196 °C [13]. An H_2 atmosphere was introduced locally to the specimen. Local specimen pressure was maintained at between 7.5×10^{-3} and 2.3×10^{-2} Torr while BF imaging video data were collected. Specimen pressure was actively throttled to prevent electron gun pressure from rising to the trip point for the gun valve to close (3.8×10^{-4} Torr). Initial experiments were conducted without an ion beam to observe potential microstructural changes associated with the formation of Pd hydride. Palladium hydride formation is characterized by a unit cell expansion [13], so electron diffraction patterns were recorded and utilized to measure the lattice strain in an H_2 environment at different temperatures. Gas pressure was maintained for 30 min at −100 °C before temperature was increased to −60 °C. Temperature was held there for approximately 20 min before being increased to −20 °C and held for an additional approximately 10 min. Temperature was then reduced back to −100 °C to maximize H_2 solubility and held for 20 min before beginning He implantation.

Helium implantation was performed using a gas-fed Colutron G-2 ion source (Colutron, Boulder, CO, USA) operating at an accelerating voltage of 10 kV. The ^{4}He ion beam had been aligned prior to Pd sample loading using a custom-built Faraday stage. Ion beam intensity was measured to be approximately 4.0×10^{13} ions/cm^2/s at the beginning of irradiation. The Pd sample was irradiated in the H_2 environment for 42 min to a final nominal ^{4}He fluence of 10^{17} ions/cm^2, all with concurrent collection of BF video data. Ion beam current was monitored throughout via a skimming cup and was observed to drop by only ~3% between the beginning and end of specimen irradiation. Once this final fluence was accomplished, BF images and diffraction patterns were collected from various locations on the specimen. The Monte-Carlo based SRIM code [14] was used to simulate material damage and ion implantation for the given irradiation conditions (Figure 1b). SRIM calculations were performed using an incident ^{4}He ion beam at an incoming angle of 18.7°, impinging on Pd metal following the procedure given by Stoller et al. [15]. Displacements per atom (dpa) was calculated using Quick Calculation mode and the phonon.txt output file. A threshold displacement energy of 34 eV [16] and a density of 11.9 g/cm^3 were used.

Bubble sizes and density were determined, where possible, using ImageJ [17] analysis. Image resolution variations can affect the bubble density analysis by up to an order of magnitude. To maintain consistency, only images in the under-focus condition were used. Bubbles were confirmed using both under- and over-focus images. A sample set of under- and over-focus images is provided in a Supplemental file. Images were all converted to a 1712 × 1712 resolution (used for in-situ video) before analysis and the same procedure was utilized on all images. Image analysis procedure was as follows: (1) Gaussian Blur with radius of 3, (2) Normalize Local Contrast with radii of 20 pixels, (3) invert to make bubbles appear dark, (4) "Mexican Hat" Filter with radius of 4 or 5, (5) threshold the entire image, and finally (6) Analyze Particles of area 0–infinity and circularity set to 0.6–1. Data were exported and bubbles with less than 1 nm diameter, the approximate TEM resolution limit, were removed from the dataset. Only average bubble size is provided because the standard deviation is small, usually less than 0.2 nm.

3. Results

3.1. Exposure to H_2 at Cryogenic Temperature

To determine the effects of an H_2 atmosphere on the Pd sample at cryogenic temperatures, a sample was subjected to an H_2 environment at temperatures between −100 °C and −20 °C. As shown in Figure 2, no significant microstructural changes were observed due to H_2 alone. Selected Area Electron Diffraction (SAED) patterns were taken after each step. Experimental error in the SAED was not explicitly quantified, but is expected to be large in these experiments due to slight variation in tilt angle and sample height with variation in H_2 flow rate or temperature. The degree of hydride phase formation, which is characterized by a 3.6% lattice expansion in $PdH_{0.57}$, was therefore unquantifiable in these experiments. Image contrast changes apparent in Figure 2 are due to the sample bending during the temperature cycles.

Figure 2. In-situ TEM images of the Pd sample during pre-implantation H_2 exposure. Images are in sequential order and show the sample (**a**) initially, (**b**) after 25 min of H_2 exposure at −100 °C, (**c**) during the 14 min of H_2 exposure at −60 °C, (**d**) after 7 min of H_2 flow at −20 °C, and (**e**) after cooling back to −100 °C in H_2 for ^{4}He implantation.

3.2. In-situ Helium Implantation and Annealing in H_2

Helium bubble evolution was observed and characterized in-situ during implantation in an H_2 environment at −100 °C. Figure 3a shows initial observation of ^{4}He bubbles after implantation to a peak concentration of 6 at.%. All reported concentration values assume no escape of He from the TEM foil. Bubbles initially had low areal-density and were approximately 1.2 nm in diameter, similar to the microstructure that was observed in the preliminary experiments at room temperature. Larger bubbles were observed in some areas, but bubble nucleation was generally homogenously distributed and uniform in size throughout the implantation. Bubble density visibly increased with increasing implantation dose as shown in Figure 3b–f. Figure 4a shows the measured bubble diameter and density changes as a function of He concentration. Bubble size remained constant during the implantation, but areal bubble density was observed to increase with implantation time, starting at 8×10^{11} bubbles/cm^2 in Figure 3a at 6 at.%, and reaching a density of 5×10^{12} bubbles/cm^2 in Figure 3f at 23 at.%. Bubble density change as a function of implantation dose was determined using a linear fit, where the slope = 2.7×10^{11} bubbles/cm^2/at.% and the intercept = -1.2×10^{12} bubbles/cm^2.

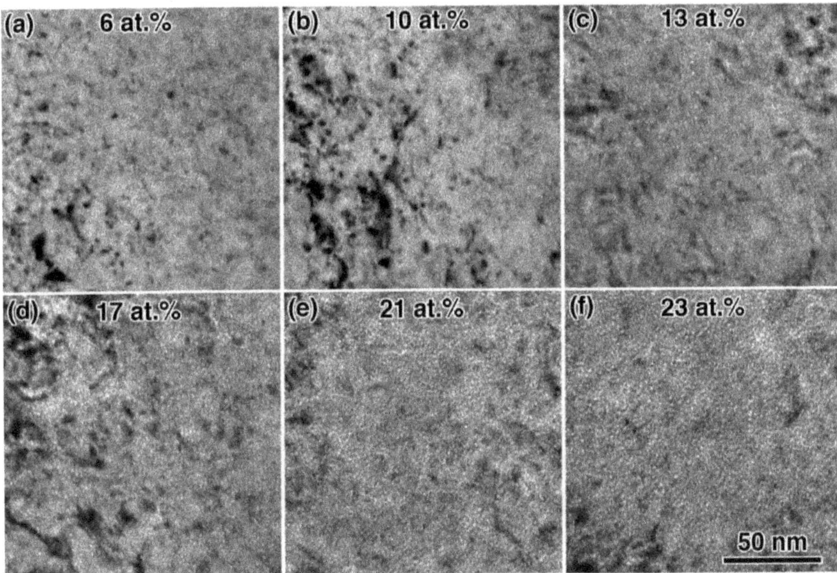

Figure 3. BF in-situ TEM images showing He bubble evolution during implantation at -100 °C under H$_2$ gas flow from peak concentrations of (**a**) 6 to (**f**) 23 at.%. The images were taken in Fresnel under-focus imaging condition, so bubbles appear as small white circles.

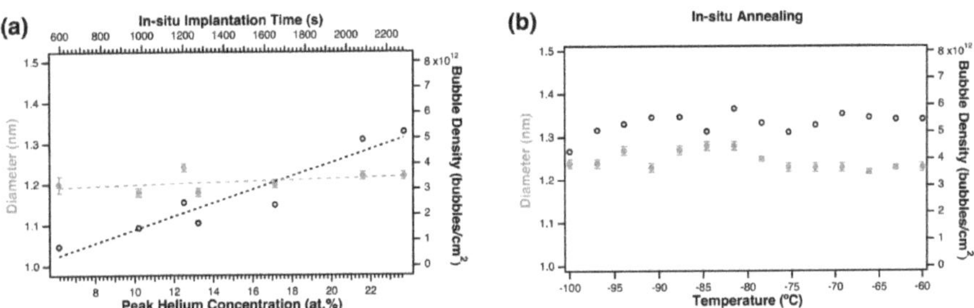

Figure 4. Bubble size and density changes (**a**) during in-situ implantation at -100 °C in H$_2$ gas, and (**b**) during annealing from -100 °C to -60 °C. Data points in (**a**) were measured from the images in Figure 3. Bubble diameter is shown as solid green circles, and bubble density is shown as empty black circles. Linear fits are shown for the data in (**a**).

After implantation, the sample was heated in-situ from -100 °C to 0 °C in H$_2$ gas. According to Figure 1a, most H$_2$ should be desorbed from the sample during this heating process. Bubble size and density were characterized during the annealing, when possible, and are summarized in Figure 4b. The sample was annealed to -60 °C in 390 s (0.10 °C/s), held at -60 °C for approximately 10 min, then ramped from -60 °C to 0 °C in approximately 3 min. Images recorded from -60 °C to 0 °C had too much dynamic contrast evolution for bubble size analysis, so two post-annealing images were analyzed and the results averaged to obtain a final density of 7×10^{12} bubbles/cm^2 and a final diameter of 1.3 nm. No significant changes in bubble diameter were observed during annealing up to 0 °C. No significant changes in bubble density were observed under annealing from -100 °C to -60 °C, but the bubble density did appear to increase from 6×10^{12} bubbles/cm^2 to 7×10^{12} bubbles/cm^2 after annealing from −60 °C to 0 °C (not shown in figure). Uncertainty is not provided for bubble density measurements because a single density value was calculated from each image.

4. Discussion

Helium bubble diameter was observed to remain constant during the cryogenic in-situ implantation in the presence of H_2, but bubble density increased as a function of time. Even though β-phase hydride formation could not be identified within experimental error, the lower concentration α-phase hydride likely formed and could impact bubble evolution. Although H_2 may influence the nucleation and growth of He bubbles in Pd, the H concentration is very low in α-phase (H/Pd = 0.03), so cryogenic implantation temperature is thought to be a more-likely dominating factor. Very little literature exists on He implantation in a H_2 environment or on cryogenic He implantation, so this section will discuss, in relation to this work, (1) a comparison of this work with previous studies on He bubble evolution in Pd, (2) theory on He bubble growth at cryogenic temperature, and (3) how H-He interactions may affect the results.

4.1. Comparison with Data on He bubble Nucleation in Pd and $PdT_{0.6}$

Although very little microscopy work has been done on fcc metals implanted with He at cryogenic temperatures, He bubble formation has been studied in aged Pd tritide, $PdT_{0.6}$. Tritium decay does not induce displacement damage in the lattice, so the vacancy concentration is limited to thermal vacancies, although H can stabilize vacancies, and formation of β-phase hydride can cause microstructural changes [18]. Bubbles are at equilibrium when the He pressure inside the bubble equals the surface tension of the host material, $p = 2\gamma/r$, where p is the He pressure, γ is the surface energy of the host material, and r is the bubble radius [19]. Bubbles formed at room temperature in $PdT_{0.6}$ are typically highly over-pressurized. He bubbles reach 1–2 nm in diameter in the first 8 months (1.34 at.% He) of storage. Bubble density estimates vary widely due to film thickness measurement error, overlapping bubbles, and the possible presence of bubble sizes near or below the resolution limit of the microscope, but values of 10^{17}–10^{19} bubbles/cm^3 are reported for aging times of 2–8 months [7–9]. If a thickness of 50 nm is assumed for the samples implanted in this work, the density varied from 10^{17}–10^{18} bubbles/cm^3 with implantation dose from 2×10^{16} to 9×10^{16} ions/cm^2. Bubble diameter did not vary with implantation dose in this work and is similar to that observed in aged $PdT_{0.6}$. Nuclear Magnetic Resonance (NMR) has been used to measure the phase and pressure of He bubbles in aged $PdT_{0.6}$. After annealing samples aged for 1 year, solid He diffusion was observed below −73 °C, and melting was observed from −73 °C to 7 °C, with corresponding bubble pressures of 6–11 GPa, and an average density of 120 He/nm^3 within a bubble [20]. Aging for 8 years, followed by multiple deuterium exchanges to remove the tritium, resulted in an average density of 90 He/nm^3 and the presumed coexistence of liquid and solid phases from −230 °C to −133 °C [21].

4.2. Comparison with Theory of He Bubble Nucleation and Growth at Cryogenic Temperatures

Several mechanisms could result in a lack of bubble growth with implantation time. Low He diffusivity at cryogenic temperature will contribute to a lower nucleation and growth rate. Low diffusivity may promote a higher bubble nucleation rate, but a lower bubble growth rate due to He becoming trapped very near its implantation site. Bubble diameter remained constant at 1.2 nm after the initial observation of He bubbles at a fluence of 2.35×10^{16} ions/cm^2 (6 at.%), but bubble density increased with implantation dose. Since the He implantation profile, shown in Figure 1b, results in lower concentrations near the surfaces, nucleation will take longer in these regions than at the center of the foil where the He concentration is highest. TEM imaging captures all the material within the thickness of the sample, which could result in a visible increase in bubble density as nucleation occurs first in the higher concentration, and subsequently in the lower concentration regions. This may contribute to an experimentally measured increase in areal bubble density.

At cryogenic temperatures, the low thermal vacancy concentration prohibits cavity growth by absorption of thermal vacancies, which cause rapid cavity growth at elevated temperatures. If the mobility of irradiation- or hydrogen-induced vacancies is low, bubble growth can only occur from dislocation loop punching. Assuming a Pd vacancy migration value of 0.63 eV [22], Pd vacancies are not mobile over the timescale of these experiments at −100 °C. Wolfer [23] has succinctly summarized He bubble growth in metals as a function of homologous temperature, T_h, and pressure. In the case of an isolated single bubble at temperatures below $T_h = 0.25$, theory indicates that the bubble pressure is high enough for loop punching only at $\mu/5$, where μ is the shear modulus of the material [24]. Below this pressure at $T_h = 0.25$, bubble growth is not expected to occur. In Pd, $\mu = 42$ GPa, making the pressure required for loop punching from an isolated single bubble 8.4 GPa. The NMR measurements discussed above [20] found bubble pressures of 6–11 GPa in the −173 °C to 77 °C range in 1 year old $PdT_{0.62}$, very close to the threshold for loop punching.

When theory is applied to the case of a He bubble array, instead of an isolated bubble, the pressure required for loop punching increases to about 50% of the shear modulus [25]. Bubbles with radii less or equal to ~$5b$, where b is the Burgers vector of a prismatic dislocation loop, require a high enough He density to create bubble pressures sufficient for loop punching, independent of the bubble density. As bubbles grow by loop punching, the accumulation of loop debris in the regions between bubbles exerts an increasing and opposing force to the subsequent formation of loops, increasing the pressure required for loop punching dramatically. Thus, during the initial He accumulation period at cryogenic temperatures, one might expect slow growth by loop punching, but as the bubble density increases and the inter-bubble spacing decreases, the pressure required for loop punching likely becomes too high for additional growth to occur. This mechanism may have caused bubble growth to cease in this experiment, while bubble areal density continued to increase.

As the bubble pressure continues to increase beyond the window where dislocation loop punching is viable, inter-bubble fracturing may occur, possibly leading to blister formation and He release [23]. Blister formation has previously been observed after implanting bulk Pd with 300 keV He to 1×10^{18} ions/cm^2 (70 at.% He at the peak) at −180 °C, a much higher He concentration that the total implanted dose in this work [1]. Theory suggests that, particularly below a bubble density of 3×10^{18} bubbles/cm^3, equilibration of the chemical potentials for gas atoms in the bubble and in interstitial solution could result in He diffusing freely throughout the solid without being trapped inside a bubble, causing rapid gas release once a critical concentration is reached [25].

4.3. Hydrogen-helium Interactions

By performing in-situ He implantation in an ETEM, this work is uniquely suited to explore the interactions of H and He in a model fcc system. Point defects are introduced in the lattice during He implantation (see Figure 1b for the dpa profile). Hydrogen is known to interact with such defects in metals [26–28], which may have influenced the bubble nucleation and growth rates observed in this work. The binding energies of H to defects in Pd are lower than many other metals [29]. Experimentally determined binding energies of H to Pd defects are about: 0.15 eV (self-interstitial), 0.23 eV (vacancy), and 0.29 eV (He bubble) [27–29]. Hydrogen is expected to have been strongly bound to these defects in the present work, which was done at −100 °C (0.015 eV), and could increase the size of He clusters. In fcc metals, up to six H atoms can occupy a monovacancy [29], which could influence the trapping kinetics of He atoms to vacancies. Additionally, the presence of H in interstitial sites may influence the diffusion of interstitial He through the lattice, and therefore influence nucleation and growth kinetics. More simulation work is needed to verify potential effects of H on He bubble nucleation and growth.

5. Conclusions

Palladium metal was implanted with 10 keV ^{4}He in-situ, at cryogenic temperature, in a H_2 environment. No lattice expansion indicating β-phase hydride formation was observed. He bubbles 1.2 nm in diameter were observed to nucleate after 6 at.% He. Bubble size did not change with implantation time, but bubble density did increase. These initial experiments highlight the strength of the MIAMI-2 facility for in-situ TEM exploration of H_2 interaction with He bubbles at various temperature extremes.

This preliminary work has highlighted the new combination of extreme environments (cryogenics, gas implantation, and reactive gas exposure) that can be explored during direct real-time observation within a TEM. Further work is needed to fully understand these initial observations. This future work would include comparison between in-situ He implantations in the presence and absence of H_2 at the same temperature, for both ambient and low temperature, to deduce the effects of H_2 and temperature on bubble formation, as well as development of methods to ensure hydride formation in the ETEM. This study points to a new multidimensional stressor approach to in-situ TEM experiments that permits greater understanding of the response to complex environments by materials.

Supplementary Materials: The following are available online at http://www.mdpi.com/1996-1944/12/16/2618/s1, Figure S1: Under-focus image showing He bubbles during implantation, Figure S2: Over-focus image showing He bubbles during implantation.

Author Contributions: C.A.T. performed preliminary experiments at Sandia, aided in experimental planning, analyzed the data, and wrote the manuscript. J.A.H., G.G., E.A., S.B., and K.H. aided in experimental planning and performed the experiments. A.M. aided in data analysis and interpretation. D.B.R. aided in experimental planning and interpretation. J.D.S. prepared TEM samples and provided interpretation.

Funding: This work was performed, in part, at the Center for Integrated Nanotechnologies, an Office of Science User Facility operated for the U.S. Department of Energy (DOE) Office of Science. Sandia National Laboratories is a multimission laboratory managed and operated by National Technology & Engineering Solutions of Sandia, LLC, a wholly owned subsidiary of Honeywell International, Inc., for the U.S. DOE's National Nuclear Security Administration under contract DE-NA-0003525. The views expressed in the article do not necessarily represent the views of the U.S. DOE or the United States Government.

Acknowledgments: The authors would like to thank Warren York for his time in preparing TEM samples, and Norm Bartelt, Doug Medlin, and Trevor Clark for thoughtful discussion.

Conflicts of Interest: The authors declare no conflict of interest.

References

1. Thomas, G.J.; Bauer, W. Helium Implantation Effects in Palladium at High Doses. *Radiat. Eff.* **1973**, *17*, 221–234. [CrossRef]
2. Bauer, W.; Thomas, G.J. Helium Release and Electron-Microscopy of Helium-Implanted Palladium. *J. Nucl. Mater.* **1972**, *42*, 96–100. [CrossRef]
3. Adams, B.D.; Chen, A. The Role of Palladium in a Hydrogen Economy. *Mater. Today* **2011**, *14*, 282–289. [CrossRef]
4. Manchester, F.D.; San-Martin, A.; Pitre, J.M. The H-Pd (Hydrogen-Palladium) System. *J. Phase Equilibria* **1994**, *15*, 62–83. [CrossRef]
5. Flanagan, T.B.; Oates, W.A. The Palladium-Hydrogen System. *Annu. Rev. Mater. Sci.* **1991**, *21*, 269–304. [CrossRef]
6. Jewell, L.L.; Davis, B.H. Review of Absorption and Adsorption in the Hydrogen-Palladium System. *Appl. Catal. A Gen.* **2006**, *310*, 1–15. [CrossRef]
7. Thiébaut, S.; Décamps, B.; Pénisson, J.M.; Limacher, B.; Percheron Guégan, A. TEM Study of the Aging of Palladium-Based Alloys During Tritium Storage. *J. Nucl. Mater.* **2000**, *277*, 217–225. [CrossRef]
8. Thomas, G.J.; Mintz, J.M. Helium bubbles in palladium tritide. *J. Nucl. Mater.* **1983**, *116*, 336–338. [CrossRef]
9. Fabre, A.; Decamps, B.; Finot, E.; Penisson, J.M.; Demoment, J.; Thiebaut, S.; Contreras, S.; Percheron-Guegan, A. On the correlation between mechanical and TEM studies of the aging of palladium during tritium storage. *J. Nucl. Mater.* **2005**, *342*, 101–107. [CrossRef]
10. Hinks, J.A. Transmission Electron Microscopy with In-situ Ion Irradiation. *J. Mater. Res.* **2015**, *30*, 1214–1221. [CrossRef]
11. Greaves, G.; Mir, A.H.; Harrison, R.W.; Tunes, M.A.; Donnelly, S.E.; Hinks, J.A. New Microscope and Ion Accelerators for Materials Investigations (MIAMI-2) system at the University of Huddersfield. *Nucl. Instrum. Methods Phys. Res. Sect. A Accel. Spectrometers Detect. Assoc. Equip.* **2019**, *931*, 37–43. [CrossRef]

12. Lasser, R.; Klatt, K.H. Solubility of Hydrogen Isotopes in Palladium. *Phys. Rev. B* **1983**, *28*, 748–758. [CrossRef]

13. Lewis, F.A. The Hydrides of Palladium and Palladium Alloys: A Review of Recent Researches. *Platin. Met. Rev.* **1960**, *4*, 132.

14. Ziegler, J.F.; Ziegler, M.D.; Biersack, J.P. SRIM-The Stopping and Range of Ions in Matter. *Nucl. Instrum. Methods Phys. Res. Sect. B Beam Interact. Mater. At.* **2010**, *268*, 1818–1823. [CrossRef]

15. Stoller, R.E.; Toloczko, M.B.; Was, G.S.; Certain, A.G.; Dwaraknath, S.; Garner, F.A. On the use of SRIM for computing radiation damage exposure. *Nucl. Instrum. Methods Phys. Res. B* **2013**, *310*, 75–80. [CrossRef]

16. Jimenez, C.M.; Lowe, L.F.; Burke, E.A.; Sherman, C.H. Radiation Damage in Pd Produced by 1–3-MeV Electrons. *Phys. Rev.* **1967**, *153*, 735–739. [CrossRef]

17. Schneider, C.A.; Rasband, W.S.; Eliceiri, K.W. NIH Image to ImageJ: 25 years of image analysis. *Nat. Methods* **2012**, *9*, 671–675. [CrossRef] [PubMed]

18. Roddatis, V.; Bongers, M.D.; Vink, R.; Burlaka, V.; Čížek, J.; Pundt, A. Insights into Hydrogen Gas Environment-Promoted Nanostructural Changes in Stressed and Relaxed Palladium by Environmental Transmission Electron Microscopy and Variable-Energy Positron Annihilation Spectroscopy. *J. Phys. Chem. Lett.* **2018**, *9*, 5246–5253. [CrossRef]

19. Trinkaus, H. Energetics and Formation Kinetics of Helium Bubbles in Metals. *Radiat. Eff.* **1983**, *78*, 1–4. [CrossRef]

20. Abell, G.C.; Attalla, A. NMR Evidence for Solid-Fluid Transition near 250 K of He-3 Bubbles in Palladium Tritide. *Phys. Rev. Lett.* **1987**, *59*, 995–997. [CrossRef]

21. Abell, G.C.; Cowgill, D.F. Low-Temperature He-3 NMR Studies in Aged Palladium Tritide. *Phys. Rev. B* **1991**, *44*, 4178–4184. [CrossRef] [PubMed]

22. Eremeev, S.V.; Lipnitskii, A.G.; Potekaev, A.I.; Chulkov, E.V. Activation Energy for Diffusion of Point Defects at the Surfaces of F.C.C. metals. *Russ. Phys. J.* **1997**, *40*, 584–589. [CrossRef]

23. Wolfer, W.G. Radiation Effects in Plutonium. *Los Alamos Sci.* **2000**, *26*, 274–285.

24. Wolfer, W.G. The Pressure for Dislocation Loop Punching by a Single Bubble. *Philos. Mag. A* **1988**, *58*, 285–297. [CrossRef]

25. Wolfer, W.G. Dislocation Loop Punching in Bubble Arrays. *Philos. Mag. A* **1989**, *59*, 87–103. [CrossRef]

26. Myers, S.M.; Baskes, M.I.; Birnbaum, H.K.; Corbett, J.W.; DeLeo, G.G.; Estreicher, S.K.; Haller, E.E.; Jena, P.; Johnson, N.M.; Kirchheim, R.; et al. Hydrogen Interactions with Defects in Crystalline Solids. *Rev. Mod. Phys.* **1992**, *64*, 559–617. [CrossRef]

27. Norskov, J.K.; Besenbacher, F.; Bottiger, J.; Nielsen, B.B.; Pisarev, A.A. Interaction of Hydrogen with Defects in Metals: Interplay between Theory and Experiment. *Phys. Rev. Lett.* **1982**, *49*, 1420–1423. [CrossRef]

28. Myers, S.M.; Wampler, W.R.; Besenbacher, F.; Robinson, S.L.; Moody, N.R. Ion Beam Studies of Hydrogen in Metals. *Mater. Sci. Eng.* **1985**, *69*, 397–409. [CrossRef]

29. Besenbacher, F.; Nielsen, B.B.; Norskov, J.K.; Myers, S.M.; Nordlander, P. Interaction of Hydrogen Isotopes with Metals: Deuterium Trapped at Lattice Defects in Palladium. *J. Fusion Energy* **1990**, *9*, 257–261. [CrossRef]

Review

Interface Effects on He Ion Irradiation in Nanostructured Materials

Wenfan Yang [1,2], **Jingyu Pang** [1,2], **Shijian Zheng** [1,3,4,*], **Jian Wang** [5], **Xinghang Zhang** [6] **and Xiuliang Ma** [1,7]

[1] Shenyang National Laboratory for Materials Science, Institute of Metal Research, Chinese Academy of Sciences, 72 Wenhua Road, Shenyang 110016, China
[2] School of Material Science and Engineering, University of Science and Technology of China, Hefei 230026, China
[3] Tianjin Key Laboratory of Materials Laminating Fabrication and Interface Control Technology, Hebei University of Technology, Tianjin 300130, China
[4] State Key Laboratory of Reliability and Intelligence of Electrical Equipment, Hebei University of Technology, Tianjin 300130, China
[5] Mechanical and Materials Engineering, University of Nebraska-Lincoln, Lincoln, NE 68588, USA
[6] School of Materials Engineering, Purdue University, West Lafayette, Indiana, IN 47907, USA
[7] School of Material Science and Engineering, Lanzhou University of Technology, Lanzhou 730050, China
* Correspondence: sjzheng@imr.ac.cn; Tel.: +86-2397-1843

Received: 20 June 2019; Accepted: 13 August 2019; Published: 19 August 2019

Abstract: In advanced fission and fusion reactors, structural materials suffer from high dose irradiation by energetic particles and are subject to severe microstructure damage. He atoms, as a byproduct of the (n, α) transmutation reaction, could accumulate to form deleterious cavities, which accelerate radiation-induced embrittlement, swelling and surface deterioration, ultimately degrade the service lifetime of reactor materials. Extensive studies have been performed to explore the strategies that can mitigate He ion irradiation damage. Recently, nanostructured materials have received broad attention because they contain abundant interfaces that are efficient sinks for radiation-induced defects. In this review, we summarize and analyze the current understandings on interface effects on He ion irradiation in nanostructured materials. Some key challenges and research directions are highlighted for studying the interface effects on radiation damage in nanostructured materials.

Keywords: He ion irradiation; cavities; interface; nanostructured materials

---

1. Introduction

1.1. Motivation and Architecture

As a reliable, sustainable and affordable energy, nuclear power provides more than 13% of electricity worldwide [1]. The next generation nuclear reactors have increasing demands for the discovery of advanced materials that can survive severe radiation environments, so that nuclear reactors can operate safely for a prolonged period of time. In nuclear fission and fusion reactors, helium (He) ions are produced by the decay of tritium and the (n, α) reaction during neutron irradiation [2,3]. He is insoluble in most nuclear materials and its diffusion activation energy is low, so it is easy to form interstitials and He bubbles [4]. He can combine with cavities and accelerate irradiation-induced swelling, hardening, embrittlement and surface deterioration [5,6]. Prior studies show that nanostructured materials possess outstanding irradiation resistance because abundant interfaces can hinder the formation of cavities [7]. However, the underlying mechanisms for interfaces enhanced irradiation resistance had not been revealed in-depth in the past. Recently, many achievements have been made to explore the interface mechanisms for designing suitable nuclear materials. To understand the interface mechanisms and provide a theoretical basis for designing nuclear materials, we summarize these research achievements in this review.

In this review, we first introduce He ion irradiation damage in materials. Then, we summarize the irradiation response in nanostructured materials, including the interactions between cavities and interfaces, and interface effects on irradiation damage. Finally, we discuss the evolutions of properties of irradiated nanostructured materials.

1.2. He Ion Irradiation Behavior

1.2.1. He Diffusion

The solubility of He in most nuclear materials is negligible [8], and the accumulation of He atoms in materials after irradiation is usually accompanied by abundant defect clusters. In addition, the way of He diffusion depends on the dose of irradiation, temperature as well as the presence of other intrinsic or irradiation-induced defects, such as point defects (interstitial and vacancy), interstitial clusters, He-vacancy clusters, cavities, acting as traps for He atoms. Regardless of the way of He diffusion, thermal activation is the most important basic process [8].

(1) Interstitial migration: He atoms at interstitial sites diffuse interstitially. Since the activation energy of the He atoms interstitial migration is only a few tenths of eV [8], the He atoms can diffuse quickly at room temperature until they are trapped by other defects. In the case of a low radiation dose, the concentration of defects introduced by irradiation is negligible low, hence He atoms diffuse effectively as interstitials at $T < 0.5T_m$ (T is irradiation temperature, T_m is melting point of materials), where both the displacement damage and the concentration of thermal vacancies are small. Therefore, interstitial migration is the most probable type of He diffusion mechanisms.

(2) Vacancy mechanism: He atoms exchange position with the neighbor vacancies. When $T > 0.5T_m$, abundant vacancies would form by thermal excitation, hence He atoms could jump from one to the other vacancy [9]. Figure 1 shows different means for He diffusion at low radiation damage level [8]:

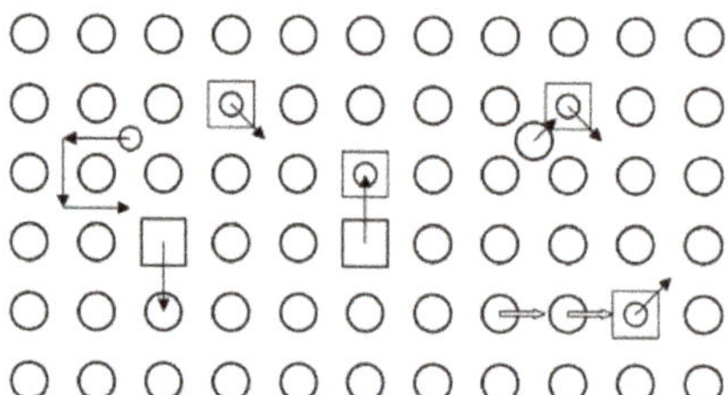

Figure 1. Schematic illustration of defect configurations and jump processes relevant for He diffusion without and with irradiation [8].

For high dose irradiation, He atoms would accumulate to form defect clusters, such as interstitial clusters, He-vacancy clusters, cavities (bubbles and voids), which display different diffusion mechanisms: when $T < 0.2T_m$, the athermal "displacement or cascade mechanism" would be the dominant mechanism for He diffusion, because vacancies are almost immobile; when $0.2T_m < T < 0.5T_m$, the "replacement mechanism" would be the dominant mechanism—at this time a He atom diffuses interstitially from a vacancy, then it can be re-trapped by another vacancy; when $T > 0.5T_m$, the "vacancy mechanism" would be dominant, because He diffusion can be assisted by numerous vacancies (formed by thermal activation) that can move easily.

1.2.2. Nucleation of He Bubbles

Before introducing the bubble nucleation, it is necessary to distinguish the forms of cavities, i.e., bubble and void. Usually, bubble is a kind of spherical cavity, possessing a high concentration of He atoms, hence the bubble has over-pressure or equilibrium-pressure. However, a void with plenty of vacancies has internal pressure much lower than equilibrium-pressure, so that the void has a faceted structure composed of close-packed planes [10]. For cavity formation, first He atoms and vacancies agglomerate into clusters, which accelerate the bubble nucleation and growth. Then with the increasing of irradiated dose or temperature, bubbles would achieve a critical radius via absorbing more He atoms and vacancies and transform into voids [11]. Consequently, bubble and void are two distinct forms of cavities, but with a strong connection.

Here, we focus on bubble nucleation. There are two types of nucleation of bubbles: homogeneous and heterogeneous nucleation [8]. Homogeneous nucleation means that bubbles nucleate in perfect lattice position, while heterogeneous nucleation preferentially occurs on defects of materials, such as dislocations and interfaces. These two ways are competing during actual nucleation process in materials suffering He ion irradiation.

At low temperatures, thermal dissociation of He atoms from their traps is negligible. Homogeneous nuclei and preexisting defects in materials are competitive for absorbing He atoms. If nuclei can capture more He atoms, in other words, the density of nuclei is higher than that of preexisting defects in the material, homogenous nucleation will dominate (usually at low temperatures or high He implantation doses).

At high temperatures, the thermal dissociation of He atoms from captured positions should be taken into account, and the relation between capture abilities of possible nucleation sites could not explain clearly which nucleation mode dominates. Some defects, such as dislocations and interfaces, have impacts on the thermodynamics of critical nuclei. Figure 2 shows the traditional understanding of how interfaces influence the thermodynamic state of a nucleus [8]. For instance, bubbles are in thermal equilibrium, and the relation among the gas pressure inside bubbles (p), the surface energy of bubbles (γ) and the radius of bubbles (r) can be written as [8]:

$$p = 2\gamma/r \tag{1}$$

With the same nucleus volume, the interfacial equilibrium between surface of a bubble and grain boundary (GB) can be seen in Figure 2b, and the equilibrium between a bubble and a GB-precipitate can be seen in Figure 2c.

Radius of the bubble at a GB is bigger than that in the grain interior, thus reducing gas pressure and equilibrium concentration of He atoms. In other words, the critical He concentration for heterogeneous nucleation at a GB is smaller than that for homogeneous nucleation. Therefore, a bubble finds it easy to nucleate around a GB where the nucleation barrier is low.

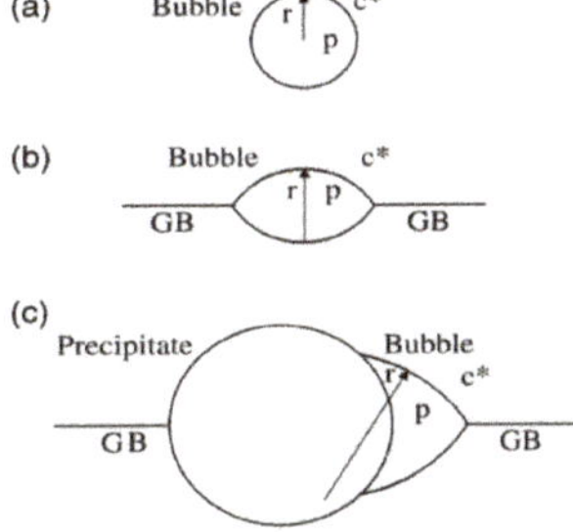

Figure 2. Illustration of interfacial effects on bubble nucleation at high temperatures; (**a**) spherical nucleus in the matrix, (**b**) lenticular nucleus at a grain boundary (GB), (**c**) truncated lenticular nucleus at a GB-precipitate, all assumed to have the same volume [8].

1.2.3. The Coarsening of Bubbles

Apparently, bubbles tend to grow into voids, namely bubble to void transformation. After nucleation, bubbles can continue to grow by capturing He atoms and vacancies. In addition, the coarsening of the bubble is influenced by several parameters, such as irradiation temperature, He concentration and irradiation time. As the temperature increases, more vacancies are activated and the probability of combining He atoms and vacancies increases, so that the rate of bubble to void transformation increases. In a similar way, higher He concentration and longer irradiation time, produce bigger voids in materials. There are two classic coarsening mechanisms as shown in Figure 3: (1) migration and coalescence (MC) [12]; (2) Ostwald ripening (OR) [13].

(1) Migration and coalescence: bubble migration is because of random rearrangements of the bubble surface via diffusion of matrix atoms. These processes can be achieved by three ways: bulk diffusion, surface diffusion and vapor transport. For the bulk diffusion process, atoms can migrate from the front of a bubble into the matrix, reaching the back of the bubble through body diffusion. Surface diffusion is due to random rearrangements of the bubble surface by diffusion of matrix atoms [8]. Vapor transport means that matrix atoms move into a bubble in the form of vapor, thereby changing its shape and causing it to migrate. Usually, the migration of bubbles is random, that is to say, Brownian motion. Small bubbles will migrate and coalesce into larger bubbles and voids by various ways. Different from bubbles, void migration depends on surface ledge nucleation significantly [14]. This mechanism has been demonstrated in vanadium by in situ hot stage transmission electron microscopy (TEM), where the voids migrate in a random walk manner at $0.55T_m$ [15].

(2) Ostwald ripening: large bubbles can coarsen by capturing and absorbing He atoms, which are dissolved by thermal activation from other small bubbles.

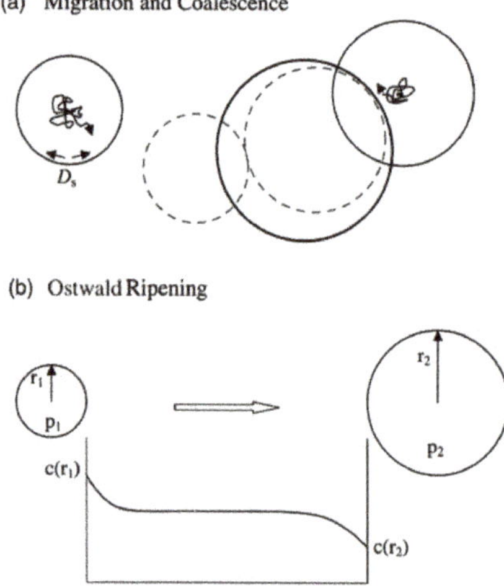

Figure 3. Schematic illustration of the two main bubble coarsening mechanisms, (**a**) migration and coalescence via surface diffusion, (**b**) Ostwald ripening due to He fluxes driven by differences in the thermal equilibrium He concentrations in the vicinity of small and large bubbles [8].

Temperature is a key parameter that determines coarsening mechanisms. The OR mechanism dominates at high temperatures, suggesting that the bubble dissociation needs higher thermal activation energy than that for driving bubble migration. The energy driving bubble dissociation is not high enough at low temperatures, therefore, the growth of bubbles at low temperatures depend strongly upon the MC mechanism, especially upon the surface diffusion [16]. Whatever mechanisms He bubbles grow by, the number of bubbles continues to decrease with the growth of bubbles. In addition, the mechanism about the growth of He bubbles is very complicated with presence of dislocations, interfaces and other defects. Under high temperature creeping, dislocations motion and GBs sliding can sweep a He bubble, thus promoting the growth of this bubble [17].

1.3. The Influence of He Ion Irradiation on Mechanical Properties

As inert gas atoms, He atoms are nearly insoluble in nuclear materials. They usually diffuse and aggregate interstitially and combine with vacancies or cavities. Cavities can continue to migrate and grow. This process can lead to hardening, swelling, embrittlement and surface deterioration, hence degrading mechanical properties.

Since He ion irradiation can induce abundant bubbles which hinder dislocation motion severely, hardening often takes place in irradiated materials. Yang et al. conducted the same irradiation experiments at room temperature for Ni, 304 stainless steel and CrMnFeCoNi, which have similar hardening behavior despite the different compositions these materials possess [18]. Similar results also appeared in nanolayered composites, for example, most multilayers suffered hardening after He ion irradiation at room temperature [19–22].

Bubbles can grow continuously by absorbing He atoms and vacancies to form large voids, thus making materials swell. Wakai et al. studied swelling behaviors in Ni alloys (Ni–Si alloys, Ni–Co alloys and so on) by high dose irradiation experiments [23]. They found the percent of swelling increases with the size and number density of voids increasing in these Ni alloys. In addition, when pure Ni was irradiated with a dose of 5×10^{17} ions/cm^2 and annealed at 973 K, the swelling was more than 6% [24].

With numerous big voids forming in materials, embrittlement and surface deterioration take place inevitably. Shinohara et al. did a series of tensile experiments at different temperatures for unirradiated and irradiated pure Fe (where samples were irradiated using 100 keV He ions with a dose of 1×10^{17} ions/cm^2) [25]. They found the ductile–brittle transition temperature for irradiated polycrystalline Fe is higher than that for unirradiated Fe. Although irradiation-induced defects are within 1 μm of the surface, bulk samples show more embrittlement after irradiation. About surface deterioration, Yoshida, et al. found the surface of W samples transformed from straight to wavy after irradiation [26].

1.4. Strategy for Irradiation Resistance: Interfaces

Generally, He ion irradiation damage results from numerous point defects (interstitials and vacancies) induced by the irradiation process. The point defects can aggregate and form cavities (bubbles and voids) which degrade materials severely [4]. Interfaces, as typical planar defects, whether homogenous GBs or heterogeneous phase interfaces, can attract, absorb and annihilate point defects efficiently [5,6].

For homogenous GBs, Bai et al. found GBs have a surprising "loading–unloading" effect during He ion irradiation [27]. During irradiation, interstitials were absorbed into GBs via the loading process, then GBs as a source emitted interstitials to annihilate vacancies in the bulk near GBs. This GB assisted process has a much lower energy barrier compared with conventional defects diffusion. Also, this mechanism is efficient for annihilating irradiation-induced defects in the bulk, resulting in self-healing of the irradiation damage.

For heterogeneous interfaces, taking Cu/Nb interfaces for example, there are numerous misfit dislocations at Cu/Nb interfaces because of lattice mismatch, and these dislocations can form misfit dislocation interactions (MDIs). Also, these MDIs can absorb and trap point defects efficiently and induce "platelet-to-bubble" transition (Figure 4) at interfaces during irradiation [28]. This transition is governed by a competition between three kinds of pressures acting on interfacial He-filled bubbles: the mechanical pressure p_{He} of the trapped He gas, the osmotic pressure p_V due to the flux of radiation-induced vacancies within the crystal to the bubble, and the capillary pressure p_C arising from the surface energy of the bubble. p_{He} and p_V tend to expand the bubble while p_C tends to shrink it. If these three pressures balance, like this [28]:

$$p_{He} + p_V = p_C \tag{2}$$

then, the bubble is in equilibrium: it neither expands nor contracts.

At the same irradiation condition, a spherical bubble, ~2 nm in diameter, forms within a crystalline solid, while a platelet-shaped He-filled bubble generates at an interface. Also, the volume of such platelets was nearly three times smaller than that of bubbles in face-centered cubic (FCC) Cu with the same number of He atoms [29]. Thus, platelets store He atoms more efficiently than spherical bubbles and leads to less He-induced hardening and swelling prior to void formation. Therefore, as platelets do not degrade materials as much as bubbles, heterogeneous interfaces can make materials possess more irradiation resistance.

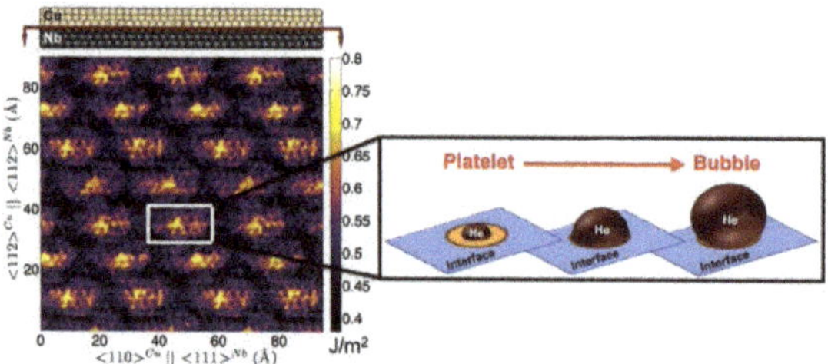

Figure 4. Left: Location-dependent energy of Cu/Nb interfaces. The bright, high-energy regions are heliophilic. Right: A He platelet transforms into a bubble once it has grown to occupy the entire heliophilic patch on which it nucleated [28].

2. Irradiation Response in Nanostructured Materials

2.1. The Interactions between Cavities and Interfaces

Although interfaces can absorb, trap and annihilate defects efficiently during He ion irradiation, cavities (bubbles and voids) form in most nanostructured materials as the implantation dose is enough. However, interfaces can still tune cavities. There are two types of interactions between interfaces and cavities. First, cavities can stay away from interfaces and generate void-denuded zones [30–32]. Second, cavities can adhere to interfaces or cross interfaces [22,32,33]. There are many factors that tailor the formation of cavities, such as implantation dose, temperature and layer constitution. Also, a bubble will transform into a void as it grows, thus bubble and void are at different stages during the growth of a cavity. Once a cavity forms, and the interaction between it and the interface does not change with the growth of this cavity. Therefore, we will take the void as an example and discuss void-interface interactions.

2.1.1. Void-Denuded Zones near Interfaces

A void-denuded zone (VDZ) can generate in accumulative roll bonding (ARB) Cu/Nb nanolayered composites when the irradiation experiment is designed like Figure 5a [31]. When He ions are implanted parallel to the interfaces, the interfaces can absorb defects induced by irradiation efficiently and make defect concentration lower near the interfaces than that within layers. The difference in defect concentration between layer interiors and interfaces leads to long-range diffusion of defects toward interfaces. Once a steady state is established, the defect concentration monotonically increases from interfaces to the interiors of layers, thus inducing the density gradient of voids which is controlled by the diffusion process [30]. In other words, the density of voids in the layers is higher than that near the interfaces. It can be proved by counting the number of voids in different areas within the Cu layer (Figure 5b,c). However, different interfaces may have different abilities for sinking defects induced by irradiation, therefore, they may form different gradients for defect concentration and generate VDZs with different widths. These abilities can be estimated by the widths of VDZs (Figure 5d). In other words, the wider VDZs are, the better sinking efficiency interfaces have. For example, Figure 5e shows widths of VDZs for different Cu/Nb interfaces in 135 nm ARB Cu/Nb, such as Cu {112}//Nb {112} interfaces, Cu {111}//Nb {110} interfaces and so on. The widths of VDZs are measured by TEM pictures and the orientation relationships of the interfaces are confirmed by precession electron diffraction (PED). According to these widths, we can find these atomic-ordered Cu {112}//Nb {112} interfaces have the best sinking abilities for defects.

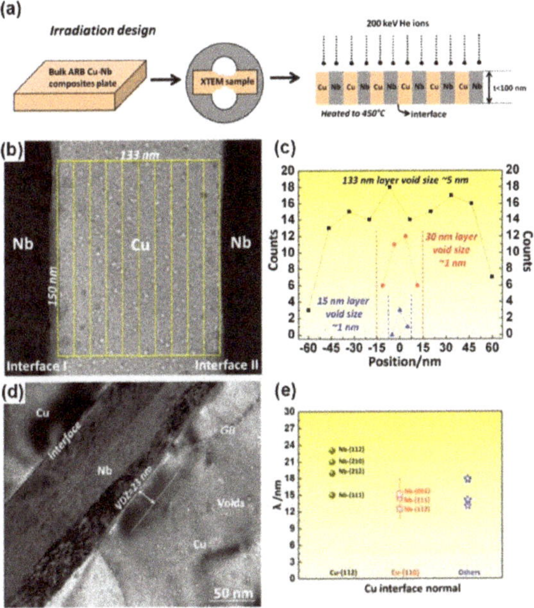

Figure 5. (**a**) Schematic illustration of the He ion irradiation experimental procedures. (**b**) Illustration of the method to determine the void number density in Cu layers. (**c**) Plot of the number density of voids as a function of distance from the center of the layer in 133 nm-, 30 nm- and 15 nm-thick Cu layers. (**d**) Under-focus TEM image showing a void-denuded zone (VDZ) near a Cu/Nb interface. (**e**) Widths of VDZs near Cu/Nb interfaces sorted by their plane orientations [31].

2.1.2. Voids Adhere to Interfaces or Cross Interfaces

Voids can adhere to interfaces or cross interfaces. In Figure 6a,b, voids adhere to interfaces in bulk Cu/Ag and ARB Cu/Nb nanolaminates irradiated using 200 keV He ions with a dose of 2×10^{17} ions/cm^2 at 450 °C. However, voids adhere to interfaces on the side of the Ag layer in bulk Cu/Ag and in the Cu layer for ARB Cu/Nb [22,33]. This asymmetrical void–interface interaction is a consequence of different surface energies of the two metals and non-uniformity in interface energy [33]. To confirm this explanation further, first, we introduce void–interface interaction on twin boundaries (TBs) as shown in Figure 6c. TBs possess uniform interface energy and the matrix and twin have the same surface energy, therefore voids cross TBs symmetrically. Second, we take bulk Cu/Ag as an example and use an interface wetting model (Figure 7a) to explain this asymmetrical interaction. The equation regarding a wetting model can be written as [33]:

$$W = \gamma_{A-C} + \gamma_{A-B} - \gamma_{B-C} \tag{3}$$

where γ_{A-C}, γ_{A-B} and γ_{B-C} are surface energy of A–C, A–B and B–C interface. In Figure 7b, if $W > 0$, voids will stay in the A phase and touch the interface because wetting is favored. In contrast, if $W < 0$, wetting is not favored, voids should have minimum energy and stay within the phase (the A phase) with the lower free surface energy entirely. In addition, there are abundant MDIs at Cu/Ag interfaces

(Figure 7c). Molecular dynamics (MD) simulation shows that W > 0 at MDIs and W < 0 at coherent parts of the interface between MDIs (Figure 7d), therefore, voids only wet MDIs during irradiation. Once a void has grown large enough to cover an entire MDI, it is not thermodynamically favorable for this void to continue to wet the interface regions where W < 0. Instead, the void prefers extending into the Ag layer with lower surface energy.

Irradiation experiment parameters, such as temperature and layer thickness, can also determine whether voids adhere to interfaces or cross interfaces [22]. For 58 nm ARB Cu/Nb, voids do not cross the interfaces independent of implantation dose at room temperature, but voids can overlap interfaces at 450 °C (Figure 8a,c,e). The increase of irradiation dose can only induce the growth of voids, and high irradiation temperature can not only increase the size of voids but also make voids cross interfaces. For 16 nm ARB Cu/Nb, voids always adhere to the interfaces from the Cu layer and never cross interfaces despite irradiation temperature (Figure 8b,d). Layer thickness is another key point for the distribution of voids. ARB Cu/Nb interfaces have excellent abilities for irradiation resistance, therefore voids are confined within Cu layers despite the temperature for 16 nm ARB Cu/Nb with a high interface density.

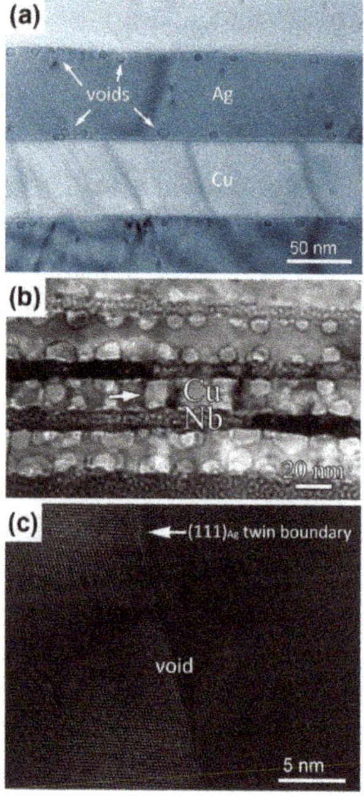

Figure 6. (a) Over-focus TEM image of the Cu/Ag composite after He ion irradiation at 450 °C [33]. (b) Under-focus TEM image of the ARB Cu/Nb nanolayered composites after He ion irradiation at 450 °C [22]. (c) High resolution transmission electron microscopy (HRTEM) image shows void wetting of a coherent (111) Ag twin boundary [33].

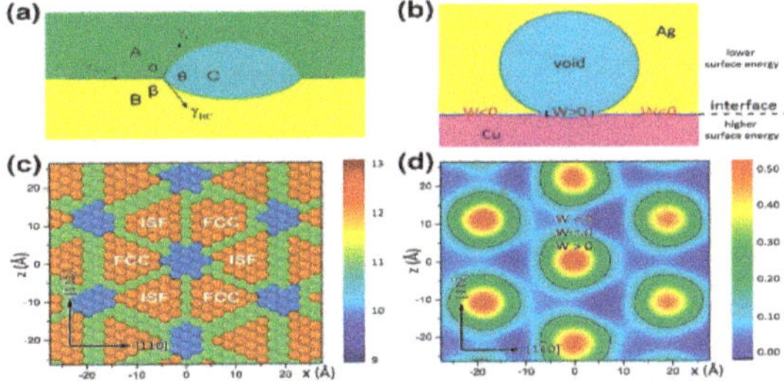

Figure 7. (**a**) Schematic of wetting model on interfaces. (**b**) Schematic of a void wetting a single misfit dislocation interaction (MDI) where W > 0 at a Cu/Ag interface. (**c**) Misfit dislocation network in the cube-on-cube Cu/Ag interface. Atoms shown are on the Cu side of the interface and colored by coordination number. (**d**) Contour plot of the location-dependent interface energy of a Cu/Ag interface. Black contours correspond to zero wetting energy [33].

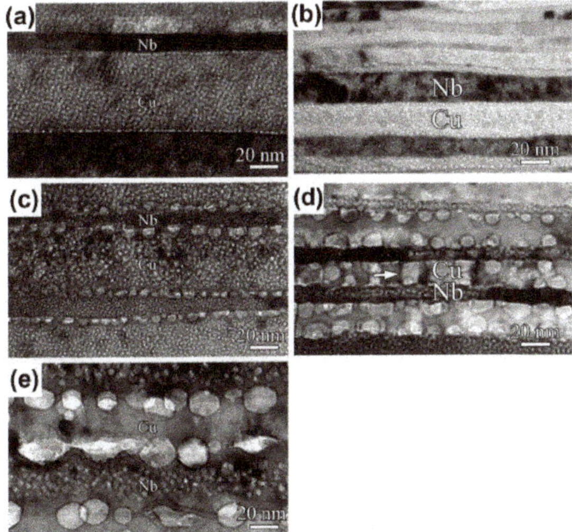

Figure 8. TEM images of accumulative roll bonding (ARB) Cu/Nb after He ion irradiation with different layer thickness, implantation dose and temperature. (**a**) 58 nm, 2×10^{17} ions/cm^2, room temperature (RT); (**b**) 16 nm, 2×10^{17} ions/cm^2, RT; (**c**) 58 nm, 6.5×10^{17} ions/cm^2, RT; (**d**) 16 nm, 2×10^{17} ions/cm^2, 450 °C; (**e**) 58 nm, 2×10^{17} ions/cm^2, 450 °C [22].

2.2. Interfaces Effects on Irradiation Damage

Cavity is a basic form of irradiation production, and its diameter, density and distribution can significantly influence irradiation damage during He ion irradiation. Also, effects of cavity on irradiation damage can be tailored by interfaces. Interfaces can be divided into incoherent interfaces, semi-coherent interfaces and coherent interfaces [34,35]. When lattice mismatch on a boundary is large, and arrays of misfit dislocations and elastic deformation on the boundary cannot accommodate this strain, the boundary is an incoherent interface. However, when arrays of misfit dislocations can mediate this strain, the boundary is a semi-coherent interface. When elastic deformation can accommodate this strain, the boundary is a coherent interface. Next, we will discuss the interface effects on cavity induced irradiation damage.

2.2.1. Incoherent Interfaces

Most GBs are typical incoherent interfaces, and they are believed to have excellent abilities for sinking defects due to the surprising "loading–unloading" effect which has been illustrated in the previous chapter. Also, this ability for sinking defects (sink strength) is related to grain size for the same materials. To describe the relationship between sink strength and grain size, Bullough et al. [36] used the cellular model and acquired this relationship:

$$k_{gb}^2 = 15/R^2, \tag{4}$$

where k_{gb}^2 is the sink strength for grain boundary and R is grain size. From this equation, we can find that sink strength increases with grain size decreasing.

Nanocrystalline (NC) materials have good irradiation resistance compared to their coarse-grained (CG) counterparts because of large sink strength for irradiation induced defects. For example, austenitic Fe–Cr–Ni ternary alloys fabricated by vacuum cast and subsequent hot isostatic pressing have coarse grains with a grain size of 700 μm (Figure 9a). The alloys display ultrafine grains (UFG) with a grain size of 400 nm after equal channel angular pressing (ECAP) up to eight passes at 500 °C (Figure 9b). In Figure 9c,d, when CG and UFG Fe–Cr–Ni alloys are irradiated together using 100 keV He ions to a dose of 6×10^{16} ions/cm^2, UFG Fe–Cr–Ni alloys have less bubbles compared to CG Fe–Cr–Ni alloys (Figure 9e). Also, the average bubbles size in UFG Fe–Cr–Ni alloys is 1.0 nm, which is smaller than that in CG Fe–Cr–Ni alloys (Figure 9f) [37].

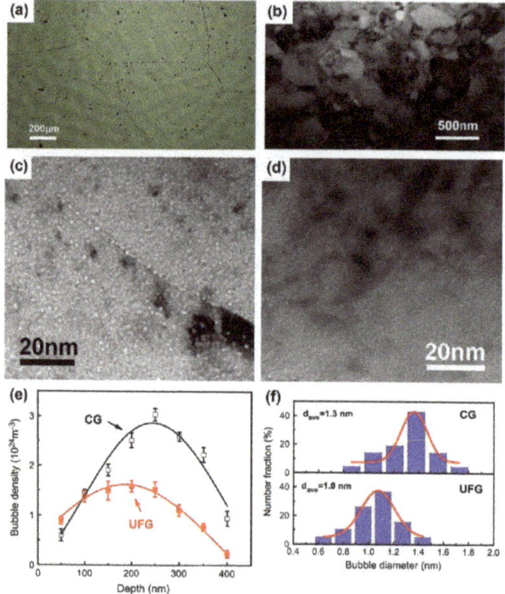

Figure 9. (**a**) An optical microscopy image of as-received coarse-grained (CG) Fe–Cr–Ni alloy. (**b**) TEM micrograph of ultrafine grain (UFG) Fe–Cr–Ni alloy after equal channel angular pressing (ECAP). (**c**,**d**) Under focused TEM micrograph of He ion irradiated CG and UFG Fe–Cr–Ni alloys. (**e**) Depth dependent bubble density in irradiated CG and UFG Fe–Cr–Ni alloy. (**f**) Bubble diameter distribution in the peak damage region of He ion irradiated CG and UFG Fe–Cr–Ni alloys [37].

In situ irradiation experiments also have been conducted in order to confirm this superior irradiation tolerance of GBs. NC Fe films fabricated by sputter deposition experienced in situ He ion irradiation with a total dose of 2.8×10^{21} ions/m^2 in TEM. Figure 10 shows TEM pictures of the samples after irradiation and the profile plotted with average void density vs. grain size (area). During this irradiation experiment, the appearance of void-denuded zones was recorded accurately in videos. Also, the smaller grain diameter resulted in less void density [38]. These results prove that GBs have superior sinking abilities and NC materials can alleviate irradiation damage effectively.

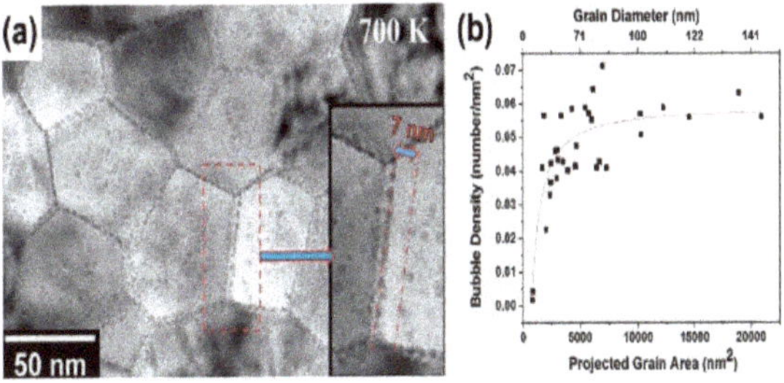

Figure 10. (**a**) Over-focused bright field TEM images of 10 keV He ion irradiation on nanocrystalline iron at calibrated temperatures of 700 K. (**b**) Areal bubble density (number/nm^2) vs. grain size (area) for 10 keV He ion irradiation on nanocrystalline iron at 700 K [38].

2.2.2. Semi-Coherent Interfaces

Most multilayers fabricated by physical vapor deposition (PVD) have semi-coherent interfaces, such as PVD Cu/Nb, Cu/V and Cu/Mo. Their interface structures are similar, for instance, PVD Cu/Nb, Cu/V and Cu/Mo have atomic straight FCC {111}//BCC {110} interfaces (where FCC is face-centered cubic and BCC is body-centered cubic). These multilayers also have superior irradiation resistance due to abundant MDIs at interfaces which can trap defects effectively and induce "platelet-to-bubble" transition during irradiation. Also, the sink strength is related to layer thickness for the same multilayer, and the relationship between sink strength and layer thickness can be described as [39]:

$$k_h^2 = 12\eta/h^2, \tag{5}$$

where k_h^2 is the layer interface sink strength, h is layer thickness and η is a constant for the multilayer. When lattice mismatch on the layer interface is large, η is also large.

There are many experimental results [7,19,40–43] which are consistent with what these theories predict. When PVD Cu/V and Cu/Mo were irradiated using He ions with a dose of 6×10^{16} ions/cm^2, the diameters of bubbles were all about 1 nm (Figure 11a,b), and the densities of bubbles all decreased with layer thickness reducing because the sink strengths increased [19,40]. Moreover, PVD Cu/Nb has the largest sink strength in these multilayers with the same layer thickness, because lattice mismatch on the Cu/Nb interface is highest according to O-lattice theory [44,45]. For PVD Cu/Nb with 100 nm layer thickness, the size of bubble was 1 nm (Figure 12a,b) when it was irradiated by He ions with a dose of 1×10^{17} ions/cm^2 [7], although this implantation dose is higher than the PVD Cu/V and Cu/Mo experienced. Also, when 2.5 nm PVD Cu/Nb experienced the same irradiation, bubbles were not observed independently of defocused conditions of TEM (Figure 12c) [7]. In other words, high density of semi-coherent interfaces, which have the superior sink strength for irradiation-induced defects, can hinder bubble formation significantly in 2.5 nm PVD Cu/Nb.

Semi-coherent interfaces also have superior irradiation resistance at high temperatures. For example, when 120 nm PVD Cu/Nb experienced He ion irradiation with a dose of 1×10^{17} ions/cm^2 at 490 °C, voids generated in the Cu layer and displayed a size range between 4.5 and 43 nm (Figure 13a). In contrast to the Cu layer, bubbles formed in the Nb layer and their diameter was 1 nm [46]. For 5 nm PVD Cu/Nb, voids in Cu layers spanned the whole Cu thickness and elongated between 7 and 11 nm, while the size of bubbles in Nb layers was still about 1 nm (Figure 13b). From these results, we can see bubbles coarsen rapidly and voids form in the Cu layer, and bubbles in the Nb layer do not change at the elevated temperatures. The reasons for this phenomenon are as follows: first, T > 0.5T$_m$ in the Cu layer, the migration of vacancies is easy, and bubbles coarsen rapidly and transform into voids by absorbing interstitials and vacancies. Second, T < 0.5 T$_m$ in the Nb layer, the migration of vacancies is not as easy as that in Cu and bubbles coarsen slowly by absorbing interstitials. Although bubbles grow rapidly and voids form inevitably in the Cu layer at high temperatures, semi-coherent interfaces can confine big voids within the layers and control the distribution of cavities. Also, the size of voids decreases with the increasing density of interfaces.

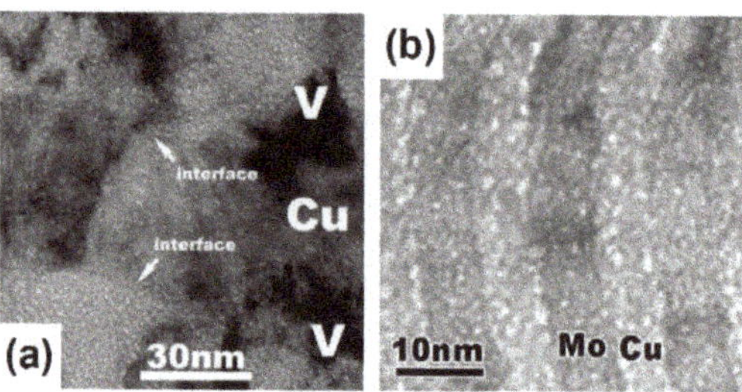

Figure 11. Under focused TEM images of He ion irradiation on multilayers with a dose of 6 × 10^{16} ions/cm^2. (**a**) Cu/V multilayers [19]; (**b**) Cu/Mo multilayers [40].

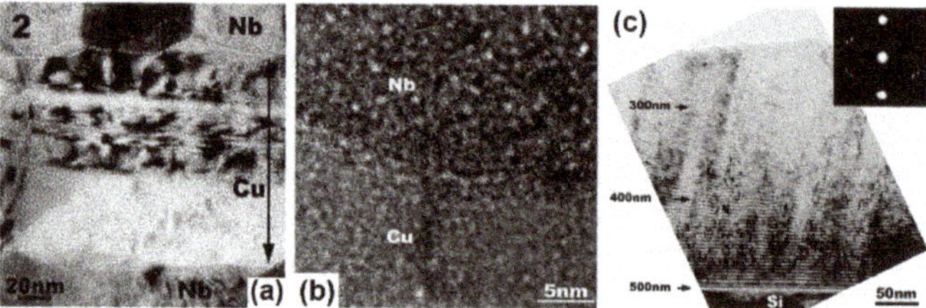

Figure 12. (**a**) Low magnification TEM micrograph of 100 nm Cu/Nb multilayer irradiated with He ions at 150 keV with a dose of 1 × 10^{17} ions/cm^2. (**b**) Defocused HRTEM micrograph of Cu/Nb interfaces indicated cavities in each constituent has 1 nm in diameter and aligned along the interface. (**c**) 2.5 nm Cu/Nb multilayers subjected to He ion irradiation at 150 keV with a dose of 1 × 10^{17} ions/cm^2 [7].

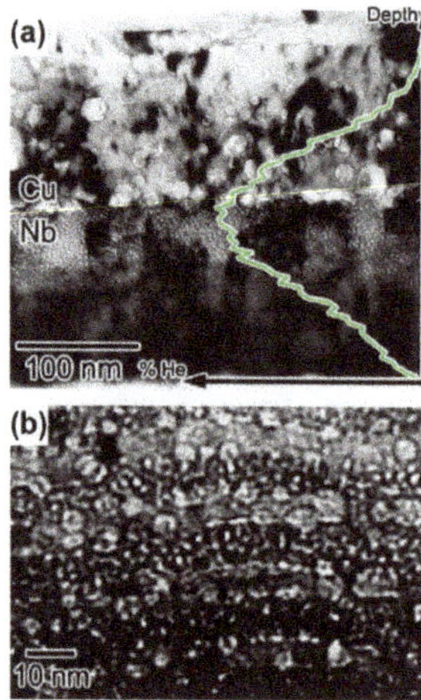

Figure 13. (**a**) Cavities in irradiated Cu/Nb multilayers with 120 nm layer thickness. (**b**) Cavities in He ion irradiated Cu/Nb multilayers with 5 nm layer thickness [46].

Besides layer interfaces in multilayers, some ferrite/precipitate interfaces (ferrite/$Y_2Ti_2O_7$ interfaces) are also semi-coherent interfaces in the nano-sized oxide dispersion strengthened (ODS) steels [47]. These interfaces can trap interstitials and vacancies efficiently and prevent abundant vacancies and He from entering ferrite. Therefore, ODS steels, which have abundant ferrite/oxide interfaces, possess excellent irradiation resistance. To explain this ability of trapping defects, Yang et al. [48] conducted structural relaxations and energetic calculations using the density functional theory (DFT) for ferrite/$Y_2Ti_2O_7$ interfaces in ODS steels. They found these interfaces have the cube-on-cube orientation relationship of {100} < 100>$_{Fe}$// {100} < 100> $_{Y2Ti2O7}$ and enrich Yi and Ti. Also, the interfacial He and vacancy formation energies are lower at the ferrite/$Y_2Ti_2O_7$ interfaces than at ferrite matrix and GBs as predicted in Reference [48].

In order to prove superior irradiation resistance of ODS steels, some irradiation experiments were also performed. For example, when 14Cr-ODS and Eurofer 97 steel experienced He ion irradiation with a dose of 1×10^{17} ions/cm^2 at 400 °C [49], the average bubble size at the peak depth was 3.6 nm in 14Cr-ODS, and 5.2 nm in Eurofer 97 (Figure 14a–c). Also, the maximum bubble volume fraction in 14Cr-ODS was 0.5%, and 1.1% in Eurofer 97 (Figure 14a,b,d). Compared to Eurofer 97, 14Cr-ODS contains abundant nano-sized $Y_2Ti_2O_7$ precipitates which are identified in Figure 11e,f. Therefore, these dominant ferrite/$Y_2Ti_2O_7$ interfaces lead to smaller bubble size and less bubble volume fraction.

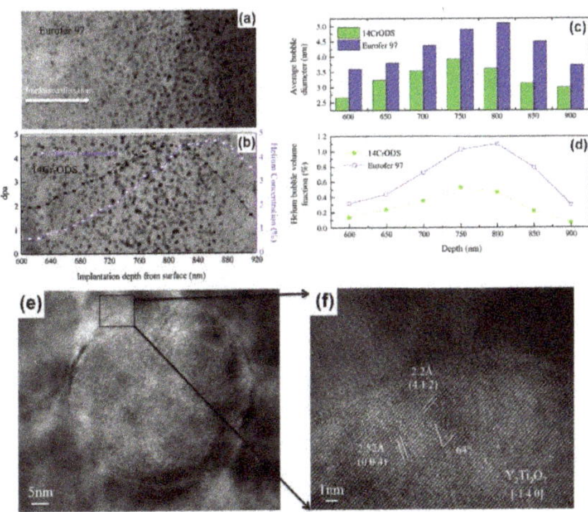

Figure 14. (**a**,**b**) High-angle annular dark field (HAADF) images of the highest helium concentration regions (600–920 nm from the surface) for Eurofer 97 and 14CrODS steel, overlaid with predicted displacement per atom (DPA) and He concentration profile; (**c**,**d**) helium bubble distribution (average bubble diameter and helium bubble volume fraction.) versus depth; (**e**) a $Y_2Ti_2O_7$ precipitate; (**f**) HRTEM of the pyrochlore structure $Y_2Ti_2O_7$ [49].

2.2.3. Coherent Interfaces

Coherent twin boundaries (CTBs) are typical fully coherent interfaces. During He ion irradiation, CTBs are believed to not possess as strong an ability for sinking irradiated defects as normal GBs, the reasons are as follows [50]: first, the excess free volume is low near CTBs; second, atomistic simulations find the vacancy and interstitial formation energies at CTBs are nearly identical to that at the interior of grains.

Nanotwined (NT) materials despite possessing high density of CTBs, however, cannot resist He ion irradiation damage efficiently. Taking NT Cu films as an example, we illustrate this weak ability of irradiation resistance for NT materials further [50]. In Figure 15, when experiencing He ion irradiation at a dose of 1×10^{17} ions/cm^2, the diameter of bubbles in NT Cu films is about 2 nm, which is larger than that in PVD Cu/Nb whose layer thickness is similar to the twin spacing. Bubbles can be observed throughout all implanted regions, even within 5 nm of the surface where He concentration is very low ($<1 \times 10^{16}$ ions/cm^2). Also, the density of bubbles in NT Cu films is similar to that in coarse-grained Cu under the same irradiation parameter.

To confirm this limited effect of CTBs on He ion irradiation resistance, some studies have compared the irradiation response of GBs and CTBs for He ion irradiation directly [51]. In their work, TEM samples of Cu which have GBs and CTBs were irradiated with a dose of 1×10^{17} ions/cm^2. Then, the distribution of cavities around GBs and CTBs was acquired. In Figure 16a, the density of cavities is lower near GBs and cavity-depleted zones (like void-denuded zones) along the GBs are readily observed. However, cavities are almost homogenously distributed across the CTBs (Figure 16b), suggesting that CTBs do not trap defects efficiently and hinder cavity formation compared to GBs. Also, some zones can be selected to evaluate the interface effects for irradiation resistance quantitatively. In Figure 16c, zone 1 and zone 2 around interfaces are chosen (GB and CTB) for counting cavities, and the statistical results are shown in Figure 16d. For GBs, the number of cavities in zone 2 is about three times of that in zone 1. However, for CTBs, the number of cavities in zone 1 and zone 2 is similar. It means that GBs can alleviate He ion irradiation resistance efficiently and lower the density of cavities locally (about two thirds) and CTBs do not have the same strong ability compared with GBs.

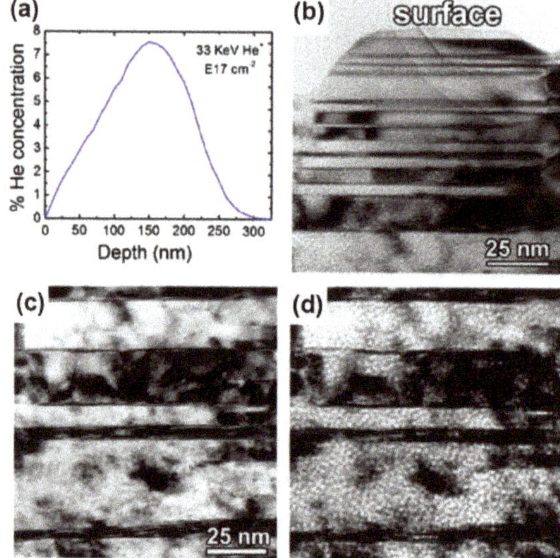

Figure 15. (**a**) Depth profile of He concentration in nanotwined (NT) Cu film implanted to a dose of 1×10^{17} ions/cm^2 with 33 keV He ions, calculated using the stopping and range of ions in matter (SRIM). (**b**) Under-focused (−2 um) bright field TEM image from near surface. (**c,d**) Near-focus and under-focused (−2 um) bright field TEM images from a depth of 150 nm (peak He concentration) [50].

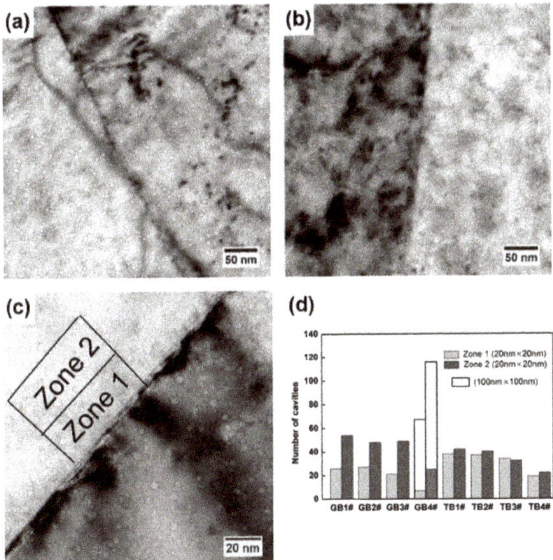

Figure 16. (**a,b**) Cavities near general a grain boundary and twin boundary in Cu after He ion irradiation. (**c**) high-magnification micrograph of the cavities near the twin boundary. (**d**) Statistical data of the cavities number adjacent to two types of the boundaries [51].

Some multilayers also have abundant coherent interfaces because lattice mismatch along their interface can be accommodated by coherency strain. Compared to twin boundaries, these heterogeneous coherent interfaces have better abilities for resisting He ion irradiation. The reasons can be summarized as follows: first, coherent interfaces have coherent stress which can promote vacancies and interstitials migration to interfaces [52]; second, coherent interfaces are interrupted by irradiation induced defects and generate disconnections which can absorb defects further; third, once bubbles nucleate on the interfaces, coherency stress can hinder bubble growth.

To prove the good sink ability for heterogeneous coherent interfaces, PVD Cu/Fe with 0.75 nm layer thickness is taken for an example [20]. PVD Cu/Fe with 0.75 nm has numerous Cu {111}//Fe {110} interfaces according to X-ray diffraction (XRD) analysis. After irradiation using 100 keV He ions with a dose of 6×10^{16} ions/cm^2 at room temperature, the diameter of bubbles is about 1 nm (Figure 17a), which is comparable to that in PVD Cu/V and Cu/Mo experienced the same He ion irradiation. Also, although these Cu/Fe interfaces are coherent, cavities are still confined by interfaces (Figure 17b) and their diameter is curtailed with layer thickness reducing (Figure 17c). In conclusion, nanostructured materials with heterogeneous coherent interfaces can also resist irradiation damage efficiently.

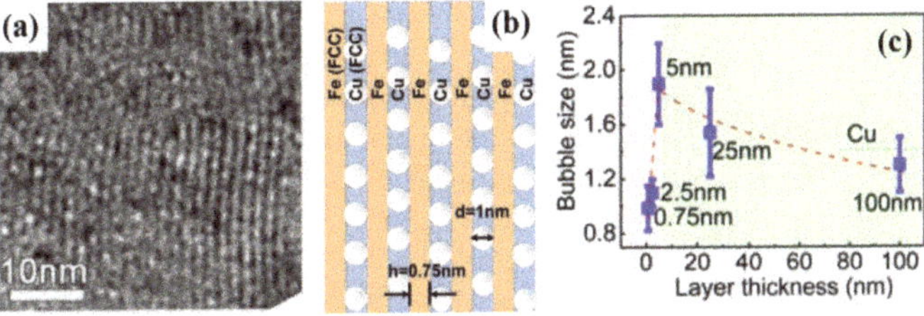

Figure 17. (**a**) TEM images of irradiated fully coherent 0.75 nm Cu/Fe multilayer. (**b**) Schematics illustrate that in a coherent 0.75 nm Cu/Fe multilayer, He bubbles prefer to nucleate in Cu layers and are constricted to reside inside Cu layers. (**c**) The variation of He bubble density with layer thickness h [20].

3. The Evolution of Mechanical Properties of Irradiated Nanostructured Materials

After irradiation, the mechanical properties of nanostructured materials change due to the formation of cavities. However, these changes can be tuned by interfaces. Next, taking hardening and softening behaviors as examples, we will discuss these interface-related changes of properties in detail.

3.1. Hardening Behavior: Small Cavities

Compared to their large-sized counterparts (where size represents grain size in NC materials, layer thickness in nanolayered composites and so on), smaller cavities can form in nanostructured materials after the same He ion irradiation at room temperature because of interface effects. These small cavities, like nanoscale precipitated phases, can block dislocation motion severely, thus contributing to hardening in nanostructured materials. Also, irradiation dose and the size of the nanostructured materials, which determine the density of small cavities, can influence the hardening rate. These influences have been investigated widely by hardness measurement before and after irradiation [22,43,53–56]. As shown in Figure 18, we can see hardening is directly proportional to irradiation dose for PVD Cu/V [53]. Also, hardening becomes weak with layer thickness decreasing for PVD Ag/V, Ag/Ni and Cu/V [43,54,55], and it becomes strong with layer thickness decreasing for PVD Cu/Co [56]. In other words, there are two types of size effects on irradiation-induced hardening in nanostructured materials. Next, taking PVD Ag/V and Cu/Co as examples, we explain these size effects.

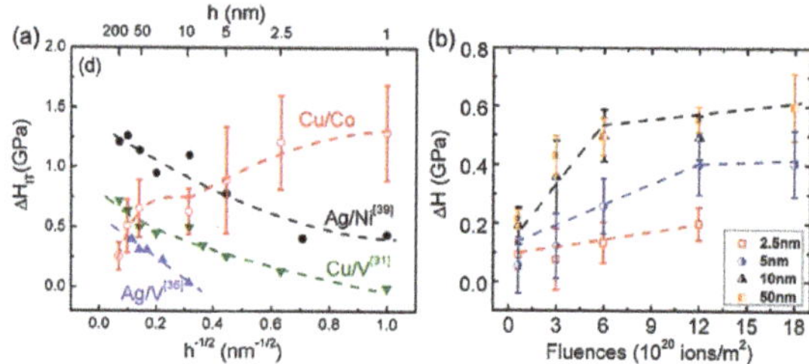

Figure 18. (**a**) Irradiation hardening in He ions irradiated Ag/V, Ag/Ni, Cu/V multilayers scaling with layer thickness. (**b**) Fluence dependence of irradiation hardening in He ion irradiated Cu/V multilayers [39].

In PVD Ag/V, the relationship between hardening ΔH and layer thickness h can be described by [43]:

$$\Delta H = 9\tau_i(1 - \frac{l}{\sqrt{2}h}),\qquad(6)$$

where τ_i is the average shear strength of bubbles, and l is the average obstacle spacing.

Also, τ_i which is related to interface can be acquired by the Orowan model and l can be gauged by TEM images. When h > 100 nm, the dislocation pile-up mechanism [57] dominates in as-deposited multilayers. At this length scale, multilayers exhibit some mechanical behavior like bulk. Therefore, h is infinite and $\Delta H \approx 9\tau_i$. When 5 nm < h < 100 nm, the confined layer slip model [58] dominates in as-deposited multilayers, and cavities appear near interfaces and inside layers which contribute to hardening. At this length scale, ΔH can be directly calculated by the equation above (Figure 19a). When h < 5 nm, the interface crossing mechanism [59] dominates in as-deposited multilayers. At this length scale, a few cavities form because the interfaces of high density can trap and annihilate numerous irradiation-induced point defects (Figure 19b), and these cavities interfere with the crossing dislocations. Under this condition l > h, ΔH is small and even negligible sometimes.

In PVD Cu/Co, the hardening increases with reducing layer thickness. This abnormal size effect is because interface structures change with decreasing layer thickness. The change of interface structure influences the strengthening mechanism for PVD Cu/Co. In order to explain this size effect in detail, we divide layer thickness into two sections.

When h < 10 nm, Cu/Co interfaces are coherent and the hardness of as-deposited multilayers is dominated by the interface barrier strength to transmission of partial dislocations. After radiation, cavities at the layer interfaces are typically over-pressurized and strong obstacles [19,54,55], thus these partials have to constrict to full dislocations (in Cu layers) before transmitting to the adjacent Co layers (Figure 20). In other words, cavities can strengthen interfaces and block dislocation transmission. At this length scale, cavities-induced hardening is predominant. When 50 nm < h < 200 nm, Cu/Co interfaces are incoherent and the hardness of as-deposited multilayers is determined primarily by high-density stacking faults (SFs) in Co. After irradiation, cavities distribute both along interfaces and within the layers. However, the average cavity spacing *l* for 100 nm PVD Cu/Co is about 9 nm, which is much larger than the average spacing between SFs (Figure 21). At this length scale, cavity-induced hardening is negligible compared to high density of SFs in multilayers.

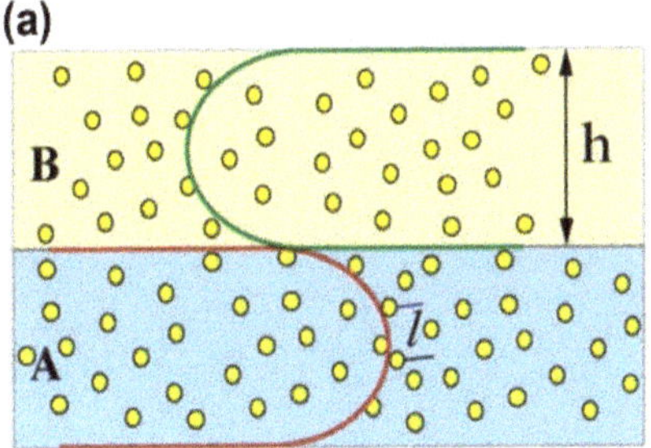

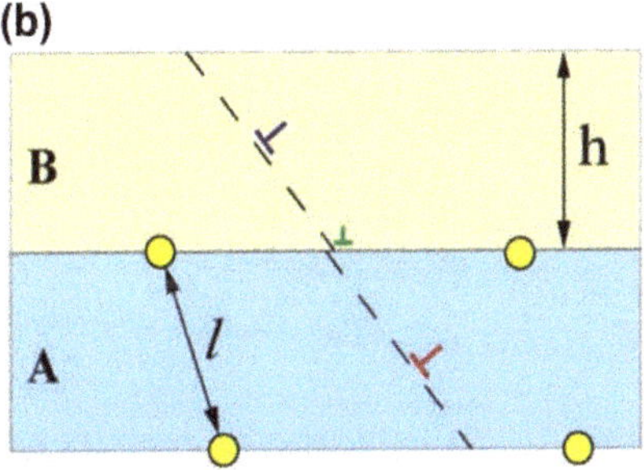

Figure 19. Schematic illustration of the bubble distribution in multilayers (the circles indicate bubbles). (**a**) When the layer thickness is a few tens of nanometers, h > *l* and the deformation is via confined layer slip. (**b**) When the layer thickness is of the order of a few nanometers, h < *l*, the yield strength is determined by the crossing of single dislocations across interfaces [43].

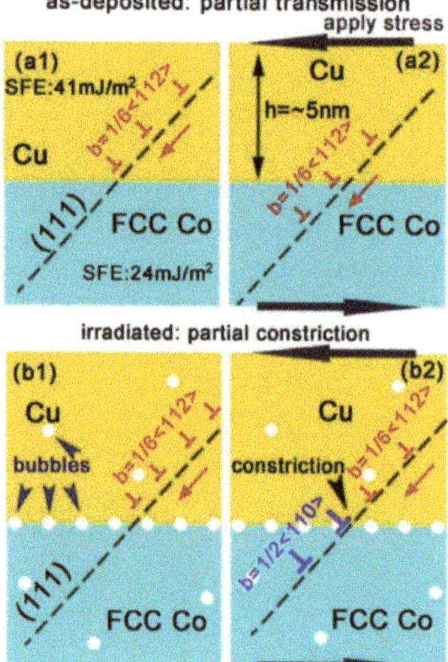

Figure 20. Hypothetical schematics compare strengthening mechanisms in as-deposited and irradiated Cu/Co (100) multilayers at small h (h = 5 nm). (**a1, a2**) In as-deposited films, partials can trespass layer interfaces owing to the low stacking fault energy of Cu and Co. (**b1, b2**) However, after radiation, bubbles at the layer interface disrupt the transmission of partials [56].

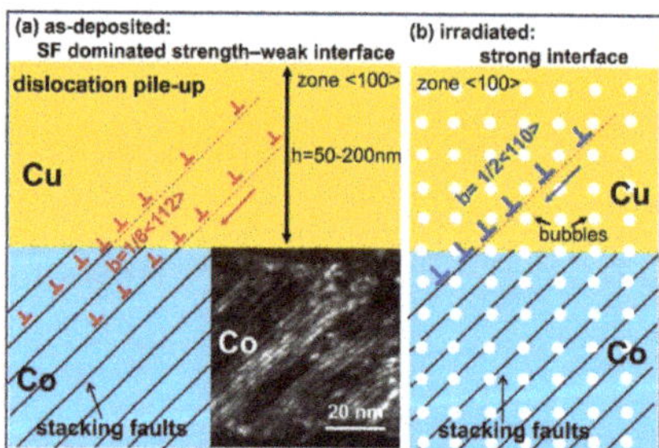

Figure 21. Hypothetical schematics illustrate different strengthening mechanisms in as-deposited and irradiated Cu/Co (100) multilayers at large h (h = 50–200 nm). (**a**) In the as-deposited state, partial dislocations can transmit across the Cu/Co interface relatively easily; (**b**) After radiation, high-density He bubbles are distributed both along the layer interface and within the layers [56].

3.2. Softening Behavior: Large Cavities

When nanostructured materials are irradiated at high temperatures (more than half of the melting point of constituent), cavities can grow rapidly by absorbing abundant interstitials and vacancies. In this case, softening would occur in nanostructured materials after irradiation. For example, hardness of 58 nm ARB Cu/Nb decreased from 3.8 GPa to 3.2 GPa [22] after He ion irradiation at 450 °C with a dose of 2×10^{17} ions/cm². Also, when NC W experienced He ion irradiation at a dose of 3.6×10^{17} ions/cm² at 950 °C, its hardness decreased severely from 8.5 GPa to 6.8 GPa [60]. According to TEM images for these irradiated samples (Figure 22), the softening behavior can be explained as follows: first, the size of cavities is large. These cavities are weak obstacles to dislocation and even are sources of dislocations owing to their large surface; second, the large cavities that cross interfaces can destroy interface structures, thus reducing the interface barrier strength to the transmission of dislocations, leading to softening.

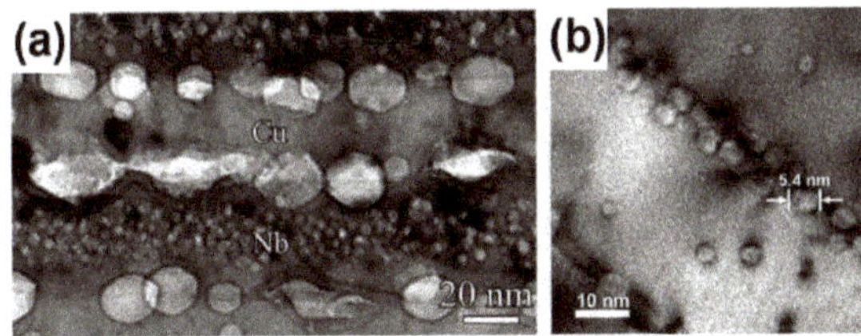

Figure 22. (**a**) Cavities in irradiated Cu/Nb multilayers with 58 nm layer thickness [22]. (**b**) Cavities in irradiated nanocrystalline W [60].

4. Summary and Outlook

He ion irradiation can induce numerous defects (interstitials and vacancies), and form He bubbles. These bubbles can induce swelling, hardening, embrittlement and surface deterioration, thus degrading mechanical properties substantially.

Nanostructured materials have excellent irradiation responses to He ion irradiation due to high density of interfaces. These interfaces can tune cavities, such as making cavities stay away from interfaces, adhere to interfaces and cross interfaces.

Different interfaces, such as GBs, oxide precipitate interfaces, heterogeneous layer interfaces and TBs, may have different sink strengths for He ion irradiated defects. The He ion irradiated results, like size and density of cavities, can be reduced effectively in nanostructured materials. As for multilayers with semi-coherent interfaces, different parameters, such as lattice mismatch on the interface, layer thickness, irradiation dose and temperature, can influence these results. The specific relationships between these factors and results are concluded in Table 1. Also, the formation of cavities can be suppressed completely. For example, when Cu/Nb multilayers with 2.5 nm layer thickness experienced He ion irradiation with a dose of 1×10^{17} ions/cm^2 at room temperature, no cavities formed.

Table 1. The relationships for multilayers between lattice mismatch, layer thickness, irradiation dose, temperature and cavity size, cavity density.

	Lattice Mismatch	Layer Thickness	Irradiation Dose	Temperature
Cavity size	−	+	+	+
Cavity density	−	+	+	−

"+" means positive relationship, for example, cavity size decreases with layer thickness reducing. "−" means negative relationship, for example, cavity density decreases with lattice mismatch increasing.

Moreover, after irradiation, both hardening and softening can occur depending on the cavity size. As for multilayers, there are opposite relationships between hardening and layer thickness due to different interface structures.

Although the interface effects on He ion irradiation have been studied extensively recently, there are still numerous subjects that require in-depth exploration.

First, studies on evolution of mechanical properties of irradiated nanostructured materials, such as tensile strength, ductility and so on, are needed. These studies can build relationships between interfaces and application-related properties for nanostructured materials.

Second, more in situ He ion irradiation experiments should be performed for understanding the interactions between interfaces and defects better for different nanostructured materials. Such as in situ TEM study on He ion irradiation can supply direct evidence for defect/cavity formation, migration and interaction with interfaces.

Third, besides interstitials, vacancies and cavities, the interface effects on other defects, like dislocation loops, should also be investigated in-depth.

Author Contributions: W.Y. wrote the manuscript; J.P. searched the stuff for the manuscript, discussed the details; S.Z. organized the structure of the manuscript; J.W. worked on the section of mechanical evolution induced by He irradiation; X.Z. worked on the section of interface effects on irradiation damage; X.M. discussed the structure of the manuscript; J.P., S.Z., J.W., X.Z. and X.M. revised the manuscript.

Funding: This research was funded by the National Natural Science Foundation of China (No. 51771201) and the Key Project of Natural Science Foundation of Liaoning Province, China (No. 20180510059).

Acknowledgments: The authors thank Amit Misra at Department of Materials Science and Engineering, College of Engineering, University of Michigan-Ann Arbor, Ann Arbor, Nathan Mara at Chemical Engineering and Materials Science, University of Minnesota-Twin Cities, Minneapolis for helpful discussions.

Conflicts of Interest: The authors declare no conflict of interest.

References

1. Zinkle, S.J.; Was, G.S. Materials challenges in nuclear energy. *Acta Mater.* **2013**, *61*, 735–758. [CrossRef]
2. Ullmaier, H. The influence of helium on the bulk properties of fusion reactor structural materials. *Nucl. Fusion* **1984**, *24*, 1039–1083. [CrossRef]
3. Zinkle, S.J.; Wolfer, W.G.; Kulcinski, G.L.; Seitzman, L.E. Effect of oxygen and helium on void formation in metals. *Philos. Mag. A* **1987**, *55*, 127–140. [CrossRef]
4. Farrell, K. Experimental effects of helium on cavity formation during irradiation—A review. *Radiat. Eff.* **1980**, *53*, 175–194. [CrossRef]
5. Braski, D.N.; Schroeder, H.; Ullmaier, H. The effect of tensile stress on the growth of helium bubbles in an austenitic stainless steel. *J. Nucl. Mater.* **1979**, *83*, 265–277. [CrossRef]
6. Schroeder, H.; Kesternich, W.; Ullmaier, H. Helium effects on the creep and fatigue resistance of austenitic stainless steels at high temperatures. *Nucl. Eng. Des. Fusion* **1985**, *2*, 65–95. [CrossRef]
7. Zhang, X.; Li, N.; Anderoglu, O.; Wang, H.; Swadener, J.G.; Höchbauer, T.; Misra, A.; Hoagland, R.G. Nanostructured Cu/Nb multilayers subjected to helium ion-irradiation. *Nucl. Instrum. Methods Phys. Res. Sect. B Beam Interact. Mater. Atoms* **2007**, *261*, 1129–1132. [CrossRef]
8. Trinkaus, H.; Singh, B.N. Helium accumulation in metals during irradiation—Where do we stand? *J. Nucl. Mater.* **2003**, *323*, 229–242. [CrossRef]
9. Ghoniem, N.M.; Sharafat, S.; Williams, J.M.; Mansur, L.K. Theory of helium transport and clustering in materials under irradiation. *J. Nucl. Mater.* **1983**, *117*, 96–105. [CrossRef]
10. Demkowicz, M.J.; Misra, A.; Caro, A. The role of interface structure in controlling high helium concentrations. *Curr. Opin. Solid State Mater. Sci.* **2012**, *16*, 101–108. [CrossRef]
11. Stoller, R.E.; Odette, G.R. Analytical solutions for helium bubble and critical radius parameters using a hard sphere equation of state. *J. Nucl. Mater.* **1985**, *131*, 118–125. [CrossRef]
12. Gruber, E.E. Calculated Size Distributions for Gas Bubble Migration and Coalescence in Solids. *J. Appl. Phys.* **1967**, *38*, 243–250. [CrossRef]
13. Greenwood, G.W.; Boltax, A. The role of fission gas re-solution during post-irradiation heat treatment. *J. Nucl. Mater.* **1962**, *5*, 234–240. [CrossRef]
14. Tyler, S.K.; Goodhew, P.J. The growth of helium bubbles in niobium and Nb-1% Zr. *J. Nucl. Mater.* **1978**, *74*, 27–33. [CrossRef]
15. Tyler, S.K.; Goodhew, P.J. Direct evidence for the Brownian motion of helium bubbles. *J. Nucl. Mater.* **1980**, *92*, 201–206. [CrossRef]
16. Evans, J.H.; van Veen, A.; Finnis, M.W. Observations and theory of thermally-induced helium bubble shrinkage in gold. *J. Nucl. Mater.* **1989**, *168*, 19–23. [CrossRef]
17. Beere, W. The growth of sub-critical bubbles on grain boundaries. *J. Nucl. Mater.* **1984**, *120*, 88–93. [CrossRef]
18. Yang, L.; Ge, H.; Zhang, J.; Xiong, T.; Jin, Q.; Zhou, Y.; Shao, X.; Zhang, B.; Zhu, Z.; Zheng, S.; et al. High He-ion irradiation resistance of CrMnFeCoNi high-entropy alloy revealed by comparison study with Ni and 304SS. *J. Mater. Sci. Technol.* **2018**, *35*, 300–305. [CrossRef]
19. Fu, E.G.; Carter, J.; Swadener, G.; Misra, A.; Shao, L.; Wang, H.; Zhang, X. Size dependent enhancement of helium ion irradiation tolerance in sputtered Cu/V nanolaminates. *J. Nucl. Mater.* **2009**, *385*, 629–632. [CrossRef]

20. Chen, Y.; Fu, E.; Yu, K.; Song, M.; Liu, Y.; Wang, Y.; Wang, H.; Zhang, X. Enhanced radiation tolerance in immiscible Cu/Fe multilayers with coherent and incoherent layer interfaces. *J. Mater. Res.* **2015**, *30*, 1300–1309. [CrossRef]

21. Li, N.; Mara, N.A.; Wang, Y.Q.; Nastasi, M.; Misra, A. Compressive flow behavior of Cu thin films and Cu/Nb multilayers containing nanometer-scale helium bubbles. *Scr. Mater.* **2011**, *64*, 974–977. [CrossRef]

22. Yang, L.X.; Zheng, S.J.; Zhou, Y.T.; Zhang, J.; Wang, Y.Q.; Jiang, C.B.; Mara, N.A.; Beyerlein, I.J.; Ma, X.L. Effects of He radiation on cavity distribution and hardness of bulk nanolayered Cu-Nb composites. *J. Nucl. Mater.* **2017**, *487*, 311–316. [CrossRef]

23. Wakai, E.; Ezawa, T.; Imamura, J.; Takenaka, T.; Tanabe, T.; Oshima, R. Effect of solute atoms on swelling in Ni alloys and pure Ni under He ion irradiation. *J. Nucl. Mater.* **2002**, *307*, 367–373. [CrossRef]

24. Chernikov, V.N.; Zakharov, A.P.; Kazansky, P.R. Relation between swelling and embrittlement during post-irradiation annealing and instability of helium-vacancy complexes in nickel. *J. Nucl. Mater.* **1988**, *155*, 1142–1145. [CrossRef]

25. Shinohara, K.; Nakamura, Y.; Kitajima, S.; Kutsuwada, M.; Kaneko, M. Low temperature embrittlement in He-implanted iron polycrystals. *J. Nucl. Mater.* **1988**, *155*, 1154–1158. [CrossRef]

26. Yoshida, N.; Iwakiri, H.; Tokunaga, K.; Baba, T. Impact of low energy helium irradiation on plasma facing metals. *J. Nucl. Mater.* **2005**, *337*, 946–950. [CrossRef]

27. Bai, X.M.; Voter, A.F.; Hoagland, R.G.; Nastasi, M.; Uberuaga, B.P. Efficient Annealing of Radiation Damage Near Grain Boundaries via Interstitial Emission. *Science* **2010**, *327*, 1631. [CrossRef]

28. Beyerlein, I.J.; Caro, A.; Demkowicz, M.J.; Mara, N.A.; Misra, A.; Uberuaga, B.P. Radiation damage tolerant nanomaterials. *Mater. Today* **2013**, *16*, 443–449. [CrossRef]

29. Kashinath, A.; Misra, A.; Demkowicz, M.J. Stable storage of helium in nanoscale platelets at semicoherent interfaces. *Phys. Rev. Lett.* **2013**, *110*, 086101. [CrossRef]

30. Han, W.Z.; Demkowicz, M.J.; Fu, E.G.; Wang, Y.Q.; Misra, A. Effect of grain boundary character on sink efficiency. *Acta Mater.* **2012**, *60*, 6341–6351. [CrossRef]

31. Han, W.; Demkowicz, M.J.; Mara, N.A.; Fu, E.; Sinha, S.; Rollett, A.D.; Wang, Y.; Carpenter, J.S.; Beyerlein, I.J.; Misra, A. Design of radiation tolerant materials via interface engineering. *Adv. Mater.* **2013**, *25*, 6975–6979. [CrossRef]

32. Wang, M.; Beyerlein, I.J.; Zhang, J.; Han, W.Z. Defect-interface interactions in irradiated Cu/Ag nanocomposites. *Acta Mater.* **2018**, *160*, 211–223. [CrossRef]

33. Zheng, S.; Shao, S.; Zhang, J.; Wang, Y.; Demkowicz, M.J.; Beyerlein, I.J.; Mara, N.A. Adhesion of voids to bimetal interfaces with non-uniform energies. *Sci. Rep.* **2015**, *5*, 15428. [CrossRef]

34. Medyanik, S.N.; Shao, S. Strengthening effects of coherent interfaces in nanoscale metallic bilayers. *Comput. Mater. Sci* **2009**, *45*, 1129–1133. [CrossRef]

35. Zherebtsov, S.; Salishchev, G.; Lee Semiatin, S. Loss of coherency of the alpha/beta interface boundary in titanium alloys during deformation. *Philos. Mag. Lett.* **2010**, *90*, 903–914. [CrossRef]

36. Bullough, R.; Hayns, M.R.; Wood, M.H. Sink strengths for thin film surfaces and grain boundaries. *J. Nucl. Mater.* **1980**, *90*, 44–59. [CrossRef]

37. Sun, C.; Yu, K.Y.; Lee, J.H.; Liu, Y.; Wang, H.; Shao, L.; Maloy, S.A.; Hartwig, K.T.; Zhang, X. Enhanced radiation tolerance of ultrafine grained Fe–Cr–Ni alloy. *J. Nucl. Mater.* **2012**, *420*, 235–240. [CrossRef]

38. El-Atwani, O.; Nathaniel, J.E.; Leff, A.C.; Baldwin, J.K.; Hattar, K.; Taheri, M.L. Evidence of a temperature transition for denuded zone formation in nanocrystalline Fe under He irradiation. *Mater. Res. Lett.* **2017**, *5*, 195–200. [CrossRef]

39. Zhang, X.; Hattar, K.; Chen, Y.; Shao, L.; Li, J.; Sun, C.; Yu, K.; Li, N.; Taheri, M.L.; Wang, H.; et al. Radiation damage in nanostructured materials. *Prog. Mater. Sci.* **2018**, *96*, 217–321. [CrossRef]

40. Li, N.; Carter, J.J.; Misra, A.; Shao, L.; Wang, H.; Zhang, X. The influence of interfaces on the formation of bubbles in He-ion-irradiated Cu/Mo nanolayers. *Philos. Mag. Lett.* **2011**, *91*, 18–28. [CrossRef]

41. Li, N.; Nastasi, M.; Misra, A. Defect structures and hardening mechanisms in high dose helium ion implanted Cu and Cu/Nb multilayer thin films. *Int. J. Plast.* **2012**, *32*, 1–16. [CrossRef]

42. Zhang, X.; Fu, E.G.; Misra, A.; Demkowicz, M.J. Interface-enabled defect reduction in He ion irradiated metallic multilayers. *Jom* **2010**, *62*, 75–78. [CrossRef]

43. Wei, Q.M.; Li, N.; Mara, N.; Nastasi, M.; Misra, A. Suppression of irradiation hardening in nanoscale V/Ag multilayers. *Acta Mater.* **2011**, *59*, 6331–6340. [CrossRef]

44. Bollmann, W. O-Lattice calculation of an FCC–BCC interface. *Phys. Status Solidi A* **1974**, *21*, 543–550. [CrossRef]

45. Beyerlein, I.J.; Demkowicz, M.J.; Misra, A.; Uberuaga, B.P. Defect-interface interactions. *Prog. Mater. Sci.* **2015**, *74*, 125–210. [CrossRef]

46. Hattar, K.; Demkowicz, M.J.; Misra, A.; Robertson, I.M.; Hoagland, R.G. Arrest of He bubble growth in Cu–Nb multilayer nanocomposites. *Scr. Mater.* **2008**, *58*, 541–544. [CrossRef]

47. Ribis, J.; De Carlan, Y. Interfacial strained structure and orientation relationships of the nanosized oxide particles deduced from elasticity-driven morphology in oxide dispersion strengthened materials. *Acta Mater.* **2012**, *60*, 238–252. [CrossRef]

48. Yang, L.; Jiang, Y.; Wu, Y.; Odette, G.R.; Zhou, Z.; Lu, Z. The ferrite/oxide interface and helium management in nano-structured ferritic alloys from the first principles. *Acta Mater.* **2016**, *103*, 474–482. [CrossRef]

49. Lu, C.; Lu, Z.; Xie, R.; Liu, C.; Wang, L. Microstructure of a 14Cr-ODS ferritic steel before and after helium ion implantation. *J. Nucl. Mater.* **2014**, *455*, 366–370. [CrossRef]

50. Demkowicz, M.J.; Anderoglu, O.; Zhang, X.; Misra, A. The influence of Σ3 twin boundaries on the formation of radiation-induced defect clusters in nanotwinned Cu. *J. Mater. Res.* **2011**, *26*, 1666–1675. [CrossRef]

51. Gao, J.; Liu, Z.J.; Wan, F.R. Limited Effect of Twin Boundaries on Radiation Damage. *Acta Metall. Sin.* **2016**, *29*, 72–78. [CrossRef]

52. Vattré, A.; Jourdan, T.; Ding, H.; Marinica, M.C.; Demkowicz, M.J. Non-random walk diffusion enhances the sink strength of semicoherent interfaces. *Nat. Commun.* **2016**, *7*, 10424. [CrossRef]

53. Fu, E.G.; Wang, H.; Carter, J.; Shao, L.; Wang, Y.Q.; Zhang, X. Fluence-dependent radiation damage in helium (He) ion-irradiated Cu/V multilayers. *Philos. Mag.* **2013**, *93*, 883–898. [CrossRef]

54. Yu, K.Y.; Liu, Y.; Fu, E.G.; Wang, Y.Q.; Myers, M.T.; Wang, H.; Shao, L.; Zhang, X. Comparisons of radiation damage in He ion and proton irradiated immiscible Ag/Ni nanolayers. *J. Nucl. Mater.* **2013**, *440*, 310–318. [CrossRef]

55. Fu, E.G.; Misra, A.; Wang, H.; Shao, L.; Zhang, X. Interface enabled defects reduction in helium ion irradiated Cu/V nanolayers. *J. Nucl. Mater.* **2010**, *407*, 178–188. [CrossRef]

56. Chen, Y.; Liu, Y.; Fu, E.G.; Sun, C.; Yu, K.Y.; Song, M.; Li, J.; Wang, Y.Q.; Wang, H.; Zhang, X. Unusual size-dependent strengthening mechanisms in helium ion-irradiated immiscible coherent Cu/Co nanolayers. *Acta Mater.* **2015**, *84*, 393–404. [CrossRef]

57. Anderson, P.M.; Li, C. Hall-Petch relations for multilayered materials. *Nanostruct. Mater.* **1995**, *5*, 349–362. [CrossRef]

58. Misra, A.; Hirth, J.P.; Kung, H. Single-dislocation-based strengthening mechanisms in nanoscale metallic multilayers. *Philos. Mag. A* **2002**, *82*, 2935–2951. [CrossRef]

59. Misra, A.; Hirth, J.P.; Hoagland, R.G. Length-scale-dependent deformation mechanisms in incoherent metallic multilayered composites. *Acta Mater.* **2005**, *53*, 4817–4824. [CrossRef]

60. Cunningham, W.S.; Gentile, J.M.; El-Atwani, O.; Taylor, C.N.; Efe, M.; Maloy, S.A.; Trelewicz, J.R. Softening due to Grain Boundary Cavity Formation and its Competition with Hardening in Helium Implanted Nanocrystalline Tungsten. *Sci. Rep.* **2018**, *8*, 2897. [CrossRef]

Letter

Comparison of Vacancy Sink Efficiency of Cu/V and Cu/Nb Interfaces by the Shared Cu Layer

Huaqiang Chen [1], Jinlong Du [1], Yanxia Liang [1], Peipei Wang [1], Jinchi Huang [2], Jian Zhang [2,*], Yunbiao Zhao [1], Xingjun Wang [3,*], Xianfeng Zhang [4], Yuehui Wang [4], George A. Stanciu [5] and Engang Fu [1,*]

[1] State Key Laboratory of Nuclear Physics and Technology, School of Physics, Peking University, Beijing 100871, China
[2] Institute for Advanced Nuclear Energy, College of Energy, Xiamen University, Xiamen 361102, China
[3] State Key Laboratory of Advanced Optical Communication Systems and Networks, Peking University, Beijing 100871, China
[4] Zhongshan Institute, University of Electronic Science and Technology, Zhongshan 528400, China
[5] Center for Microcopy-Microanalysis and Information Processing, University Politehnica of Bucharest, 313 Splaiul Independentei, 060042 Bucharest, Romania
* Correspondence: zhangjian@xmu.edu.cn (J.Z.); xjwang@pku.edu.cn (X.W.); efu@pku.edu.cn (E.F.)

Received: 18 June 2019; Accepted: 8 August 2019; Published: 18 August 2019

Abstract: This paper provides a new method to compare and then reveal the vacancy sink efficiencies quantitively between different hetero-interfaces with a shared Cu layer in one sample, in contrast to previous studies, which have compared the vacancy sink efficiencies of interfaces in different samples. Cu-Nb-Cu-V nanoscale metallic multilayer composites (NMMCs) containing Cu/V and Cu/Nb interfaces periodically were prepared as research samples and bombarded with helium ions to create vacancies which were filled by helium bubbles. A special Cu layer shared by adjoining Cu/V and Cu/Nb interfaces exists, in which the implanted helium concentration reaches its maximum and remains nearly constant with a well-designed incident energy. The results show that bubble-denuded zones (BDZ) close to interfaces exist, and that the width of the BDZ close to the Cu/V interface is less than that of Cu/Nb interface. This result is explained by one-dimensional diffusion theory, and the ratio of vacancy sink efficiency between Cu/V and Cu/Nb interfaces is calculated. Conclusively, Cu/Nb interfaces are more efficient than Cu/V interfaces in eliminating vacancies induced by radiation.

Keywords: radiation damage; sink efficiency; multilayer composite; interface; vacancy

1. Introduction

Nanoscale metallic multilayer composites (NMMCs) have been widely studied due to their ultra-high strengths and enhanced radiation damage tolerance. During the last decade, a number of NMMCs were fabricated and researched, including Cu/Nb [1–3], Cu/V [4–9], Cu/Co [10], Cu/W [11], Ag/Ni [12] Al/Nb [13] etc. Hetero-interfaces, which separate adjacent layers with different structures or chemistry elements in these NMMCs, are found to act as sinks for point defects [14–16] during ion irradiation. In order to characterize the ability of a specific interface in absorbing point defects, the quantity called "sink efficiency" was firstly proposed by Sutton and Balluffi [17]. Experimental and computational studies suggest that sink efficiency can depend strongly on the interface crystallography and chemistry [18]. Cu/Nb and Cu/V NMMCs are usually chosen as the research targets to investigate how hetero-interfaces eliminate radiation-induced point defects efficiently. However, few experiments [19] provide a quantitative measurement of sink efficiency up to now, in which Shimin Mao et al. successfully compared the sink efficiencies of three kinds of interfaces, Cu/Nb, Cu/V, and Cu/Ni, based on the effects of different interfaces on radiation-enhanced diffusion. In this paper, a new method, based on observing the distribution of radiation-induced defect clusters, is proposed to conveniently and visually compare and then reveal the sink efficiencies of different interfaces in the same sample.

Cu/Nb and Cu/V NMMCs have two things in common. The first is that the Kurdjumov-Sachs (K-S) orientation relationship [20] exists in both Cu/Nb and Cu/V interfaces, i.e., FCC (face centered cubic) (111) // BCC (body centered cubic) (110) // interface, and FCC <1 1 0> // BCC <1 1 1>. The other is the commonly used Cu element. Thus, we propose to take Cu as the intermediate layer connecting the Cu/V interface and Cu/Nb interface. If the two interfaces behave in the same way, the defects in Cu layer are expected to be distributed uniformly. However, interfaces play a big role in the distribution of defect clusters in their vicinity zones [20,21]. By means of observing the distribution of radiation-induced defect clusters in this intermediate Cu layer, we attempt to investigate the difference between Cu/V and Cu/Nb interfaces, and to determine the extent of the difference.

2. Experiments

Cu-Nb-Cu-V NMMCs were prepared using the DC magnetron sputtering technique at room temperature on Si (100) substrates with individual layer thickness of 48 nm and total thickness of 576 nm (referred to as Cu-Nb-Cu-V 48 nm thereafter). In the deposition process, a base pressure of 4×10^{-5} Pa was reached prior to deposition and argon partial pressure during sputtering was kept at 0.5 Pa. The deposition rate was approximately 0.2 nm/s.

The Stopping and Range of Ions in Matter (SRIM) [22] computer program developed by J. Ziegler and based on the Monte Carlo method was used to calculate the depth profile of He concentration in Cu-Nb-Cu-V NMMCs irradiated by He ions at an energy of 119 keV and a total dose of 6×10^{16} ions/cm^2. Reasonably, the target model in SRIM simulation for Cu-Nb-Cu-V 48 nm NMMCs is the multilayered model, as Figure 1 shows, exactly corresponding to the layered morphology of which the multilayered composite consists. This model is used in the SRIM simulations of multilayered composites, differently from the compound model used in previous studies [1–13] in which multilayered composites are regarded as compounds. Compared to the compound model, the multilayered model highlights the presence of every individual layer with a corresponding chemical element. Therefore, the profile of the He concentration in every individual layer can be obtained. The threshold displacement energies of 30 eV for Cu, 40 eV for V and 60 eV for Nb [23] are chosen in the SRIM simulation.

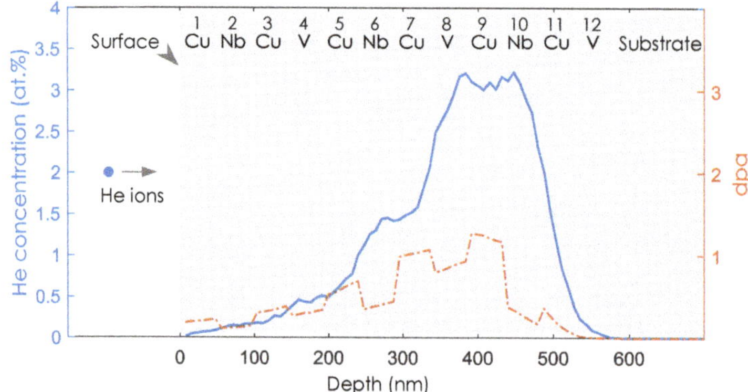

Figure 1. The depth distribution of helium concentration and dpa distribution in Cu-Nb-Cu-V 48 nm NMMCs simulated by SRIM. The helium ions are injected perpendicularly to the surface layer which is made of Cu and numbered as the 1st layer. The NMMCs are repeated of Cu-Nb-Cu-V periodically and the bottom layer is numbered as the 12th layer. The energy of the incident helium ions is 119 keV, and the total dose is 6×10^{16} ions/cm^2.

Helium ions were chosen to radiate samples to create point defects inside with an incident energy of 119 keV. The injection dose of 6×10^{16} ions/cm^2 was conducted at room temperature to make sure that the He concentration exceeded a critical value so that the defect clusters could be observed in TEM, as described in previous studies [6]. The temperature was almost constant during the He ion irradiation, with a negligible fluctuation of less than 2 °C. The cross-sectional transmission electron microscopy (XTEM) specimens were prepared in a sequential manner of grinding, Ar ion milling and low energy Ar ion polishing. Then, the microstructures of Cu-Nb-Cu-V NMMCs before and after irradiation were characterized using a FEI Tecnai F30 transmission electron microscope.

3. Results

The depth distributions of helium concentration (DHC) and displacements per atom (dpa) in Cu-Nb-Cu-V 48 nm NMMCs are predicted by SRIM, as Figure 1 shows. The He ions are injected perpendicularly to the surface layer, which is made of Cu and numbered as the 1st layer. The simulation predicts that the He concentration will initially increase with the increase of the penetration depth, before reaching a peak value of ~3.1 at.% at the beginning of the 9th Cu layer. Importantly, it remains nearly constant in the whole of the 9th Cu layer with a fluctuation of only ±0.1 at.%. This is due to the well-selected incident helium energy. Then, it decays over the Cu/Nb interface. Generally, the helium concentration is symmetrically distributed in the region composed of the 8th V layer, the 9th Cu layer and 10th Nb layer. The helium concentration reduces to zero at the interface between the bottom vanadium layer and silicon substrate, meaning that all of these implanted helium atoms stay in the nanolayered composite.

The in-focused XTEM micrographs of as-deposited Cu-Nb-Cu-V 48 nm NMMC are presented in Figure 2a. The sample was composed of 12 layers in a periodical sequence of Cu-Nb-Cu-V from surface to substrate. In order to conveniently identify the layers in the nanolayered composite, each layer is numbered sequentially. The layer below the surface is numbered as the first layer, and that attached to the substrate is given as the 12th layer. The surface fluctuates because of some wear and tear during sample preparation, due to cutting and polishing. The thicknesses of these layers is almost uniform, at around 48 nm. The layered morphology and the selected area electron diffraction (SAED) image posted in the top-right corner collectively suggest that a strong texture structure with K-S orientation relationships exits, i.e., Cu (111)//V (110)//Nb (110), Cu <110>//V <111>//Nb <111>.

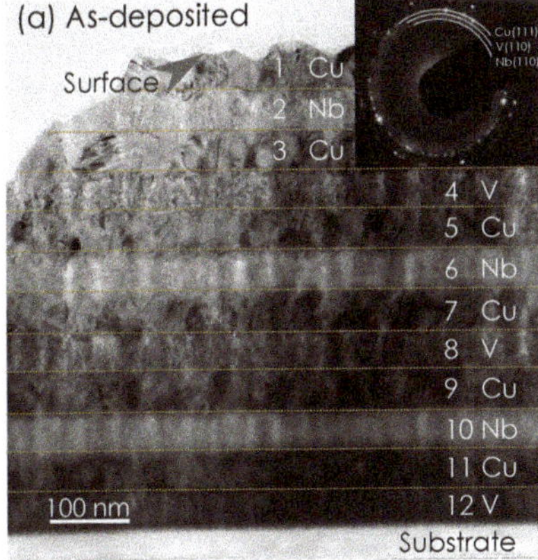

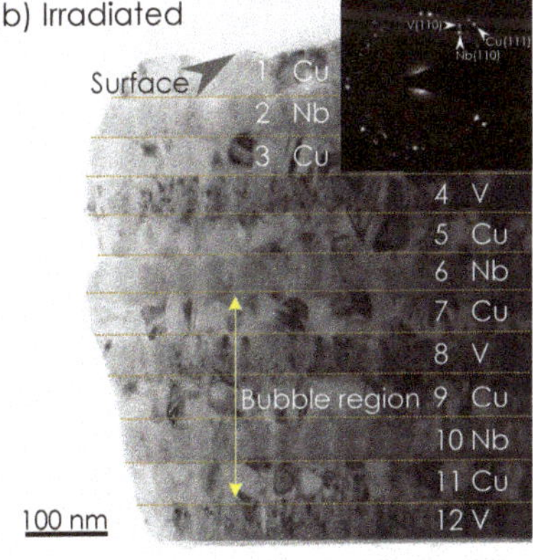

Figure 2. The bright-field XTEM images and corresponding SAED images of Cu-Nb-Cu-V 48 nm NMMCs in conditions of (**a**) as-deposited, (**b**) irradiated at room temperature with a dose of 6×10^{16} ions/cm^2 at under-focused condition. In (**b**), bubbles are observed in the region of 7th to 11th layers and called as "Bubble region".

Figure 2b displays the under-focused TEM image of Cu-Nb-Cu-V 48 nm NMMCs irradiated at room temperature with a total dose of 6×10^{16} ions/cm^2. Bubbles are observed in the region of 7th to 11th layers. These bubbles filled by helium atoms may be imaged by structure factor contrast under dynamical or bright-field kinematical imaging conditions [24]. The contrast mechanism is similar to that for disordered zones in ordered alloys or amorphous zones in crystalline matrices. Changing the focus from over focus to under focus, bubbles appear dark to bright, respectively. In this way, the existence of bubbles is confirmed. Setting the focus under a proper value, bubbles can be observed in an optimal contrast, i.e., they are too small to be seen at this magnification. Therefore, we didn't indicate any bubbles individually in Figure 2b as typical examples. Alternatively, we have indicated the region where bubbles mainly exist in Figure 2b, and called it the "Bubble region". The morphologies of both Cu/V and Cu/Nb interfaces in Cu-Nb-Cu-V 48 nm NMMCs remain immiscible after irradiation. The SAED images suggest that the K-S orientation relationships still exits after He ion irradiation, but it is slightly weaker than that of the as-deposited material.

Figure 3 shows the typical microstructures of Cu/V and Cu/Nb interfaces adjoining the 9th Cu layer in Cu-Nb-Cu-V 48 nm which was irradiated at room temperature with a dose of 6×10^{16} ions/cm^2. The existence of the bubbles is confirmed and indicated by arrows. As we can see, bubbles are observed in Layer 9-Cu, but not in the 8-V and 9-Nb layers. Meanwhile, two zones free of bubbles in Layer 9-Cu close to the Cu/Nb and Cu/V interfaces exist. These zones, depleted of defect clusters near the interfaces, are referred to as "bubble denuded zones" (BDZs) [20,21]. The BDZ boundaries were checked based on their definition and followed by the reported study [21]. By undertaking five measurements at different positions, the average width of the BDZ in the Layer 9-Cu close to Cu/Nb interface was determined to be 3.5 ± 0.3 nm, and 2.5 ± 0.2 nm for Cu/V interface.

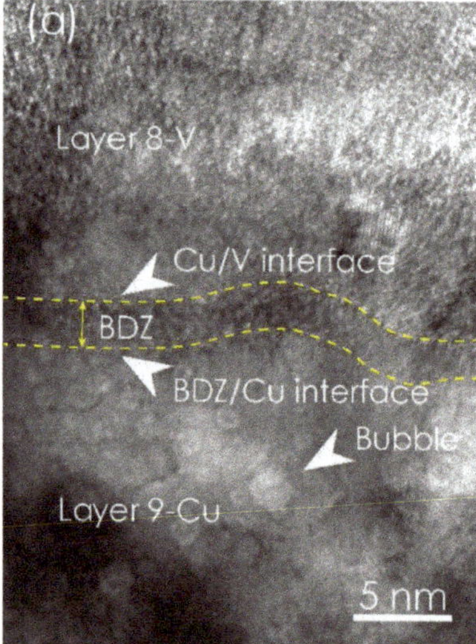

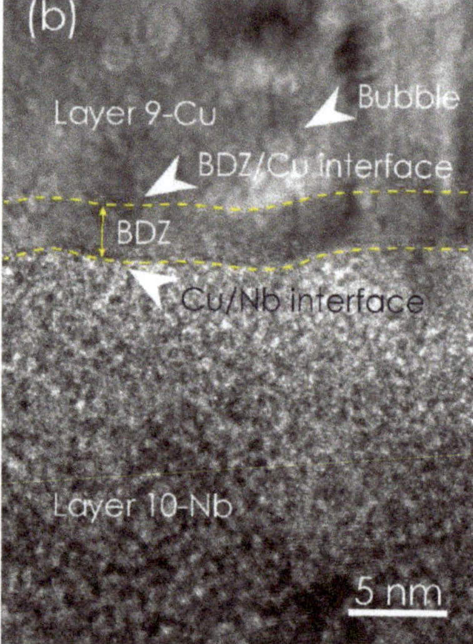

Figure 3. The typical HRTEM images of (**a**) the Cu/V interface sandwiched by the 8th V layer and the 9th Cu layer and (**b**) the Cu/Nb interface sandwiched by the 9th Cu layer and the 10th Nb layer in irradiated Cu-Nb-Cu-V 48 nm NMMC at room temperature with a dose of 6×10^{16} ions/cm^2. The average width of the bubble denuded zone near the Cu/V interface in the 9th Cu layer is 2.5 ± 0.2 nm, while 3.5 ± 0.3 nm near the Cu/Nb interface in the 9th Cu layer. Bubbles are indicated by arrows.

4. Discussion

In order to compare the sink efficiencies of Cu/V and Cu/Nb interfaces in the same sample for the first time, Cu-Nb-Cu-V NMMCs containing Cu/V and Cu/Nb interfaces are composed in a single sample for this paper. Previous studies have stated that the interfaces play a big role in the distribution of defect clusters in their vicinity zones [20,21]. Therefore, the Cu layer sandwiched by the V and Nb layers in Cu-Nb-Cu-V NMMC can be treated as an indication layer. The distribution of defect clusters in this Cu indication layer indicates the difference, and its extent difference, between the Cu/V and Cu/Nb interfaces.

The most primary form of atomic damage sustained by an irradiated material is point defects, including vacancies and interstitials. These interstitials within the intermediate Cu layer after irradiation include helium atoms and disordered Cu atoms. As helium ion irradiation occurs over time, point defects in NMMCs gradually evolve into defect clusters. For Cu-Nb-Cu-V 48 nm NMMCs, bubbles are observed at room temperature, as Figure 2b shows. As a result of the constantly distributed helium concentration shown in Figure 1, the shared 9th Cu layer can behave as a stage by which display the strength of neighboring two Cu/V and Cu/Nb interfaces in eliminating point defects. Following the idea mentioned earlier, we focus on the bubbles in the 9th Cu layer. As Figure 3 shows, the bubbles are small and crowded at room temperature. The BDZ in the 9th Cu layer can be noticed and the average width of BDZ in the 9th Cu layer close to Cu/V interface is $\lambda_{Cu/V} = 2.5 \pm 0.2$ nm, while $\lambda_{Cu/Nb} = 3.5 \pm 0.3$ nm for Cu/Nb interface. The difference in BDZ width between these two interfaces may originate from the different interaction degrees of these two interfaces with radiation defects produced during helium bombardment; a detailed discussion is provided in the following paragraphs.

In the formation stage of helium bubbles, vacancies are firstly formed and then combined with helium atoms into a stable cluster. These clusters may build up by capturing vacancies and helium atoms, and gradually become detectable large bubbles. Hence, the difference in width of these two BDZs originates from different vacancy sink efficiencies of interfaces. Following a previous study [16], the vacancy concentration in the vicinity of an interface with arbitrary vacancy sink efficiency η may be written as:

$$c(x) = c_{eq} + \frac{K_0}{K_s}\left(1 - \eta e^{-x\sqrt{\frac{K_s}{D}}}\right) \tag{1}$$

where x is the position along the direction normal to the interface plane, c is the vacancy concentration, c_{eq} is the vacancy concentration under thermal equilibrium conditions and D is the vacancy diffusivity. K_0 is the rate of vacancy generation under radiation, which can be obtained using the Norgett-Ronbinson-Torrens model [25] with the help of SRIM. K_s is the vacancy reduction rate in some kind of medium. In Figure 3a, the interface at $x = 0$ is exactly referred to the Cu/V interface. Similarly, in Figure 3b, the interface at $x = 0$ is exactly referred to the Cu/Nb interface. As is well known, a critical vacancy concentration is necessary when supersaturated vacancies in matrix precipitate into bubbles. Thus, the vacancy concentration in the BDZ/Cu interface near Cu/V is the critical vacancy concentration to form bubbles; we denote its value as c_b. From Equation (1), we can see that the distance corresponding to the critical vacancy concentration is exactly the width of the BDZ, i.e., $c(x = \lambda_{Cu/V}) = c_b$. Substituting c_b and $\lambda_{Cu/V}$ into Equation (1), and rewriting the equation, we get the vacancy sink efficiency for the Cu/V interface:

$$\eta_{Cu/V} = \frac{1 - \left(c_b - c_{eq}\right)K_s/K_0}{e^{-\lambda_{Cu/V}\sqrt{\frac{K_s}{D}}}} \tag{2}$$

Because the BDZ close to Cu/Nb interface and the BDZ close to Cu/V interface exist in the same Cu layer, the vacancy concentration in the BDZ/Cu interface near Cu/Nb interface takes the same value of c_b. So, for Cu/Nb interface, the sink efficiency is:

$$\eta_{Cu/Nb} = \frac{1 - \left(c_b - c_{eq}\right)K_s/K_0}{e^{-\lambda_{Cu/Nb}\sqrt{\frac{K_s}{D}}}} \tag{3}$$

Note that the numerators of these two sink efficiencies are exactly same. So, the ratio of sink efficiencies of these two interfaces is:

$$\alpha = \frac{\eta_{Cu/V}}{\eta_{Cu/Nb}} = e^{\sqrt{\frac{K_s}{D}}(\lambda_{Cu/V} - \lambda_{Cu/Nb})} \tag{4}$$

The rate of recombination K_s takes the value of 5×10^6 s^{-1} [26]. The vacancy diffusivity D_v is computed as $D = 1.938 \times 10^{-9}$ m^2 s^{-1} [14]. Taking all the values of these listed parameters in the equation above, we get $\alpha = 0.95$, i.e., the vacancy sink efficiency of Cu/V interface is only 95% of Cu/Nb.

Conclusively, the Cu/Nb interface had a higher sink efficiency than that of the Cu/V interface. The difference in vacancy sink efficiency for Cu/V and Cu/Nb interfaces may be attributed to defect trapping sites in them. Misfit dislocation interactions (MDIs) have been shown to serve as preferential vacancy trapping sites at some hetero-interfaces in previous studies. And the Cu/Nb interface contains a 5 times greater density of MDIs than Cu/V [16]. Therefore, The Cu/Nb interface prevails over the Cu/V interface in eliminating vacancies, suppressing the formation of helium-vacancy clusters and making bubble-denuded zones slightly wider in their vicinities.

5. Conclusions

In this paper, a new method to quantitively compare the vacancy sink efficiencies between different interfaces is proposed. This method is the first to integrate the comparison in the same sample, in contrast to the previous studies which have compared two interfaces existing in two individual samples. The average width of bubble-denuded zones was measured to be 2.5 ± 0.2 nm for the Cu/V interface and 3.5 ± 0.3 nm for the Cu/Nb in our designed experiment. Based on one-dimensional diffusion theory, the ratio of vacancy sink efficiencies of Cu/V and Cu/Nb interfaces was shown to be approximately 95% of Cu/Nb, indicating that Cu/Nb interfaces are slightly more efficient than Cu/V interfaces in eliminating point defects and associated He bubbles induced by radiation.

Author Contributions: Conceptualization, H.C., J.D. and E.F.; methodology, H.C., J.D., Y.L., P.W., Y.Z., J.H., X.Z. and Y.W.; software, H.C.; validation, J.D., G.A.S. and E.F.; formal analysis, H.C., J.D. and E.F.; investigation, H.C.; resources, J.Z., X.W. and E.F.; data curation, H.C.; writing—original draft preparation, H.C.; writing—review and editing, J.D., J.Z.; X.W. and E.F.; visualization, H.C.; supervision, E.F.; project administration, E.F.; funding acquisition, E.F.

Funding: This work was supported by National Magnetic Confinement Fusion Energy Research Project with numbers of 2015GB121004, 2017YFE0302500 and 2018YFE0307100 from Ministry of Science and Technology of China and by grant with numbers of 11975034, 11921006, 11375018 and 11528508 from National Natural Science and Foundation of China (NSFC). E.F. appreciates the support from China-Romania Science and Technology Cooperation Committee 43rd Regular Meeting Exchange Program, the Ion Beam Materials Lab at Peking University and the Instrumental Analysis Fund of Peking University. X.W. acknowledges financial support from the Program for New Century Excellent Talents in University.

Acknowledgments: The authors would like to gratefully acknowledge the technical support from Xiumei Ma and Chuan Xu.

Conflicts of Interest: The authors declare no conflict of interest.

Data Availability Statement: All data included in this study are available upon request by contact with the corresponding author.

References

1. Höchbauer, A.T.; Misra, A.; Hattar, K.; Hoagland, R.G. Influence of interfaces on the storage of ion-implanted He in multilayered metallic composites. *J. Appl. Phys.* **2005**, *98*, 123516. [CrossRef]
2. Demkowicz, M.J.; Hoagland, R.G.; Hirth, J.P. Interface Structure and Radiation Damage Resistance in Cu-Nb Multilayer Nanocomposites. *Phys. Rev. Lett.* **2008**, *100*, 136102. [CrossRef] [PubMed]
3. Hattar, K.; Demkowicz, M.; Misra, A.; Robertson, I.; Hoagland, R. Arrest of He bubble growth in Cu–Nb multilayer nanocomposites. *Scr. Mater.* **2008**, *58*, 541–544. [CrossRef]

4. Fu, E.G.; Carter, J.; Swadener, G.; Misra, A.; Shao, L.; Wang, H.; Zhang, X.; Swadener, J.G. Size dependent enhancement of helium ion irradiation tolerance in sputtered Cu/V nanolaminates. *J. Nucl. Mater.* **2009**, *385*, 629–632. [CrossRef]

5. Fu, E.G.; Misra, A.; Wang, H.; Shao, L.; Zhang, X. Interface enabled defects reduction in helium ion irradiated Cu/V nanolayers. *J. Nucl. Mater.* **2010**, *407*, 178–188. [CrossRef]

6. Fu, E.G.; Wang, H.; Carter, J.; Shao, L.; Wang, Y.; Zhang, X. Fluence-dependent radiation damage in helium (He) ion-irradiated Cu/V multilayers. *Philos. Mag.* **2013**, *93*, 883–898. [CrossRef]

7. Liu, Y.; Yang, K.; Hay, J.; Fu, E.G.; Zhang, X. The effect of coherent interface on strain-rate sensitivity of highly textured Cu/Ni and Cu/V multilayers. *Scr. Mater.* **2019**, *162*, 33–37. [CrossRef]

8. Wang, P.P.; Wang, X.J.; Du, J.L.; Ren, F.; Zhang, Y.; Zhang, X.; Fu, E.G. The temperature and size effect on the electrical resistivity of Cu/V multilayer films. *Acta Mater.* **2017**, *126*, 294–301. [CrossRef]

9. Wang, P.P.; Xu, C.; Fu, E.G.; Du, J.L.; Gao, Y.; Wang, X.J.; Qiu, Y.H. The study on the electrical resistivity of Cu/V multilayer films subjected to helium (He) ion irradiation. *Appl. Surf. Sci.* **2018**, *440*, 396–402. [CrossRef]

10. Chen, Y.X.; Liu, Y.; Fu, E.G.; Sun, C.; Yu, K.; Song, M.; Li, J.; Wang, Y.; Wang, H.; Zhang, X. Unusual size-dependent strengthening mechanisms in helium ion-irradiated immiscible coherent Cu/Co nanolayers. *Acta Mater.* **2015**, *84*, 393–404. [CrossRef]

11. Gao, Y.; Yang, T.F.; Xue, J.M.; Yan, S.; Zhou, S.; Wang, Y.; Kwok, D.T.; Chu, P.K.; Zhang, Y. Radiation tolerance of Cu/W multilayered nanocomposites. *J. Nucl. Mater.* **2011**, *413*, 11–15. [CrossRef]

12. Yu, K.Y.; Liu, Y.; Fu, E.G.; Wang, Y.; Myers, M.; Wang, H.; Shao, L.; Zhang, X. Comparisons of radiation damage in He ion and proton irradiated immiscible Ag/Ni nanolayers. *J. Nucl. Mater.* **2013**, *440*, 310–318. [CrossRef]

13. Zhang, X.; Fu, E.G.; Li, N.; Misra, A.; Wang, Y.-Q.; Shao, L.; Wang, H. Design of Radiation Tolerant Nanostructured Metallic Multilayers. *J. Eng. Mater. Technol.* **2012**, *134*, 041010. [CrossRef]

14. Demkowicz, M.J.; Hoagland, R.G.; Uberuaga, B.P.; Misra, A. Influence of interface sink strength on the reduction of radiation-induced defect concentrations and fluxes in materials with large interface area per unit volume. *Phys. Rev. B* **2011**, *84*, 104102. [CrossRef]

15. Zhang, X.; Fu, E.G.; Misra, A.; Demkowicz, M.J. Interface-enabled defect reduction in He ion irradiated metallic multilayers. *JOM* **2010**, *62*, 75–78. [CrossRef]

16. Demkowicz, M.; Misra, A.; Caro, A. The role of interface structure in controlling high helium concentrations. *Curr. Opin. Solid State Mater. Sci.* **2012**, *16*, 101–108. [CrossRef]

17. Sutton, A.P.; Balluffi, R.W. *Interfaces in Crystalline Materials*; Clarendon Press: Oxford, UK, 2009.

18. Beyerlein, I.J.; Demkowicz, M.J.; Misra, A.; Uberuaga, B.P. Defect-interface interactions. *Prog. Mater. Sci.* **2015**, *74*, 125–210. [CrossRef]

19. Mao, S.; Shu, S.; Zhou, J.; Averback, R.S.; Dillon, S.J. Quantitative comparison of sink efficiency of Cu–Nb, Cu–V and Cu–Ni interfaces for point defects. *Acta Mater.* **2015**, *82*, 328–335. [CrossRef]

20. Han, W.Z.; Demkowicz, M.J.; Mara, N.A.; Fu, E.G.; Sinha, S.; Rollett, A.D.; Wang, Y.Q.; Carpenter, J.S.; Beyerlein, I.J.; Misra, A. Design of Radiation Tolerant Materials Via Interface Engineering. *Adv. Mater.* **2013**, *25*, 6975–6979. [CrossRef]

21. Han, W.Z.; Demkowicz, M.; Fu, E.G.; Wang, Y.Q.; Misra, A. Effect of grain boundary character on sink efficiency. *Acta Mater.* **2012**, *60*, 6341–6351. [CrossRef]

22. Ziegler, J.F.; Ziegler, M.D.; Biersack, J.P. SRIM—The stopping and range of ions in matter (2010). *Nucl. Instrum. Methods Phys. Res. B* **2010**, *268*, 1818–1823. [CrossRef]

23. Was, G.S. *Fundamentals of Radiation Materials Science: Metals and Alloys*; Springer: New York, NY, USA, 2018.

24. Williams, D.B.; Carter, B.C. *Transmission Electron Microscopy*; Springer: New York, NY, USA, 2009.

25. Norgett, M. A proposed method of calculating displacement dose rates. *Nucl. Eng. Des.* **1975**, *33*, 50–54. [CrossRef]

26. Ehrhart, P.; Ullmaier, H. *Atomic Defects in Metals*; Springer: Berlin, Germany, 2006.

Communication

Resistance to Helium Bubble Formation in Amorphous SiOC/Crystalline Fe Nanocomposite

Qing Su [1,*], Tianyao Wang [2], Jonathan Gigax [2], Lin Shao [2] and Michael Nastasi [1,3,4]

[1] Department of Mechanical and Materials Engineering, University of Nebraska-Lincoln, Lincoln, NE 68583-0857, USA; mnastasi2@unl.edu

[2] Department of Nuclear Engineering, Texas A&M University, College Station, TX 77843-3128, USA; wty19920822@tamu.edu (T.W.); gigaxj@tamu.edu (J.G.); lshao@tamu.edu (L.S.)

[3] Nebraska Center for Energy Sciences Research, University of Nebraska-Lincoln, Lincoln, NE 68583-0857, USA

[4] Nebraska Center for Materials and Nanoscience, University of Nebraska-Lincoln, Lincoln, NE 68588-0298, USA

* Correspondence: qsu3@unl.edu; Tel.: +1-402-472-1685

Received: 29 November 2018; Accepted: 25 December 2018; Published: 28 December 2018

Abstract: The management of radiation defects and insoluble He atoms represent key challenges for structural materials in existing fission reactors and advanced reactor systems. To examine how crystalline/amorphous interface, together with the amorphous constituents affects radiation tolerance and He management, we studied helium bubble formation in helium ion implanted amorphous silicon oxycarbide (SiOC) and crystalline Fe composites by transmission electron microscopy (TEM). The SiOC/Fe composites were grown via magnetron sputtering with controlled length scale on a surface oxidized Si (100) substrate. These composites were subjected to 50 keV He+ implantation with ion doses chosen to produce a 5 at% peak He concentration. TEM characterization shows no sign of helium bubbles in SiOC layers nor an indication of secondary phase formation after irradiation. Compared to pure Fe films, helium bubble density in Fe layers of SiOC/Fe composite is less and it decreases as the amorphous/crystalline SiOC/Fe interface density increases. Our findings suggest that the crystalline/amorphous interface can help to mitigate helium defect generated during implantation, and therefore enhance the resistance to helium bubble formation.

Keywords: radiation tolerant materials; amorphous silicon oxycarbide; nanocrystalline Fe; composite; interface

1. Introduction

The combination of irradiation defects and helium (He) lead to a microstructural evolution of bubbles, cavities and voids, which ultimately lead to the degradation of mechanical properties in first-wall materials as well as fuel cladding in fission nuclear reactors [1,2]. For example, formation of He bubbles at grain boundaries of austenitic stainless steel has been found to occur even at very low overall He concentrations, causing deleterious effects such as swelling and embrittlement [3,4].

Over past decades, extensive researches have been conducted to understand the behavior of inert gases such as helium in pure elemental metals [5,6]. The development of radiation tolerant composite materials via the introduction of interfaces, phase boundaries, and grain boundaries has also been discussed in a number of investigations [7–10]. For instance, oxide dispersion strengthened (ODS) steels, which contain a high volume fraction of metal/nanoscale oxides interfaces, has shown that nanoscale precipitates can promote the recombination of radiation-induced point defects, and therefore mitigate He bubbles formation [1,11,12]. Similar to the interface effect due to precipitates in ODS steels, the introduction of well-controlled nanoscale metallic interfaces (e.g., interfaces between face-centered cubic and body-centered cubic materials) have also been shown to be efficient for trapping He and mitigating the onset of He bubble formation [9,13,14].

While the above discussion on interface design strategies have shown that it is possible to delay the deleterious effects of He, recent studies have shown that in some materials it is possible to avert helium bubble formation entirely by continually removing it as it is implanted [15]. Amorphous SiOC, a new class of superior radiation tolerant materials, has shown very good steady-state irradiation properties [16]. Previous studies have demonstrated that amorphous SiOC alloys are stable under irradiation, sustaining their glassy states over a wide range of irradiation conditions [17–19]. More interestingly, implanted He atoms were found to diffuse out of the SiOC matrix as fast as it was implanted, even at liquid nitrogen temperatures, resulting in time-invariant structure and properties. In addition, amorphous SiOC can be paired with a crystalline metal component such as Fe to form a composite with enhanced thermal, mechanical and irradiation properties [20,21]. However, at present the properties of SiOC/Fe composites under helium implantation, remain virtually uncharted. Similar to what has been observed in metal/metal nano-composites and ODS steels, we hypothesize that the crystalline/amorphous interfaces in the SiOC/Fe composite films are able to facilitate vacancy and interstitial recombination, therefore, resulting in enhanced helium bubble formation resistance [10,22,23]. In this work, we investigated and compared the He implantation responses of pure Fe films and SiOC/Fe composites with controlled length scale for the first time. The results serve to better understand the role of SiOC/Fe amorphous/crystalline interfaces on helium management and defect mitigation in harsh environments.

2. Materials and Methods

Magnetron sputtering was used to synthesize SiOC/Fe multilayer films with controlled individual layer thicknesses. Direct current (DC) magnetron sputtering was used to deposit α-Fe layers, while radio frequency (RF) sputtering was used to synthesize amorphous SiOC layers from co-sputtering SiO_2 and SiC targets. Prior to depositions, a base pressure of 9.2×10^{-6} Pa or lower was obtained and the typical argon partial pressure during sputtering was ~0.65 Pa. The thickness of the pure α-Fe film and two kinds of SiOC/Fe multilayered films was ~1 µm. In thick SiOC/Fe multilayer films, the thickness of Fe and SiOC layers were 80 and 60 nm, respectively, while the thickness of Fe and SiOC layers for thin SiOC/Fe multilayer films were 16 and 12 nm. The roughness of typical SiOC and Fe layer ranges from 3 to 5 nm, as suggested by previous X-ray reflectivity experiment [24]. All targets including SiO_2 (purity 99.995%), SiC (purity 99.5%) and Fe (purity 99.95%), were obtained from AJA International, Inc. (North Scituate, MA, USA)

The pure SiOC film, pure Fe film and SiOC/Fe multilayers were subject to 50 keV He ions implantation at room temperature. Stopping and Range of Ions in Matter (SRIM)-2008 software was used to simulate depth profiles of implanted ion concentration and irradiation damage using the ion distribution and quick calculation of damage option [25]. The SiOC/Fe nanolaminate was treated as a uniformly distributed amorphous target material for the purpose of the simulations. The nominal composition for thick and thin SiOC/Fe nanolaminates are $Fe_{13.3}Si_3O_4C_3$. The assumed displacement energies for Si, O, C and Fe are 15, 28, 28 and 40 eV, respectively. Rutherford backscattering spectrometry and X-ray reflectivity results suggest the SiOC films possess chemical composition of Si-30%, O-40%, C-30% and density of 2.2 g/cm^3 [24]. The density of Fe layers is 6.92 g/cm^3, approximately 14% lower than that of pure Fe target due to shadowing effects during sputtering process [26]. The base pressure during He implantation was better than 5×10^{-4} Pa. To obtain a 5 at% He peak concentration, fluences of 6.8×10^{16}, 6.5×10^{16} and 7.0×10^{16} ion/cm^2 was implanted into pure SiOC film, pure Fe film and SiOC/Fe multilayers, respectively. The 5 at% He peak was chosen in order to visualize the He bubbles in Fe layers. Because the He concentration profile is near-Gaussian and therefore, the implant concentration varied between 0 and 5 at% as a function of depth. It allowed to investigate He bubble formation in this range of implant concentration. The beam spot size was 8 mm × 10 mm. The fluence variation within the beam spot was typically within ±10%. The fluence was measured by monitoring the charge collection on the target. The target was biased during the irradiation to suppress the error caused by secondary electrons. Such a setup has shown good accuracy in fluence determination, of uncertainty of <15%, based on previous testing from secondary ion mass spectrometry analysis of various implants in Si and Fe substrates. The cross-sectional microstructure of SiOC/Fe multilayers and pure Fe films before and after implantation was characterized by TEM. The cross-sectional TEM specimen was prepared by conventional dimple and grinding followed by ion-milling. Low energy (3.5 keV) and low angle (5°) were selected to reduce the ion milling damage. A FEI Tecnai G2 F20 TEM (FEI, Hillsboro, OR, USA) was used to investigate the microstructure of these films before and after He implantation. The typical TEM operation voltage was 200 kV. The thickness of TEM foils of all specimens were determined by Electron energy loss spectroscopy (EELS) log-ratio technique, as described in the literature [27]. The thicknesses of analyzed pure Fe film, thin and thick SiOC/Fe multilayers were 68.2 ± 6.1, 74.1 ± 8.5 and 52.8 ± 4.2 nm. The detailed microstructure analysis of as-deposited SiOC film, Fe film and SiOC/Fe multilayers films can be found in previous references [28,29].

3. Results

3.1. SRIM Simulation

To obtain He doping within these films, 50 keV He ions were selected for ion implantation. The depth profiles of radiation damage in units of displacement per atom (dpa) and helium concentration in the SiOC, Fe and Fe/SiOC films are shown in Figure 1a–c, respectively. The simulations, as shown in Figure 1, implies that all films were subjected an implantation that would result in a 5 at% peak helium concentration, assuming all the implanted He was retained. In addition to helium implantation, the He$^+$ irradiation results in maximum 2.6 dpa in the pure SiOC films and ~3 dpa in both the Fe and the SiOC/Fe composite films.

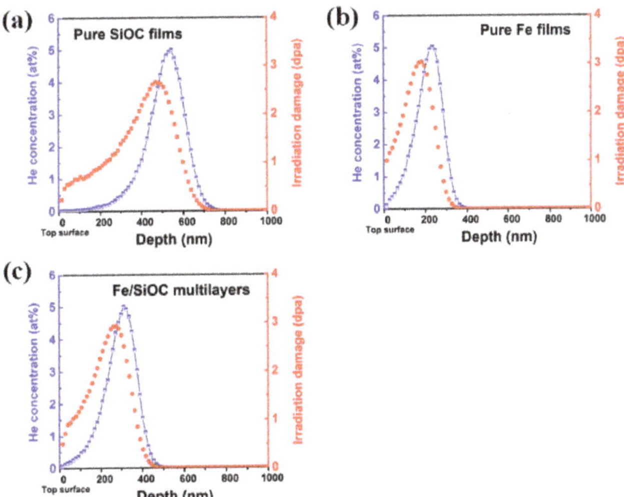

Figure 1. The simulated depth profile of radiation damage and helium concentration in (a) pure SiOC film, (b) pure Fe film and (c) Fe/SiOC multilayers. The peak concentration for all films are 5 at%.

3.2. He Implantation in Fe and SiOC Films

The cross-sectional TEM images of the pure Fe film after 5 at% He implantation are shown in Figure 2a,b. Similar to the as-deposited film, the implanted Fe film exhibits a columnar structure. The corresponding selective area diffraction (SAD) pattern shown in the inset of Figure 2a suggests the implanted Fe film still retained a bcc structure. An under-focused cross-sectional TEM micrograph of the implanted Fe film is shown in Figure 2b. The micrograph was collected in the He peak region, which is 233 nm underneath the surface. A high density of He bubbles is observed within the columnar grains as well as along grain boundaries. To estimate the average He bubble size, cross-sectional TEM micrographs were taken at an under-focus distance of 400 nm. The average bubble diameter for the Fe film is 1.2 ± 0.1 nm, in comparison to a 1.1 ± 0.1 nm bubble size in bulk Fe [8]. In contrast to the high density of He bubbles in implanted Fe films, no helium bubbles (>1 nm) were observed in pure SiOC film after 5 at% He implantation, Figure 2c,d. This result was consistent with previous finding that He atoms in SiOC remain in solution and are able to outgas from the material via atomic-scale diffusion [15,28]. In addition, the irradiation does not lead to any void formation, element segregation or crystallization throughout the SiOC film.

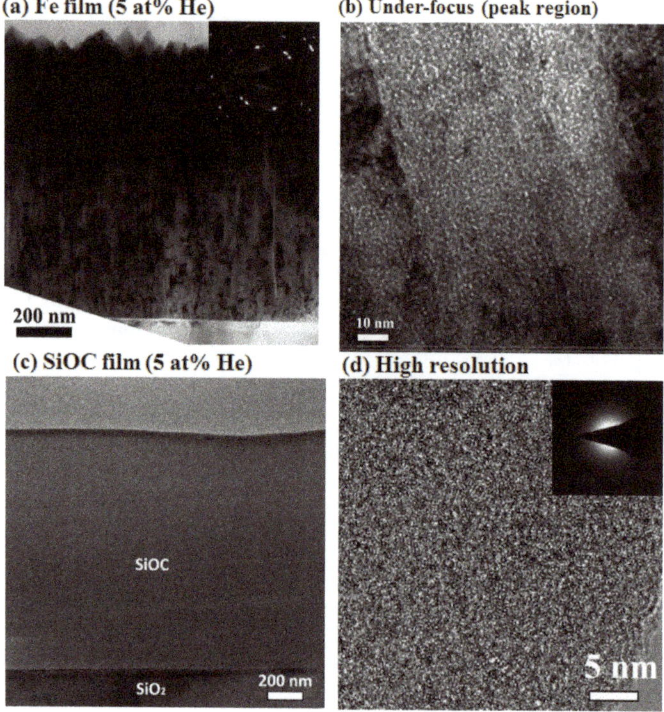

Figure 2. The typical cross-sectional TEM images of (**a**) pure Fe film and (**c**) pure SiOC film after 5 at% He implantation. The high-resolution TEM images of He peak regions are shown as (**b,d**), respectively.

3.3. He Implantation in SiOC/Fe Multilayers

To investigate the amorphous/crystalline SiOC/Fe interface effect on helium bubble formation in SiOC/Fe composite films, He$^+$ implantations were also conducted for these composite films at room temperature. Figure 3a,b shows low magnification TEM micrographs from thick and thin Fe/SiOC films after 5 at% peak He implantation. Fresnel contrast is used to examine the helium formation which are presented as dark dots surrounded by a bright fringe for the over-focus condition and bright dots surrounded by a dark fringe in the under-focus condition. The cross-sectional micrographs of under-focused and over-focused thin Fe/SiOC films at the He peak concentration regions are shown at Figure 3c,d, respectively. In order to compare the helium behavior between thin and thick Fe/SiOC films, Figure 3e,f presents typical under-focus and over-focus images for thick multilayer specimen. One of most important observations is the presence of helium bubbles in Fe layers but not in SiOC layers. For example, these helium bubbles are revealed via Fresnel contrast in both the under-focus condition (Figure 3c,e) and the over-focus condition (Figure 3d,f) in Fe layers. It is also interesting to note no helium bubbles were observed along the SiOC/Fe interface. Scanning transmission electron microscopy (STEM) was utilized under a high-angle annular dark field (HAADF) condition to further examine the composition profile of thin and thick SiOC/Fe multilayers after 5 at% He implantation. The corresponding STEM images of Figure 3a,b are shown as Figure 3g,h, respectively. These data indicate that the layer structure is maintained and SiOC/Fe interface remain quite sharp after He implantation.

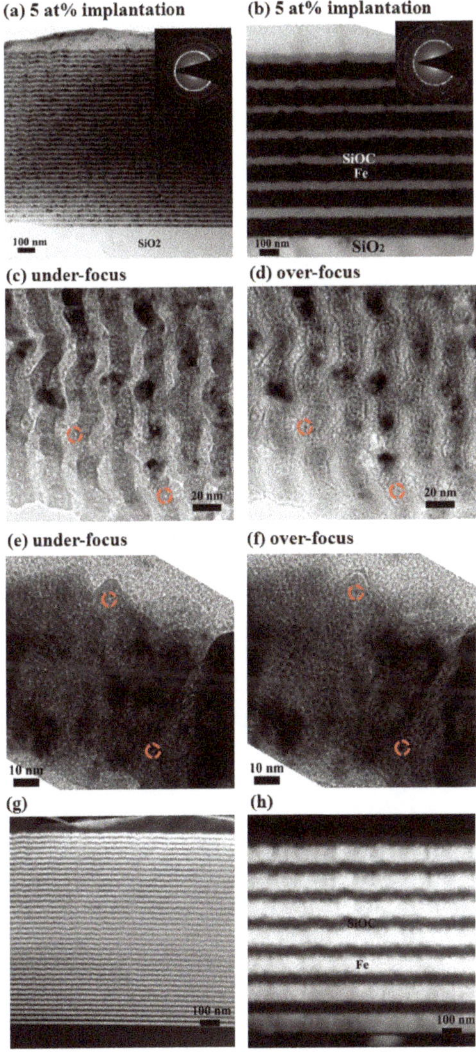

Figure 3. The typical cross-sectional TEM images of (**a**) thin, (**b**) thick Fe/SiOC multilayers after 5 at% He implantation. The He peak damage regions of thin and thick Fe/SiOC multilayers are taken at under-focus (−400 nm) and over-focus (+400 nm) condition, respectively, shown as (**c**–**f**). The corresponding scanning transmission electron microscopy (STEM) images are present as (**g**,**h**), respectively.

3.4. Depth Profile of He Bubble Density

The density of He bubbles at different depths in SiOC/Fe multilayers and Fe films was measured via extensive cross-sectional TEM analysis. Figure 4a,b summarizes that the bubble density profile in SiOC/Fe multilayers and in pure Fe film, respectively. It is interesting to note that the bubble density depth profiles in all films coincided with simulated He concentration profiles (solid line) superimposed in the same figure. In SiOC/Fe multilayers, the helium bubble density in Fe layers increases continuously and approaches a maximum at an implantation depth of ~310 nm, while the helium bubble density in pure Fe film reaches maximum at a depth of ~240 nm. The average bubble diameters for the Fe layers in thick and thin composite were 1.2 ± 0.1 and 1.2 ± 0.2 nm, respectively. No or very few change in bubble diameters was observed with depth. It can be seen that, under the same effective He implanted concentration, the helium bubble density in pure Fe films is approximately twice as that in thick SiOC/Fe multilayer sample. In addition, a lower helium bubble density profile is observed in thin SiOC/Fe multilayer sample. We also attempted to estimate the threshold concentration needed for the formation of detectable He bubbles in pure Fe films and SiOC/Fe multilayers. As shown in Figure 4a, a minimum He concentration to visualize bubbles is obtained from the intersections of the vertical lines with the SRIM simulation. Compared to the low threshold concentration in pure Fe film (<0.1 at%), the threshold concentration is approximately 0.82 at% and 1.95 at% in thick and thin SiOC/Fe multilayer sample, respectively, which represents an order of magnitude improvement for the composites.

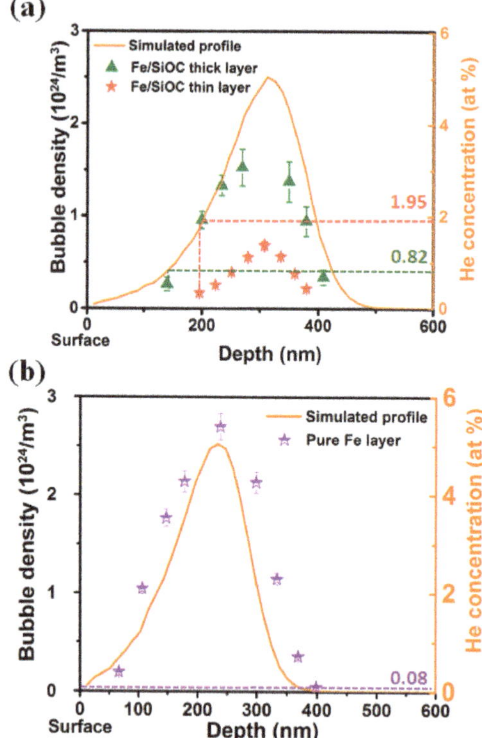

Figure 4. He bubble density (scattered points) as a function radiation depth in (**a**) Fe/SiOC multilayer and (**b**) pure Fe films. Solid line in each figure is simulated helium depth profile. The dashed line was drawn to help estimate the threshold concentration for the formation of detectable He bubbles. Bubble density in Fe layers of Fe/SiOC multilayers is significant lower than that in pure Fe film.

4. Discussion

Previous work has shown that radiation induced defects prefer to migrate to the interfacial regions that act as effective defect sinks [30,31]. The SiOC/Fe amorphous/crystalline interface density in SiOC/Fe 12/16 nm nanolayers is 10 times higher than that in SiOC/Fe 60/80 nm nanolayers. Therefore, it appears that a high density of amorphous/crystalline interfaces also facilitate defect removal. Recent molecular dynamic (MD) simulation studies of helium defect properties in bcc iron showed that helium interstitials as well as He interstitial clusters are quite mobile [32,33]. The mobility of helium interstitials or He interstitial clusters will decrease if they are trapped by vacancies. These vacancies could be intrinsic vacancies, irradiation induced vacancies, or vacancies created via kick-out of a self-interstitial atom by He clusters. The smaller helium bubble density observed in the SiOC/Fe multilayer system suggests that there is a lower concentration of vacancies in the Fe layers of SiOC/Fe composite specimen compared to that in the pure Fe layer during implantation. In both implanted Fe films and SiOC/Fe multilayers, He bubbles have an average diameter of ~1 nm and the bubble size distributions is quite narrow. As suggested by cross-section TEM images, the average bubble size depends very little on Fe layer thickness. The reduction in He bubble density in thin SiOC/Fe multilayers compared to that in thick SiOC/Fe multilayers and Fe films indicates that the vacancy concentration must have been dramatically reduced. All these results imply that the amorphous/crystalline interfaces act as efficient defect sinks to promote interstitials and vacancies recombination and to lessen helium bubble formation in the Fe layers.

Another interesting observation is that no helium bubbles were observed at the SiOC/Fe interfaces. Considering the fast helium diffusivity in SiOC, the helium atoms most likely eventually diffuse out of the system once they reach the SiOC/Fe interfaces. Therefore, the amount of helium that could diffuse out the Fe layers depends on whether helium atoms or helium clusters are able to reach the interface region. This suggests that by refining the layer thickness, it may be possible to obtain a composite free of helium bubbles.

5. Conclusions

In conclusion, we examined the effect of helium implantation on the Fe/SiOC composite system after 5 at% implantation at room temperature. By introducing amorphous/crystalline SiOC/Fe interfaces, the onset concentration needed to visualize helium bubbles in the Fe layers of SiOC/Fe composites is ten times lower than needed in pure Fe films. In addition, the helium bubble density decreases as interface density increases. All these observations suggest that the Fe/SiOC crystalline/amorphous interfaces act as efficient defects sinks, promoting interstitials and vacancies recombination and mitigating helium bubble accumulation.

Author Contributions: Q.S. prepared all specimens and designed the experiment with M.N.; T.W. and J.G. carried out He ion implantation under the supervision of L.S.; Q.S. wrote the original draft of the paper and all authors discussed the results and commented on the manuscript.

Funding: This research was funded by DoE Office of Nuclear Energy, Nuclear Energy Enabling Technologies, award DE-NE0000533. This work was supported by the U.S. Department of Energy, Office of Nuclear Energy under DOE Idaho Operations Office Contract DE-AC07- 051D14517 as part of a Nuclear Science User Facilities experiment.

Acknowledgments: The research was performed in part in the Nebraska Nanoscale Facility: National Nanotechnology Coordinated Infrastructure and the Nebraska Center for Materials and Nanoscience, which are supported by the National Science Foundation under Award ECCS: 1542182, and the Nebraska Research Initiative. The author thank D. Y. Xie for his help on EELS thickness measurement.

Conflicts of Interest: The authors declare no conflict of interest.

References

1. Odette, G.R.; Hoelzer, D.T. Irradiation-tolerant Nanostructured Ferritic Alloys: Transforming Helium from a Liability to an Asset. *JOM* **2010**, *62*, 84–92. [CrossRef]
2. Zinkle, S.J.; Busby, J.T. Structural materials for fission & fusion energy. *Mater. Today* **2009**, *12*, 12–19.
3. Schroeder, H.; Kesternich, W.; Ullmaier, H. Helium effects on the creep and fatigue resistance of austenitic stainless steels at high temperatures. *Nucl. Eng. Des. Fusion* **1985**, *2*, 65–95. [CrossRef]
4. Braski, D.N.; Schroeder, H.; Ullmaier, H. The effect of tensile stress on the growth of helium bubbles in an austenitic stainless steel. *J. Nucl. Mater.* **1979**, *83*, 265–277. [CrossRef]
5. Evans, J.H.; Vanveen, A.; Caspers, L.M. Direct Evidence for Helium Bubble-Growth in Molybdenum by the Mechanism of Loop Punching. *Scr. Metall.* **1981**, *15*, 323–326. [CrossRef]
6. Thomas, G.J. Experimental Studies of Helium in Metals. *Radiat. Eff. Defects Solids* **1983**, *78*, 37–51. [CrossRef]
7. Odette, G.R.; Miao, P.; Edwards, D.J.; Yamamoto, T.; Kurtz, R.J.; Tanigawa, H. Helium transport, fate and management in nanostructured ferritic alloys: In situ helium implanter studies. *J. Nucl. Mater.* **2011**, *417*, 1001–1004. [CrossRef]
8. Yu, K.Y.; Liu, Y.; Sun, C.; Wang, H.; Shao, L.; Fu, E.G.; Zhang, X. Radiation damage in helium ion irradiated nanocrystalline Fe. *J. Nucl. Mater.* **2012**, *425*, 140–146. [CrossRef]
9. Fu, E.G.; Misra, A.; Wang, H.; Shao, L.; Zhang, X. Interface enabled defects reduction in helium ion irradiated Cu/V nanolayers. *J. Nucl. Mater.* **2010**, *407*, 178–188. [CrossRef]
10. Di, Z.F.; Bai, X.M.; Wei, Q.M.; Won, J.; Hoagland, R.G.; Wang, Y.Q.; Misra, A.; Uberuaga, B.P.; Nastasi, M. Tunable helium bubble superlattice ordered by screw dislocation network. *Phys. Rev. B* **2011**, *84*, 052101. [CrossRef]
11. Wong, C.P.C.; Nygren, R.E.; Baxi, C.B.; Fogarty, P.; Ghoniem, N.; Khater, H.; McCarthy, K.; Merrill, B.; Nelson, B.; Reis, E.E.; et al. Helium-cooled refractory alloys first wall and blanket evaluation. *Fusion Eng. Des.* **2000**, *49–50*, 709–717. [CrossRef]
12. Li, Q.; Parish, C.M.; Powers, K.A.; Miller, M.K. Helium solubility and bubble formation in a nanostructured ferritic alloy. *J. Nucl. Mater.* **2014**, *445*, 165–174. [CrossRef]
13. Misra, A.; Demkowicz, M.J.; Zhang, X.; Hoagland, R.G. The radiation damage tolerance of ultra-high strength nanolayered composites. *JOM* **2007**, *59*, 62–65. [CrossRef]
14. Li, N.; Fu, E.G.; Wang, H.; Carter, J.J.; Shao, L.; Maloy, S.A.; Misra, A.; Zhang, X. He ion irradiation damage in Fe/W nanolayer films. *J. Nucl. Mater.* **2009**, *389*, 233–238. [CrossRef]
15. Su, Q.; Ding, H.; Price, L.; Shao, L.; Hinks, J.A.; Greaves, G.; Donnelly, S.E.; Demkowicz, M.J.; Nastasi, M. Rapid and damage-free outgassing of implanted helium from amorphous silicon oxycarbide. *Sci. Rep.* **2018**, *8*, 5009. [CrossRef] [PubMed]
16. Nastasi, M.; Su, Q.; Price, L.; Colón Santana, J.A.; Chen, T.; Balerio, R.; Shao, L. Superior radiation tolerant materials: Amorphous silicon oxycarbide. *J. Nucl. Mater.* **2015**, *461*, 200–205. [CrossRef]
17. Su, Q.; Cui, B.; Kirk, M.A.; Nastasi, M. In-situ observation of radiation damage in nano-structured amorphous SiOC/crystalline Fe composite. *Scr. Mater.* **2016**, *113*, 79–83. [CrossRef]
18. Su, Q.; Cui, B.; Kirk, M.A.; Nastasi, M. Cascade effects on the irradiation stability of amorphous SiOC. *Philos. Mag. Lett.* **2016**, *96*, 60–66. [CrossRef]
19. Su, Q.; Wang, T.; Gigax, J.; Shao, L.; Lanford, W.A.; Nastasi, M.; Li, L.; Bhattarai, G.; Paquette, M.M.; King, S.W. Influence of Topological Constraints on Ion Damage Resistance of Amorphous Hydrogenated Silicon Carbide. *Acta Mater.* **2018**, *165*, 587–602. [CrossRef]
20. Su, Q.; Price, L.; Shao, L.; Nastasi, M. Dose dependence of radiation damage in nano-structured amorphous SiOC/crystalline Fe composite. *Mater. Res. Lett.* **2016**, *4*, 48–54. [CrossRef]
21. Su, Q.; Price, L.; Shao, L.; Nastasi, M. High temperature radiation responses of amorphous SiOC/crystalline Fe nanocomposite. *J. Nucl. Mater.* **2016**, *479*, 411–417. [CrossRef]
22. Zhang, X.; Li, N.; Anderoglu, O.; Wang, H.; Swadener, J.G.; Hochbauer, T.; Misra, A.; Hoagland, R.G. Nanostructured Cu/Nb multilayers subjected to helium ion-irradiation. *Nucl. Instrum. Methods Phys. Res. Sect. B Beam Interact. Mater. Atoms* **2007**, *261*, 1129–1132. [CrossRef]
23. Demkowicz, M.J.; Hoagland, R.G.; Hirth, J.P. Interface structure and radiation damage resistance in Cu-Nb multilayer nanocomposites. *Phys. Rev. Lett.* **2008**, *100*, 136102. [CrossRef] [PubMed]

24. Su, Q.; Zhernenkov, M.; Ding, H.; Price, L.; Haskel, D.; Watkins, E.B.; Majewski, J.; Shao, L.; Demkowicz, M.J.; Nastasi, M. Reaction of amorphous/crystalline SiOC/Fe interfaces by thermal annealing. *Acta Mater.* **2017**, *135*, 61–67. [CrossRef]

25. Ziegler, J.F.; Biersack, J.P. *The Stopping and Range of Ions in Solids*; Pergamon Press: New York, NY, USA, 1985.

26. Nastasi, M.; Mayer, J.W.; Hirvonen, J.K. *Ion-Solid Interactions: Fundamentals and Applications*; Cambridge University Press: Cambridge, UK, 1996.

27. Malis, T.; Cheng, S.C.; Egerton, R.F. EELS log-ratio technique for specimen-thickness measurement in the TEM. *J. Electron. Microsc. Tech.* **1988**, *8*, 193–200. [CrossRef] [PubMed]

28. Su, Q.; Inoue, S.; Ishimaru, M.; Gigax, J.; Wang, T.Y.; Ding, H.P.; Demkowicz, M.J.; Shao, L.; Nastasi, M. Helium Irradiation and Implantation Effects on the Structure of Amorphous Silicon Oxycarbide. *Sci. Rep.* **2017**, *7*, 3900. [CrossRef] [PubMed]

29. Su, Q.; Price, L.; Colon Santana, J.A.; Shao, L.; Nastasi, M. Irradiation tolerance of amorphous SiOC/crystalline Fe composite. *Mater. Lett.* **2015**, *155*, 138–141. [CrossRef]

30. Beyerlein, I.J.; Demkowicz, M.J.; Misra, A.; Uberuaga, B.P. Defect-interface interactions. *Prog. Mater. Sci.* **2015**, *74*, 125–210. [CrossRef]

31. Demkowicz, M.J.; Misra, A.; Caro, A. The role of interface structure in controlling high helium concentrations. *Curr. Opin. Solid State Mater. Sci.* **2012**, *16*, 101–108. [CrossRef]

32. Stewart, D.M.; Osetsky, Y.N.; Stoller, R.E.; Golubov, S.I.; Seletskaia, T.; Kamenski, P.J. Atomistic studies of helium defect properties in bcc iron: Comparison of He-Fe potentials. *Philos. Mag.* **2010**, *90*, 935–944. [CrossRef]

33. Deng, H.Q.; Hu, W.Y.; Gao, F.; Heinisch, H.L.; Hu, S.Y.; Li, Y.L.; Kurt, R.J. Diffusion of small He clusters in bulk and grain boundaries in alpha-Fe. *J. Nucl. Mater.* **2013**, *442*, S667–S673. [CrossRef]

Article

The Role of Helium on Ejecta Production in Copper

Saryu Fensin [1,*], David Jones [1], Daniel Martinez [1], Calvin Lear [1] and and Jeremy Payton [2]

[1] MST-8, Los Alamos National Laboratory, Los Alamos, NM 87545, USA; djones@lanl.gov (D.J.); daniel_t@lanl.gov (D.M.); crlear@lanl.gov (C.L.)
[2] P-23, Los Alamos National Laboratory, Los Alamos, NM 87545, USA; payton@lanl.gov
* Correspondence: saryuj@lanl.gov

Received: 21 December 2019; Accepted: 24 February 2020; Published: 11 March 2020

Abstract: The effect of helium (He) concentration on ejecta production in OFHC-Copper was investigated using Richtmyer–Meshkov Instability (RMI) experiments. The experiments involved complex samples with periodic surface perturbations machined onto the surface. Each of the four target was implanted with a unique helium concentration that varied from 0 to 4000 appm. The perturbation's wavelengths were $\lambda \approx 65$ μm, and their amplitudes h_0 were varied to determine the wavenumber $(2\pi/\lambda)$ amplitude product kh_0 at which ejecta production beganfor Cu with and without He. The velocity and mass of the ejecta produced was quantified using Photon Doppler Velocimetry (PDV) and Lithium-Niobate (LN) pins, respectively. Our results show that there was an increase of 30% in the velocity at which the ejecta cloud was traveling in Copper with 4000 appm as compared to its unimplanted counterpart. Our work also shows that there was a finer cloud of ejecta particles that was not detected by the PDV probes but was detected by the early arrival of a "signal" at the LN pins. While the LN pins were not able to successfully quantify the mass produced due to it being in the solid state, they did provide information on timing. Our results show that ejecta was produced for a longer time in the 4000 appm copper.

Keywords: high strain rate strength; metals; radiation damage

---

1. Introduction

It is well known that when a shock wave in a material reaches a free surface, it can lead to ejection of particles from the surface [1–3]; this process and the launched material is referred to as ejecta. Over the last few decades there has been a body of research to understand the relationship between the total ejected mass and parameters associated with the free surface itself. Previous work has shown that the roughness of the free surface is the single most important parameter that impacts ejection of mass from surfaces [4]. Work by Asay et al. [2] also showed that defects on the surface in the form of cavities and scratches can create a surface roughness which leads to ejecta formation. Other works have also shown that impurities along with grain boundaries and slip bands can affect this process [5]. In addition, the total amount of ejected mass has been linked to the volume of surface defects, shape of the shock wave, yield strength of the material, and the phase of the material on release (solid or liquid). Specifically, Androit et al. [6] studied the effect of surface finish on tantalum (Ta) and Tin (Sn), and showed that the presence of grooves on the surfaces leads to higher ejected mass in comparison to polished surfaces. The specific roughness of the machined grooves had the most influence on total ejected mass, even for materials like Sn that melt on release at low pressures. More recent work by Zellner et al. [7] to study the role of surfaces prepared with different processes and final finishes on ejecta production in aluminum 1100 and Sn also showed a similar sensitivity of total ejected mass and the density distribution of ejected fragments to the final finish of the surface. This has also been confirmed with numerous molecular dynamics simulations that suggest surface roughness to be the determining factor in ejecta production [8]. Although, all the studies consistently agree that for a given surface finish, the maximum amount of ejecta is produced when the material is in liquid phase rather than solid. An additional effect of shock-wave shape—square to Taylor—on ejecta production was also documented, showing higher and constantly increasing ejecta production during unsupported shock generated using Taylor waves in materials like Sn [9]. The authors refer readers to Reference [10] for a thorough review of studies that investigate the role of all these factors on total ejecta production. A common theme in all these studies is the focus of these works on single-phase materials.

Although, there do exist a handful of studies that have attempted to investigate ejecta production in multicomponent materials. Androit et al. [6] investigated the effect of density inhomogeneities by using SnPb with 14 wt % Pb and 38 wt% Pb and showed that under the same shock pressure, the amount of ejected mass significantly increased with increasing density inhomogeneities with addition of Pb even in comparison to Sn. This difference was attributed to the impedance mismatch between pure Sn and SnPb eutectic especially because the microstructure of the SnPb alloy consisted of pure Sn grains included in a eutectic SnPb matrix of higher density. This work was also extended to two CuPb alloys with 15 wt% and 36 wt% Pb. Once again, an increase in ejected mass was observed in comparison to pure copper but was attributed to melting of Pb on release. More recently, the authors also investigated the affect of addition of 1–2 wt% lead to copper on both ejecta production [11]. Our results showed that addition of small amounts of lead causes the formation of background ejected mass that is absent in pure Copper. Our work agrees with the initial observations of Androit et al. [6]. However, studies of this nature remain rare and there are still open questions regarding the importance of material microstructure to ejecta production, especially as it pertains to the presence of inhomogeneities in the material.

In this work, we investigate the effect of helium concentration on ejecta production through the use of Richtmyer–Meshkov Instability experiments. The rest of the paper is organized as follows. The next section discusses the experimental techniques employed in this study. Section 3 discusses the results, followed by a brief conclusion.

2. Experimental Methodology

In this section, we outline the materials along with the experimental approach utilized in this investigation.

2.1. Material and Sample Design

The experiments used a total of four targets; out of which targets 2, 3, and 4 were implanted with 1000, 2000, and 4000 appm of helium, respectively. One of the targets was not implanted with Helium to serve as a reference. The RMI targets were machined as 2-mm thick by 35-mm diameter disks with five different perturbation regions. The perturbations zones were machined to be 3-mm wide separated by 2-mm flat regions, as shown in Figure 1. The perturbations were nominally sine-wave-like features with a wavelength λ of 65 μm. The amplitudes h for the five regions were varied such that the final finishes spanned the kh (where $k = 2\Pi/\lambda$ products of 0.5, 0.6, 0.7, 0.8, and 0.9).

These samples were implanted with helium at the Michigan Ion Beam Laboratory (MIBL) at the University of Michigan in a region that was 25 by 6 mm in size, as shown by the red rectangle in Figure 1. Implantation of each sample was carried out using He^+ or He^{++} ions at seven energies: 0.8, 1.4, 2.0, 2.6, 3.2, 3.8, and 4.4 MeV. These sequential, room-temperature implantations were designed to produce partially overlapping peaks in helium distribution from the sample surface to a depth of about 9 μm. Thermal imaging of the samples implanted at MIBL indicate that beam heating was less than 10 °C, while beam current densities were consistently below 1 μA/cm^2. All implantations were performed at room temperature and the target chamber vacuum was maintained below 5×10^{-7} torr during the implantation.

Since we were interested in studying the effect of He on ejecta production in copper, it was essential to verify that the initial microstructure (grain size, texture) for copper was not altered during the implantation process. This was essential to ensure quantitative comparison between data obtained from unimplanted and implanted copper. The change in the overall microstructure was quantified before and after irradiation using electron back-scattered diffraction (EBSD) with a FEI Inspect. The morphology of the implanted helium was investigated through the use of transmission electron microscopy (TEM) with a FEI Tecnai F20 (Hillsboro, OR, USA).

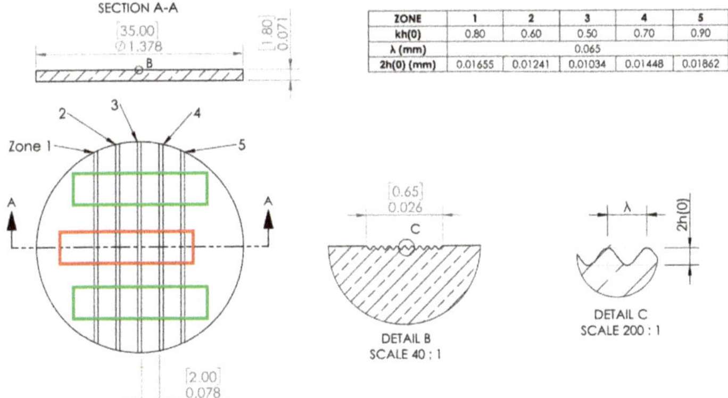

Figure 1. Target geometry with multiple surface perturbations on the copper sample used for the gun-drive experiments. The red region represents the area implanted with helium, whereas the green area shows the region used to extract data corresponding to unimplanted copper. Please note that the dimensions on the actual drawing are in inches with mm in square brackets.

To preserve the reflecting surfaces of the RMI targets, characterization specimens were collected from surrogate copper samples prepared from the same plate and implanted under the same conditions as the targets. Standard metallographic techniques were used to prepare samples for EBSD analysis in a FEI Inspect scanning electron microscope (SEM) (Hillsboro, OR, USA). A final polish of μm Al_2O_3 was followed by an electrochemical polish in a 2:1 solution of phosphoric acid and water at 1.9 V for 30–40 s. The samples were then lightly etched using a solution of $FeCl_3$ and HCl. Surface preparations were completed prior to any implantations. Post-implantation grain sizes (60 μm) and textures (random) were found to be similar to unimplanted copper [12], as shown in Figure 2.

TEM foils were then extracted from these same surrogates and thinned to electron transparency according to conventional "lift-out" techniques in a FEI Helios 600 focused ion beam (FIB) system. Final polishing was performed to minimize preparation-induced damage artifacts, using 2 kV Ga+ ions in the same FIB. Specimen microstructures were primarily examined using bright-field TEM techniques in a 300 kV FEI Tecnai F20, as shown in Figure 3.

The formation of helium bubbles in these samples is to be expected, given the favorable trapping of excess point defects at dissolved helium atoms and the suppressive effect of helium on the critical radii for cavity growth [13]. The authors acknowledge, however, that helium was not introduced homogeneously by the implantation method used here, instead concentrating in bands around the peak ion ranges of each implantation energy. Implantation profiles for each ion energy were simulated using Stopping and Range of Ions in Matter (SRIM) [14] and are shown in Figure 4. With the low diffusivity of helium in copper near room temperature, this leads to a nonuniform distribution of helium bubbles with depth. Future studies on the effects of helium distribution will incorporate ion-energy-degradation techniques to allow for more uniform implantation of helium at elevated temperature.

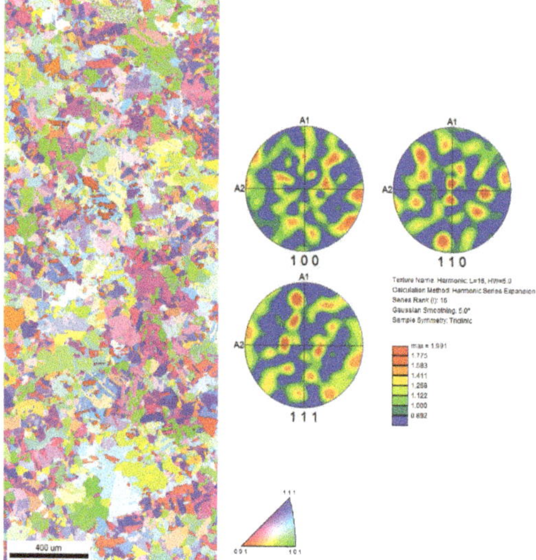

Figure 2. Microstructure of Cu after Helium implantation, characterized using Electron Back Scatter Diffraction along with the texture plot. This shows that the grain size and texture of copper were not altered due to the implantation process.

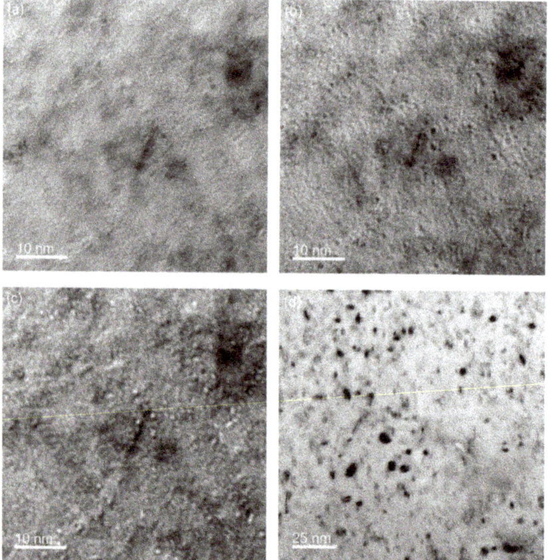

Figure 3. Small helium bubble (2-nm diameter) were observed in the implanted samples using through-focus imaging (± 1 μm). Bubbles are nearly invisible in the (**a**) focused image, but appear dark and light in the (**b**) overfocused and (**c**) underfocused conditions, respectively. Additional displacement related defects (e.g., dislocation loops) were observed, as seen in the (**d**) 4000 appm implanted sample.

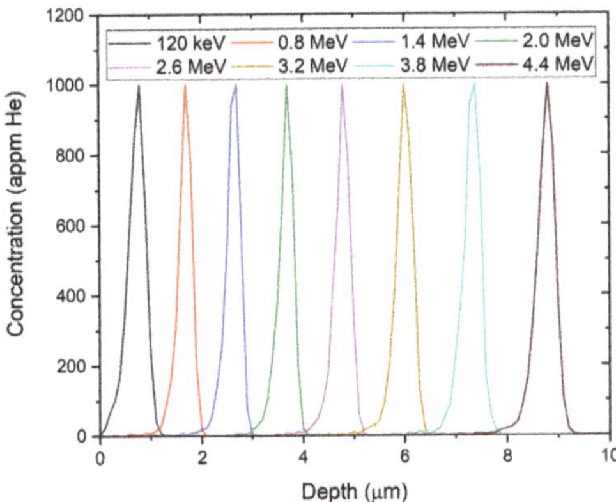

Figure 4. Helium concentration with depth into a copper target, as simulated using SRIM for each energy used here.

2.2. Impact Experiments

Plate impact experiments were performed on these targets using the 80-mm bore single-stage light gas-gun in MST-8 at LANL. The copper targets were bonded into a lexan plate, 152.4-mm wide by 127-mm high, and 12.7 mm thick, to allow for mounting and alignment in the gas-gun target tank (Figure 5, left). The impact face of the samples, i.e., the side with no perturbations, was 0.4-mm proud of the lexan plate. To ensure a planar impact between the projectile and the target, a mirror was affixed to the front of the target, and a laser was used to align a spot down and back along the 9.2 m barrel. With this method, the deviation from parallel at impact is typically sub-milli-radian. The flyer-plates were tantalum, 50-mm diameter by 2-mm thick, affixed to the front of a lexan projectile. With a projectile velocity of 1.1 km/s, this generated an approximately 1 µs duration square or flat-topped shock in the sample with a peak stress of 30 GPa.

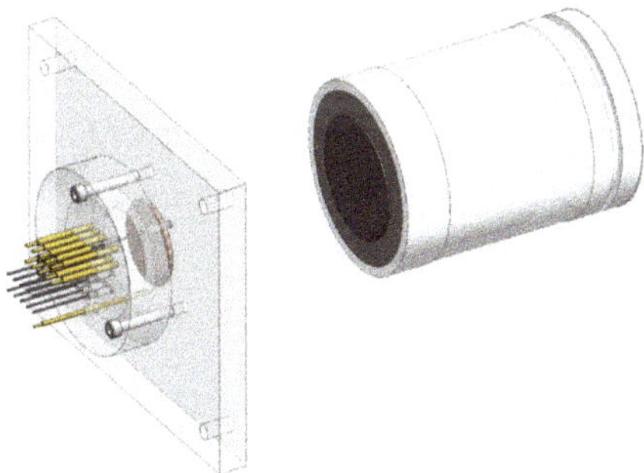

Figure 5. Left: Image showing the sample in the lexan plate that is used to mount the target in the gun along and right along with full the diagnostic assembly. **Right**: the projectile with the tantalum impactor.

The diagnostics package was mounted on the back of the lexan plate, such that it could cover the rear of the sample where the perturbations were located. This consisted of a round disc with rows of holes to mount either photon Doppler velocimetry (PDV) probes or lithium niobate (LN) pins. The former are used to measure velocity through a noncontact-laser-based interferometric technique. A narrow linewidth laser is used to illuminate the target, and the reflected light is collected by the probe. If the target surface is moving, this reflected light is Doppler shifted. The collected light is mixed with a reference source, producing a beat frequency that is proportional to the surface velocity. Fourier transform techniques are used to extract the velocity-time history. With a sufficiently high-bandwidth oscilloscope, PDV can resolve velocities from rest to 104 m/s with excellent dynamic range. PDV is able to resolve multiple velocities in the field of view at once, hence it is ideally suited to RMI work, where we want to measure the velocities of the ejecta, bubble, and bulk surface. The probes were supplied by AC photonics, part number 1CL15P020LCD01. These are collimating probes with a spot size of approximately 300 μm at a 20-mm distance, as such, each probe captures multiple wavelengths of the perturbation region it is interrogating.

The LN pins were supplied by Dynasen, part number CA-1136-Li-1. They consisted of a 2.4-mm diameter lithium niobate crystal at the end of a 25.4-mm-long brass tube. Lithium niobate is a piezoelectric material, in that it will output a voltage when strained. This voltage can be used to infer (through a series of assumptions such as conservation of momentum) the mass impinging on the LN crystal, providing an estimate of the mass ejected from the perturbed regions of the sample.

A diagram of the diagnostics package is shown in Figure 6. For each perturbation, there was one PDV probe and one LN pin in both the helium-implanted and -unimplanted regions. There were also two PDV probes on the flat regions, one between kh values of 0.6 and 0.5 and the other between kh values of 0.5 and 0.7, to provide a breakout time and the jump-off free surface velocity. A PDV probe was placed looking through the entire target assembly (magenta circle in Figure 6) to measure the projectile velocity. Similarly, another piezoelectric probe (orange circle, Figure 6) was glued onto the lexan plate to provide a voltage spike at target impact to trigger all of the diagnostics. The alignment of the PDV probes to the perturbations was checked with a red laser, to allow visible confirmation of the spot on the desired region. The standoff height—the distance from the rear of the target to the front face of the LN pins and PDV probes—was measured to be 22 mm for all targets. Images of the rear of the mounted target, the diagnostics package, and the full target assembly mounted in the gas gun are shown in Figure 7.

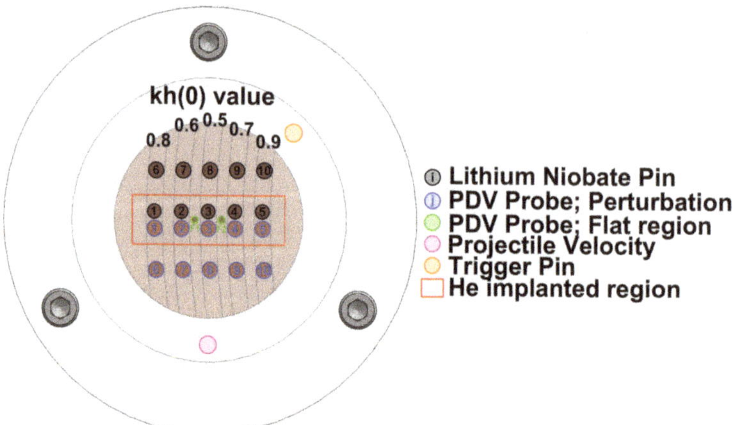

Figure 6. Schematic of the rear of the diagnostics package and target.

Figure 7. Images showing **left**—rear of the target mounted in the lexan plate, **center**—back of the diagnostics package showing the LN pins (top two rows) and PDV probes (bottom two rows), **right**—rear of the target and diagnostics assembly mounted in the gas gun.

3. Results

The four experiments were performed; three at an impact velocity of 1.1 mm/µs (0, 1000, and 2000 appm) and one at 1.0 mm/µs (4000 appm). The velocity–time histories corresponding to these experiments are shown in Figure 8. This data shows the breakout velocities to be ˜1.3 km/s for the three samples (0, 1000, and 2000 appm) and 1.2 km/s for the 4000-appm copper.

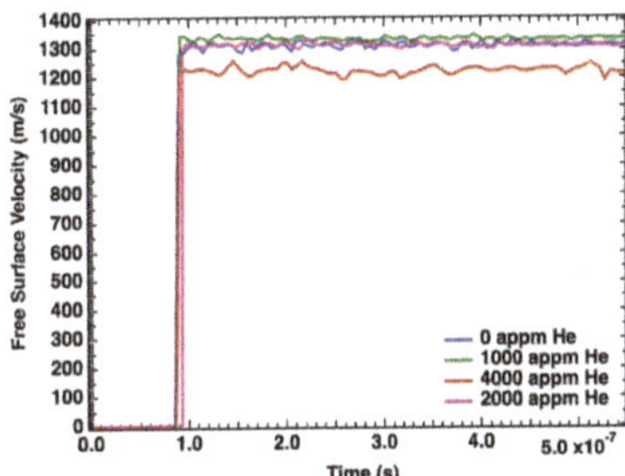

Figure 8. Velocity–time history from one of the flat regions for each of the four experiments.

This difference in shock-breakout velocities is important to note when trying to compare the four experiments to each other. Our results also show that perturbations with kh values of 0.8 and 0.9 went unstable and produced ejecta in all the samples, irrespective of helium concentration. This is consistent with previous work performed by Buttler et al. [15]. Figure 9 shows the velocity–time history corresponding to the perturbation with a kh of 0.9 on the four copper samples. We decided to focus on discussing results from a kh of 0.9 in this paper, as the focus of this work is on the effect of helium on ejecta production rather than the effect of kh on ejecta production.

This data shows the formation of a spike which is consistent with the perturbations inverting and growing. There were only minor variations measured in the peak spike velocity as a function of helium concentration. Specifically, the spike velocity was measured to be 2.48, 2.50, 2.48, 2.31 mm/µs with the helium concentration increasing from 0 to 4000 appm. The fact that there is no change in the velocity of the spike, which is related to the high strain rate strength, suggests that at these high He/Dpa ratios, there is almost no change in the material strength. This is in contrast to observations at low strain rate, where an increase in flow stress is observed with helium implantation, especially due to the formation of a gas-bubble superlattice [16,17]. This contrast could be because, in general, the strengthening in materials with helium is due to the helium acting as obstacles to dislocation motion. However, at high strain rates (in this experiment, the strain rate is 10^7/s), dislocations do not have time to actually move and interact with the bubbles. So, the instantaneous yield strength is generally due to the shear modulus and the dislocation density of the material [18]. While altering the helium concentration changes the dislocation density somewhat, we postulate that it is not enough to cause a measurable difference in the strength of the material. It is important to note that the decrease in the spike velocity for 4000 appm could be due to the lower shock-breakout velocity and not due to the actual helium-concentration increase.

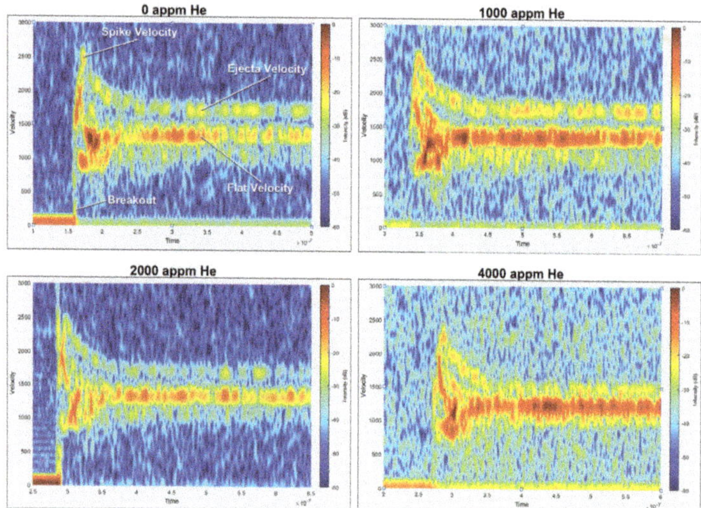

Figure 9. Spectrograms corresponding to copper with (**top-left**) 0 appm, (**top-right**) 1000 appm, (**bottom-left**) 2000 appm, and (**bottom-right**) 4000 appm helium. The spectrograms are colored by intensity, which varies between experiments depending on the signal to noise ratio for that particular experiment.

In addition to the spike velocity, data from Figure 9 can be integrated to extract position–time history, which when coupled with the standoff distance of the LN probes, can be extrapolated to extract an "expected" timing for the impact of ejecta cloud on the LN pins. The velocity-time data shows that there is a slight change in the velocity at which these ejected particles were traveling as a function of helium concentration. Specifically, the ejecta velocity was measured to be 1.72, 1.755, 1.715, and 1.460 mm/μs for the 0-, 1000-, 2000-, and 4000-appm copper, respectively. These velocities coupled with the LN pin standoff distance of 21.8 mm are used to calculate the timing at which ejecta cloud should arrive at the LN pins. This time was calculated to be 12.7, 12.4, 12.7, and 14.9 μs for the 0-, 1000-, 2000-, and 4000-appm copper, respectively. The increase in the timing for the 4000-appm copper is attributed to the lower shock-breakout velocity.

This timing should be consistent with the timing when the LN pins are activated, unless there is material present that is not measured by the PDV probes due to its extremely small size. Figure 10 shows the voltage–time history for the four shots obtained from the LN pins.

These plots show that the arrival time for ejecta on the pins was actually 10 μs for the 0 appm, 12.5 μs for 1000 appm, and 10.5 μs for 2000 appm copper instead of the times obtained from the PDV data. This is an approximately 3 μs difference in the arrival time of ejecta at the LN pins. We hypothesize that this suggests the presence of a fine, atomic-size ejecta cloud that is traveling at a much higher velocity and is too small to reflect light and be captured by the PDV probes. The most curious result is the arrival of ejecta at the LN pins at 7 μs instead of 14 μs for the 4000-appm copper. This is especially interesting because the shock-breakout velocity was lower for the 4000-appm copper compared to the others, which means the ejecta should have been traveling at a lower velocity. This could be due to the presence of helium in the sample, which might be making the material brittle and leading to the formation of finer ejecta that is then moving at a higher velocity. This is a question that is actively being researched right now. It is important to note that ejecta in this work refers to copper atoms only. The helium atoms are too light to be detected by the LN pins. In addition, with the discrepancy in timing it is interesting to note the differences in the total time that signal is produced on the LN pins. This difference in the total time, especially for the 4000-appm copper, suggests the presence of higher amounts of ejecta in the form of copper atoms, leading to the conclusion that ejecta mass is increased as a function of helium concentration. This is somewhat consistent with our recent results from molecular dynamics simulations that show a large difference in total ejected mass from copper with and without helium [19]. Through our modeling work, we had hypothesized that the presence of helium bubbles can lead to an increased ejected mass due to two reasons: (1) The bubbles alter the shock velocity in their vicinity due to differences in their density as compared to copper. This leads to the formation of a nonplanar shock front that can increase ejecta production. (2) The shock wave can compress the bubble and lead to internal jetting, similar to shape-charge phenomenon. So scientifically, an increase in ejecta mass is plausible and is supported by the increase in the time for which an increase in voltage is observed with increasing Helium concentration.

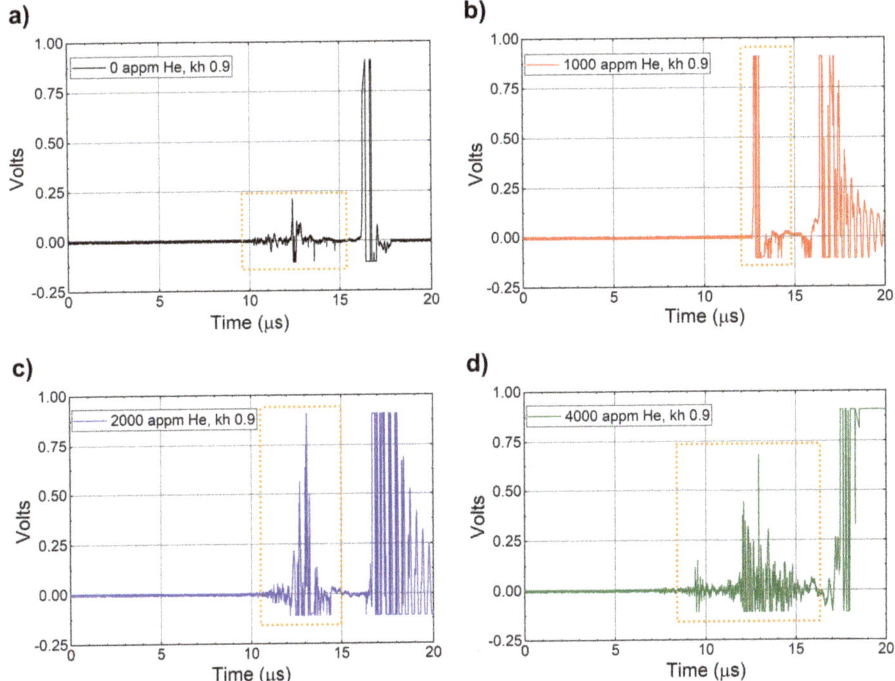

Figure 10. Voltage–time history corresponding to (**a**) 0 ppm, (**b**) 1000 ppm, (**c**) 2000 ppm, and (**d**) 4000 ppm for the four shots obtained from LN pins placed at a kh of 0.9. The dotted boxes outline the beginning and start times associated with ejecta at the LN pins. The late time signal is attributed to the free surface of the copper impacting the LN pins.

4. Conclusions

The goal of this work was to understand the effect of heterogeneities like helium on ejecta production in metals. OFHC copper was used as a model material for this study. Gas-gun driven RMI experiments, which required special perturbations be machined onto the copper targets, were used to investigate the effect of helium concentration on ejecta production. Four different concentrations of helium—0, 1000, 2000, and 4000 appm—were implanted into the copper sample at the Michigan Ion Beam Laboratory. These samples were then subjected to shock loading using a gas-gun, and diagnostics were used to measure the velocity–time history and mass–time history as the perturbations inverted, grew, and eventually went unstable to produce ejecta. Our results show that kh values of 0.8 and 0.9 produced ejecta in all targets regardless of the presence of helium. Further analysis of the velocity and mass data associated with kh of 0.9 showed (1) an increase in the production of finer ejecta traveling at higher velocities as a function of helium concentration, and (2) longer times associated with a finite voltage, suggesting an increase in mass as a function of Helium concentration. These results are significant and show that heterogeneities like Helium can actually alter important dynamic properties like ejecta production from metals.

Author Contributions: Conceptualization, S.F.; methodology, S.F., D.J., D.M., and C.L.; formal analysis, D.J. and S.F.; investigation, D.J., D.M., J.P., and C.L.; resources, S.F.; writing—original draft preparation, S.F.; project administration, S.F.; funding acquisition, S.F. All authors have read and agreed to the published version of the manuscript.

Funding: Los Alamos National Laboratory is operated by LANS, LLC, for the NNSA and the U.S Department of Energy under contract DE-AC52-06NA25396. Funding was provided by Office of Experimental Science.

Acknowledgments: We would also like to thank Bill Blumenthal, Rusty Gray and Peter Hosemann for helpful discussions. We also wish to acknowledge Derek Schmidt and John Martinez for machining the targets. The authors would like to acknowledge the assistance of O. Toader, F. Naab, and T. Kubley of Michigan Ion Beam Laboratory

Conflicts of Interest: The authors declare no conflict of interest.

References

1. Asay, J.R.;Mix, L.P.; Perry F.C. Ejection of material from shocked surfaces. *Appl. Phys. Lett.* **1976**, *29*, 284. [CrossRef]

2. Asay76. *Material Ejection from Shock-Loaded Free Surfaces of Aluminum and Leas*; SAND-76-0542; Sandia National Laboratories: Albuquerque, NM, USA, 1976.

3. Buttler, W.T.; Williams, R.J.R.; Najjar, F.M. Foreword to the Special Issue on Ejecta. *J. Dyn. Behav. Mater.* **2017**, *3*, 151–155. [CrossRef]

4. Asay, J.R.;Bertholf, L.D. *A Model for Estimating the Effects of Surface Roughness on Mass Ejection from Shocked Surfaces*; SAND-78-1256; Sandia National Laboratories: Albuquerque, NM, USA, 1978

5. Asay, J.R. *Effect of Shock Wave Risetime on Material Ejection from Aluminum Surfaces*; SAND-77-0731, Sandia National Laboratories: Albuquerque, NM, USA, 1977.

6. Andriot, P.; Chapron, P.; Olive, F. Ejection of materials from shocked surfaces of tin, tantalum and lead-alloys. *AIP Conf. Proc.* **1982**, *78*, 505.

7. Zellner, M.B.; McNeil, W.V.; Gray, D.T., III; Huerta, D.C.; King, N.S.; Neal, G.E.; Valentine, S.J.; Payton, J.R.; Rubin, J.; Stevens, G.D.; et al. Surface preparation methods to enhance dynamic surface propery measurements of shocked metal surfaces. *J. Appl. Phys.* **2008**, *103*, 083521. [CrossRef]

8. Germann, T.C.; Hammerberg, J.E.; Holian, B.L. Large Scale Molecular Dynamics Simulations of Ejecta Formation in copper. *AIP Conf. Proc.* **2004**, *706*, 285.

9. Zellner, M.B.; Dimonte, G.; Germann, T.C.; Hammerberg, J.E.; Rigg, P.A.; Stevens, G.D.; Turley, W.D.; Buttler, W.T. Influence of shockwave profile on ejecta. *AIP Conf. Proc.* **2009**, *1195*, 1047.

10. Buttler, W.T.; Oró, D.M.; Olson, R.T.; Cherne, F.J.; Hammerberg, J.E.; Hixson, R.S.; Monfared, S.K.; Pack, C.L.; Rigg, P.A.; Stone, J.B.; et al. Second shock ejecta measurements with an explosively driven two-shockwave drive. *J. Appl. Phys.* **2014**, *116*, 103519. [CrossRef]

11. Buttler, W.T.; Gray, G.T., III; Fensin, S.J.; Grover, M.; Prime, M.B.; Stevens, G.D.; Stone, J.B.; Turley, W.D. Yield strength of Cu and CuPb alloy (1%Pb). *AIP Conf. Proc.* **2017**, *1793*, 110005.

12. Fensin, S.J.; Escobido, J.P.; Gray, G.T., III; Patterson, B.M.; Trujillo, C.P.; Cerreta, E.K. Dynamic damage nucleation and evolution in multiphase materials. *J. Appl. Phys.* **2014**, *115*, 203516. [CrossRef]

13. Mansur, L.K.; Lee, E.H.; Maziasz, P.J.; Rowcliffe, A.P. Control of helium effects in irradiated materials based on theory and experiment. *J. Nucl. Mater.* **1986**, *141*, 633 [CrossRef]

14. Stopping and Range of Ions in Matter (SRIM). Available online: http://www.srim.org (accessed on 12 September 2019).

15. Buttler, W.T.; Oró, D.M.; Preston, D.L.; Mikaelian, K.O.; Cherne, F.J.; Hixson, R.S.; Mariam, F.G.; Morris, C.; Stone, J.B.; Terrones, G.; et al. The study of high-speed surface dynamics using a pulsed proton beam. In *Shock Compression of Condensed Matter-2011*; Elert, M.L., Buttler, W.T., Eds.; American Institute of Physics: Melville, NY, USA, 2012; Volume 1426, pp. 999–1002.

16. Wang, Z.J.; Allen, F.I.; Shan, Z.W.; Hosemann, P. Mechanical Behavior of Copper Containing a gas-bubble superlattice. *Acta Mater.* **2016**, *121*, 78. [CrossRef]

17. Klener, D.; Hosemann, P.; Maloy, S.A.; Minor, A.M. In-situ nanocompression testing of irradiated copper. *Nat. Mater.* **2011**, *10*, 608–613.

18. Gray, G.T., III. High strain rate deformation: Mechanical beahvior and deformation substructures induced. *Annu. Rev. Mater. Res.* **2012**, *42*, 285–303. [CrossRef]

19. Flanagan, R.M.; Hahn, E.N.; Germann, T.C.; Meyers, M.A.; Fensin, S.J. Molecular dynamics simulations of ejecta formation in helium-implanted copper. *Scripta Mat.* **2020**, *178*, 114–118. [CrossRef]

MDPI
St. Alban-Anlage 66
4052 Basel
Switzerland
Tel. +41 61 683 77 34
Fax +41 61 302 89 18
www.mdpi.com

Materials Editorial Office
E-mail: materials@mdpi.com
www.mdpi.com/journal/materials